Explanation in Geography

To my father and mother
who sacrificed so much
for my early education

Explanation in Geography

DAVID HARVEY

Professor of Geography and Environmental Engineering,
The Johns Hopkins University

 Edward Arnold

© David Harvey 1969, 1973

First published 1969 by
Edward Arnold (Publishers) Ltd
25 Hill Street, London W1X 8LL
Reprinted 1970, 1971
First published in paperback 1973
Reprinted 1973

Boards edition ISBN: 0 7131 5464 0

Paper edition ISBN: 0 7131 5693 7

Reproduced photo-litho in Great Britain by
J. W. Arrowsmith Ltd., Bristol 3.

Preface

Let me begin by explaining how this book on explanation in geography came about. It is often the case that the inner justification for writing a book is rather different from the external justification for publishing it. I wrote this book mainly to educate myself. I sought to publish it because I feel sure that there are many geographers, both young and old, who are in a similar state of ignorance to that which I was in before I commenced to write. If they can obtain from reading this book only a small fraction of the understanding and insight which I gained from writing it, then they will have benefited materially. Let me explain the nature of my own ignorance as it existed prior to setting pen to paper.

The so-called 'quantitative revolution' with its innovation centre in the University of Washington diffused slowly throughout the geographic community and by the early 1960s it became fashionable among the avant-garde to compute correlation coefficients, run 't' tests, and the like. Not wishing to be left behind, I naturally indulged in this fashion, but found to my consternation that I only managed to accumulate a drawer full of unpublished and unpublishable papers. I have to acknowledge my debt to several perceptive (or perhaps they were prejudiced) editors who by refusing to publish these papers undoubtedly saved my academic reputation from premature destruction! I also found to my further consternation that often I could not interpret the results of my own analyses. Initially I put this down to my lack of command of statistics and mathematics (a lamentable state of affairs that stemmed from a strong 'Arts' background at school and University). This lack of an adequate training undoubtedly did account for the many technical blemishes on my work (the most celebrated published example being a regression equation estimated the wrong way round—I did not realise that if X was regressed on Y it yielded a different result from Y regressed on X!). But the more I brushed up on my techniques (a never-ending process it seems), the more I became convinced that something more was involved. I therefore decided to devote some time to a systematic investigation of the quantitative revolution and its implications. I was fortunate to be allowed to teach an under-

graduate course in geographical methodology at Bristol University
and I would like to acknowledge the patience of the successive undergraduate audiences who over a period of five or six years listened to
my fumbling attempts to sort out rather complicated conceptual and
methodological problems. I must likewise thank a graduate seminar
class at Penn State who also bore with me in the session 1965–6.
Out of this investigation came a central and to me vital conclusion.
The quantitative revolution implied a philosophical revolution. If I
did not adjust my philosophy, the process of quantification would
simply lead me into a cul-de-sac. My own lack of success with the
new methodology was simply the result of trying to pour new wine
into old bottles. I was forcing my philosophical attitudes into an
alien methodology. I then had to decide whether to abandon my
philosophical attitudes (steadily accumulated from six years of indoc-
trination in what I can only call 'traditional' geography at Cambridge), or whether to abandon quantification. I examined this
question very carefully and found, to my surprise, that I could retain
much of the philosophical outlook which I valued, while venturing
down the path of quantification. Those aspects of my geographical
philosophy which had to be abandoned turned out to be those which
I could most easily dispense with on other grounds. The assumption
(often hidden and vague as it was) that things really are unique, or
that human behaviour cannot be measured, and so on, turned out
to be inhibiting and not very meaningful when subjected to critical
thought. I also found that I was often misinterpreting the assumptions upon which statistical methods were necessarily based and that
once these misinterpretations were removed the conflict between
much of my geographical philosophy and the new methodology was
also removed. When I sought to bring together the positive aspects
of traditional geographical thought and the philosophy implied by
quantification, I was amazed to observe how much more vigorous
and vital the whole philosophy of geography became. It opened up
a whole new world of thought in which we were not afraid to think
theoretically and analytically, in which we could talk of individuals
and populations in the same sentence, in which we could generalise
about pattern and particularise about locations in the same context.
There seemed to me to be nothing wrong with the aims and objectives of traditional geography (indeed they are to be prized and
cherished), but as an academic enterprise it had managed somehow
or other to hedge itself about with so many inhibiting taboos and
restrictions that it could not hope to realise the aims and objectives
it had set itself. In particular, geographers were failing, by and large,
to take advantage of the fantastic power of the scientific method.

And it was the philosophy of the scientific method which was implicit in quantification.

Some people may flinch at the term 'scientific method', so let me make it clear that I interpret this in a very broad sense to mean the setting up and observing of decent intellectual standards for rational argument. Now it is obvious that we can observe these standards without indulging in quantification. Good geographers have always observed them. But the curious thing was that it took quantification to demonstrate to me how extraordinarily lax my own standards were—hence all those unpublishable papers. I believe that the most important effect of quantification has been to force us to think logically and consistently where we had not done so before. This conclusion led me to change the emphasis of my approach. Although it was no accident that quantification was forcing us to up-grade our standards of argument, we could, if we so wished, up-grade those standards without any mention of quantification. The issue of quantification *per se* therefore faded into the background and I became much more interested in the general issue of the standards and norms of logical argument and inference which geographers ought to accept in the course of research. These standards could not be divorced from those of science as a whole. In short, I became interested in the role of scientific method (however conceived) in geography. Now there are many who have been nurtured so long in the ways of science that they appear to need no formal instruction in its method. To instruct such people seems like formalising what they already intuitively know. But there are many geographers who need formal instruction because, like me, they were not raised in the ways of science. But even those geographers who possess a fair intuitive grasp of scientific method cannot afford to ignore the formal analysis of it. An intuitive grasp arises from teaching by precept and example. Such a grasp is usually sufficient to handle routine work (and most of science is routine). But it cannot always handle the new questions, the problems for which there are no precedents. At this point it is often necessary to understand the philosophical underpinnings of scientific method as a whole.

Science provides us with very sharp tools. But as any craftsman will tell you it is the sharp tools which can do most damage when misapplied. The sharpest tools are those provided by mathematics and statistics. The former provides us with a means for formulating arguments rigorously and simply, while the latter provides us with the tools for data analysis and hypothesis testing with respect to data. I believe that these tools have often been misapplied or misunderstood in geography. I certainly plead guilty in this respect. If we are to control the use of these sharp tools in research we must understand the philosophical and methodological assumptions upon which their use necessarily rests. These assumptions are, of course, built up explicitly through an analysis of scientific method. But we have to ensure that the assumptions which we accept with respect to these particular tools of science do not conflict with the broader assumptions which we employ in setting up standards for rational argument and inference. The problem of adequate method is therefore doubly emphasised at the point where quantitative techniques and ordinary rational argument and inference come together. Hence the importance of quantification. We must, therefore, reconcile our assumptions at all levels in geographical research. What started for me as a quest to understand the nature of certain powerful tools of science, thus ended as a quest for an understanding of the totality of the process which leads to the acquisition and codification of geographical understanding.

This book is therefore about the ways in which geographical understanding and knowledge can be acquired and the standards of rational argument and inference that are necessary to ensure that this process is reasonable. I have sought to show that we can formulate criteria to judge whether or not an argument is sound, a technique properly used, or an explanation reasonable. I do not claim to have specified these criteria correctly. Ignorance is a relative thing. Compared to my situation five years ago I now feel much more learned and wise, but relative to what I still have to learn I feel more ignorant than ever. Indeed, since completing this manuscript in June 1968 I have changed several opinions and I can already identify errors and shortcomings in the analysis. This is therefore very much an interim report—one person's view at a particular point in time. I do not wish it to form the basis for some new kind of orthodoxy and I for one will certainly not defend it in those terms. My aim is to open up the field of play rather than to close it off to future development.

In constructing this interim report, I have had a good deal of help. I spent the year 1960–1 at the University of Uppsala on a Leverhulme scholarship and I would like to acknowledge this financial help for it gave me a year after the completion of my doctorate in which to think about all kinds of things which I had not had time for previously. During my stay in Uppsala, I formed a lasting friendship with Gunnar Olsson, and in the early stages we gave each other much mutual support. In the summer of 1964 I received help from the National Science Foundation to attend a conference on spatial statistics at Northwestern University and I must say the experience was a traumatic one. Michael Dacey, who provided considerable stimulus to my thought at that conference, has since provided me with unpublished materials and I am grateful to him for allowing me to quote from these. Waldo Tobler has similarly sent me unpublished materials, and I am also grateful to him for allowing me to quote from them. I must also thank my colleagues in the University of Bristol, particularly Allan Frey and Barry Garner, for providing a convivial and stimulating atmosphere in which to think and work. Peter Gould, during my year at Penn State, also provided me with plenty to think about. Various people have nibbled at the manuscript and the ideas contained therein. Allan Frey, Art Getis, Les King, Allen Scott, Roger Downs, Bob Colenutt, Rod White, Keith Bassett, Conrad Strack, and several others, made suggestions some of which I have incorporated. No acknowledgement would be complete without a mention of the 'terrible twins' of British geography, Dick Chorley and Peter Haggett. The former first introduced me to statistical methods just before I left Cambridge in 1960, and since that time has continued to throw ideas around in a most stimulating fashion. Peter Haggett has also been extremely kind to me and, particularly since he became professor at Bristol, he has been a never-ending source of advice and encouragement. I owe these two, and I believe British geography owes them also, an enormous debt.

I also want to acknowledge those motley folk who helped to keep me sane during the writing of this book. Marcia, Miles Davis, John Coltrane, Dionne Warwick, the Beatles and Shostakovitch, Titus, Phinneas T. Bluster and Jake, have all helped to maintain my inner equilibrium when things looked bleak. I think Marcia may help with the indexing too. To each and every one of this jolly crew my heartfelt thanks.

D. H.   Clifton, Bristol

March 1969

Acknowledgements

The author and publishers wish to thank the following for permission to use copyright material: George Allen and Unwin Ltd. and Columbia University Press for table 2.3.1, p. 56, from *Theory and methods of social research* by Johan Galtung, 1967; the American Statistical Association and the author for two maps, pp. 385–8, 'Maps based on probabilities' by M. Choynowski, *Journal of the American Statistical Association* 54, 1959; the Association of American Geographers and the author for map, p. 320, 'The market as a factor in the localization of industry in the United States' by C. D. Harris, *Annals* 44, 1954; the Association of American Geographers for extracts from 'The spatial structure of agricultural activities' by W. L. Garrison and D. F. Marble, *Annals* 47, 137–44, 1957, and from 'The nature of geography' by R. Hartshorne, *Annals* 29, 1939; University of California Press for extracts from *The direction of time* by H. Reichenbach, 1956, and from *Land and life—A selection from the writings of Carl Ortwin Sauer* edited by John Leighly, 1963; Cambridge University Press and the authors for figure 142, p. 294, *On growth and form* by W. D'Arcy Thompson, 1961 (abridged edition), and for extracts from *Scientific explanation* by R. B. Braithwaite, 1953; Chandler Publishing Company for extracts from *The conduct of inquiry* by A. Kaplan, 1964; University of Chicago Press and the author for extracts from *The structure of scientific evolutions* by T. S. Kuhn, 1962; Dover Publications Inc. for extracts from *An introduction to symbolic logic and its applications* by R. Carnap, 1958 (English edition, translated by W. H. Meyer and J. Wilkinson); The Free Press of Glencoe Inc. for extracts from *Aspects of scientific explanation and other essays in the philosophy of science* by C. G. Hempel, 1965; C. W. K. Gleerup Publishers and the authors for figure 38, p. 73, *Innovationsförloppet ur korologisk synpunkt* by T. Hägerstrand, 1953, and for figures 9.30 & 9.31, pp. 278–9, 'Theoretical geography', by W. Bunge, *Lund Studies in Geography, Series C*, 1, 1966 (second edition); Charles Griffin & Co. Ltd. and the authors for figure 22.1, *Advanced theory of statistics*, volume II, by M. G. Kendall and A. Stuart, 1967 (second edition); P. Haggett for figure 10.10, p. 300, *Locational analysis in human geography*, 1965 (Edward Arnold Publishers Ltd.); Harcourt, Brace and World Inc. for extracts from *The structure of science* by E. Nagel, 1961; J. W. House for extracts from 'Chance and landscape' by L. Curry in *Northern geographical essays* edited by J. W. House, Oriel Press Ltd., 1966; the Institute of British Geographers for figure 2, 'Central Europe—Mitteleuropa—Europe Centrale', by K. A. Sinnhuber, *Transactions* 20, 1954; Methuen Publishers for figure 3.1, 'Models in geomorphology' by R. Chorley in *Models in geography* edited by R. Chorley and P. Haggett, 1967; the authors and Michigan Inter-University Community of Mathematical Geographers for figure 5, p. 17, 'A note on surfaces and paths and applications to geographical problems' by W. Warntz, *Discussion Papers* 6, 1965, and for figure 14.6, p. 220, 'Numerical map generalization' by W. Tobler, *Discussion Papers* 8, 1966; the University of North Carolina Press for extracts from *Causal inferences in nonexperimental research* by H. M. Blalock, 1964 edition; Aldine Publishing Co. and Routledge and Kegan Paul Ltd. for extracts from *Explanation in social science* by R. B. Brown, 1963; John Wiley & Sons Inc. for extracts from *Scientific method: optimizing applied research decisions* by R. L. Ackoff, 1962, from *Decision and value theory* by P. C. Fishburn, 1964, and from *The foundations of statistics* by L. J. Savage, 1954; and Yale University Press for extracts from *The philosophy of symbolic forms*, Volume 2: *The phenomenology of knowledge* by E. Cassirer, 1957 (English edition, translated by R. Manheim and C. W. Hendel).

Contents

List of Figures

PART ONE
Philosophy, Methodology and Explanation

Chapter 1
Philosophy and Methodology in Geography

Consider the following statement:

Geography is concerned with the description and explanation of the areal differentiation of the earth's surface.

Such a statement might be regarded by some as an adequate definition of the field of geography. Others might demur and propose some alternative definition. It is not my aim to argue with this statement. I wish merely to analyse its form. A cursory examination of the statement thus reveals that it can be divided into two halves. The first half is concerned with *how* we should go about studying phenomena, and in particular it is concerned with the two operations of description and explanation. The second half of the statement is concerned with *what* we should study; it identifies a domain of objects and events to which the operations of description and explanation should be applied. Since this difference is fundamental to this book it is important that we begin by considering it in some detail. For convenience, we shall refer to the second half of the statement as the goals or substantive *objectives* of geographical study. The first half will be referred to as the *method* of study.

The choice of the 'areal differentiation of the earth's surface' as the objective of geographical investigation is an arguable one. It is of interest to ask, however, how we could dispute it. We cannot show, for example, that it is logically unsound, logically incoherent, or logically inconceivable. Some tautological objectives might be dealt with in this way, but the particular objective we are here considering does not contain any inherent contradiction. We might, however, argue that it is not worth while as an objective, that it is too vague to be of very much use to us, or that it does not fit the objectives which most geographers set up when they undertake some substantive investigation. We might even argue that the objective logically entails a programme of study that is not likely to be realisable in the near future and that, thus, the objective is unsound. Whatever logical argument we may produce, however, it is clear that the only grounds upon which we may ultimately dispute the objective are grounds

3

of belief. As individuals we possess values. These values, it is true, are not independent of the society in which we live and work, and in a narrower context they are not likely to be independent of other geographers with whom we have contact and interact. These values guide us to objectives that we feel are worth while. Given our own values we may dispute 'the areal differentiation of the earth's surface' as a worth-while objective for study. We may even refer to social values at large and show that this particular objective is at odds with the prevailing values which exist in our own society at the present time. A social geographer, for example, wholeheartedly committed to the study of planning problems, might prefer 'the spatial organisation of human activity' as the objective for geographical study, to 'the areal differentiation of the earth's surface'.

Different geographers and groups of geographers thus tend to have rather different objectives, depending upon their own particular sets of values. If we wish to convert someone to our own view of the objectives of geography, we can only do so *via* his beliefs. We might play upon his social conscience, for example, and point to the starvation and misery in the streets of Calcutta and seek thereby to convert him to a view of geography that is highly committed to doing something which will be useful in alleviating that starvation and misery. Or we might play upon his aesthetic feeling, take him slowly through the ruins of Rome and convert him to a standpoint that encompasses the 'feel' of landscape through time. We cannot, however, destroy his beliefs by logical argument, any more than we can support our own by such argument.

The beliefs upon which we rest the objectives of our study form our philosophy, our own individual view of life and living. It is convenient, therefore, to designate the manifestation of these beliefs in geographical work as the *philosophy of geography*. There are many such philosophies. Each provides us with a distinctive view of the *nature* of geography. Such philosophies vary from country to country, from group to group, and over time. There are those who have regarded these varied philosophies as faint manifestations of some hidden pervasive attitude in man that forms the essence of geography. Substantive geographical work is thus viewed as some shadowy representation of the *essential* geography, in rather the same way that Plato regarded perceptual experience as some kind of shadowy representation of the 'essences' that lay beyond. It is not my purpose, however, to discuss the varied philosophies of geography or to examine how they might be synthesised. The essential point for our present purpose is to demonstrate that such philosophies are dependent upon beliefs, and that although we may analyse them to make certain of their consistency and coherence, we cannot analyse away their very foundations.

Having indicated the subjective foundation of such philosophies of geography, we ought to consider the utility of such philosophies. The plain fact is that all analysis is barren unless there is some objective. The objective may not be plainly stated, it may be implied rather than explicit, it may even be extraordinarily fuzzy. But without some notion as to what is to be studied there can be no geography or, indeed, any knowledge, save the empty analytic understanding provided by mathematical systems and logically constructed calculi. We cannot, therefore, proceed without some objective, and defining an objective amounts, however temporarily, to assuming a certain philosophical position with respect to geography itself. Beliefs regarding the philosophy or nature of geography are therefore crucial to the prosecution of substantive geographical work. The second half of the statement with which we began this chapter thus provides us with one set of objectives, one set of beliefs, which we may use as the basis for our investigations. We either accept it or find some alternative objectives which we regard as being in some way more worthwhile.

The first half of the statement makes reference to description and explanation. When we use the term description we usually mean some sort of *cognitive* description. We are not content to describe events in a random manner. We seek, rather, to impose some coherence upon our descriptions, to make them rational and realistic, to try to bring out what we understand of a situation by patterning our descriptive remarks in a particular way. It proves difficult to differentiate, therefore, between the cognitive descriptions which we may provide and explanations. Modern analysis suggests, in fact, that cognitive description and explanation are different merely in degree, not in kind. The latter thus places a more explicit emphasis upon the analysis of interconnections, whereas the former tends to imply the necessity of such connections. Whether or not we accept that description and explanation are essentially similar does not really matter, for the considerations that follow can equally well be applied to them separately or jointly. For convenience, therefore, we shall merely lump them together under the loose heading of explanation.

It is of interest to ask, in the manner that we did with respect to objectives, how we might dispute a particular explanation. With respect to explanation it is clear that we can dispute it on logical grounds. We can protest that it is logically unsound, that the conclusions cannot logically be derived from the premises, or that there is some logical inconsistency in the argument which renders the explanation meaningless. Explanation is, therefore, very much a logical procedure and can be subjected to deep logical analysis. There are, it is true, some issues regarding explanation which cannot be resolved independently of philosophical beliefs—the problem of verification and confirmation being a prime example. Nevertheless, we may rightly insist that an explanation should be logically sound, before we even bother to consider its philosophical underpinnings. Now it has been the concern of logicians and philosophers (particularly the logical positivists and philosophers of science) to deepen our understanding of the logic of sound explanation. They have not, it is true, concerned themselves solely with this, but much of the work that has been done during the last fifty years or so has been intended to set up criteria by which we may judge whether or not a particular explanatory argument is or is not sound. In rendering explanations we must take account of such criteria and endeavour to show at the minimum that the explanations we offer are consistent with them.

It seems to me to be the task of the *methodologist* of geography to consider the application of such criteria to the explanation of geographical phenomena. The methodologist, therefore, is concerned with 'the logic of justification' rather than with the philosophical underpinnings of our beliefs with respect to the nature of geography. The philosopher and methodologist therefore have rather different tasks. The former is concerned with speculation, with value judgements, with inner questioning regarding what is or is not worth while. The latter is concerned primarily with the logic of explanation, with ensuring that our arguments are rigorous, that our inferences are reasonable, that our method is internally coherent.

This distinction between the activities of the philosopher and the methodologist is absolutely vital to this book. This is, in short, a book concerned with methodology rather than with philosophy. The basic concern is, therefore, to elaborate the criteria that can be developed with respect to explanation in geography, and to analyse the various ways in which we can ensure sound and consistent explanations. This emphasis upon methodology is, of course, partly a personal predilection. But it is also partly a reaction against the prevailing bias in geographical discourse towards discussing philosophical issues with scarcely a glance at methodological problems, or, at best, towards so mixing up philosophy and methodology that it is scarcely possible to discern which is what. My feeling is that a good eighty per cent or so of the literature concerned with the foundations of geographical thought has been speculative and philosophical in style. There is, of course, nothing inherently wrong in that. But in many instances it seems as if the ability to speculate is somehow constrained by particular methodological views which, on analysis, turn out to be unnecessary or simply unsound. Now there can be no doubt that methodology and philosophy interact. They are not independent of each other. But by and large we have misunderstood the nature of the interactions because we have failed to understand the differences between them.

This difference can perhaps best be demonstrated by considering a difference to which we shall have cause to refer at various points in coming chapters. This difference is that between assuming a philosophical position with respect to some mode of analysis and a methodological position with respect to it. The methodologist may adopt some mode of analysis because it is convenient and effective, because it does the job he requires of it in an efficient manner. He may, for example, adopt a deterministic or a stochastic model to examine the interaction between certain sets of phenomena. The philosopher, on the other hand, may adopt a mode of analysis because he believes it is the only appropriate way of examining a particular set of interactions. Believing in free will, for example, he may eschew deterministic models, deterministic cause-and-effect language, and insist that the only language appropriate, given the basic premiss of free will, is a basic language that incorporates indeterminacy (e.g. probability theory). It is important to recognise, however, that the adoption of a methodological position does not entail the adoption of a corresponding philosophical position. Laplace, for example, believed that the world of phenomena was a determined world, and thus he was a philosophical determinist. Yet he could develop the calculus of probability as a convenient method, arguing that our own ignorance and inability was such that we required such an approximation in order to get anywhere with the analysis of certain types of phenomena. Because we use a probabilistic model, we do not need to adopt a position of philosophical indeterminacy. In this respect methodological and philosophical positions are very different from each other. This difference has not always been appreciated in the geographical literature. It has sometimes been argued, for example, that the use of a particular method, such as cause-and-effect analysis, entails a particular philosophical position, such as philosophical determinism. We cannot argue, therefore, from a methodological position in support of a philosophical position. In the other direction the relationship is rather closer. Assuming a philosophical position thus entails a methodological position, unless we can find a good excuse for not adopting such a position. Philosophical determinists thus tend to use deterministic modes of analysis exclusively, unless, like Laplace, they admit that some other mode of analysis is a convenient approximation. Even here, therefore, it is possible to adopt a flexible methodology while holding to a rigid philosophical position.

This difference between methodology and philosophy is extremely useful to us. It allows us, for example, to adopt varying strategies in our substantive investigations without necessarily committing ourselves to some alien philosophy; it allows us to hold a particular philosophy without constraining our methods of investigation. The clear separation between methodology and philosophy thus provides us with a strategy of the utmost flexibility in tackling substantive problems. It means that every method is open for us to use, provided we can show that its use is reasonable under the circumstances. It is thus one of the major tasks of methodological analysis to show how and under what circumstances a particular mode of analysis is appropriate, to specify the assumptions necessary for the employment of a particular technique, and to demonstrate the form of analysis which must be followed if the analysis itself is to be rigorous and logically sound. All this can be accomplished independently of philosophy.

Having said this, however, we should clearly recognise the points where philosophy and methodology are likely to interact. We have already suggested that geography without some objective is barren. We shall see in later chapters that most methods cannot be evaluated independently of objectives and purpose. We shall also see that the construction of theory, itself a key element in the whole procedure of explaining, is highly dependent upon the speculative objectives which geographers have defined for themselves. There are some philosophers, logical positivists of the extreme variety, who have held that all knowledge and understanding can be developed independently of philosophical presuppositions. Such a view is not now generally held, for logical positivism in such an extreme form has turned out to be barren. Methodology without philosophy is thus meaningless. Our ultimate view of geography must therefore take both methodology and philosophy into account. Such an ambitious synthesis will not be attempted here, for before we can hope to achieve it, we need a much better understanding of methodological problems alone. But although the emphasis in this book is primarily upon methodological problems, we will have cause on several occasions to refer to important philosophical issues concerning the nature of geography. The main aim, however, will be to sort out those aspects of analysis which are a matter of logic and those aspects that are contingent upon philosophical presupposition. It is my own belief that only by separating out these different aspects shall we be capable of building a sound methodology and a sound philosophy of geography.

Chapter 2
The Meaning of Explanation

Separating out methodological and philosophical issues provides a convenient way for approaching many of the problems involved in explanation. Braithwaite (*1960*) and Rudner (1966), for example, both make considerable use of such a distinction in their analyses of the form and nature of explanation. They do so because it allows discussion to concentrate on the more limited (and more tractable) problem of explanatory form without necessarily presupposing a philosophical position with respect to the nature of experience, the nature of perception, or what Bertrand Russell (1914) calls *Our Knowledge of the External World*. These latter questions have, of course, been the subject of philosophical speculation through the ages, and in their modern form they are usually posed as questions regarding the nature of human communication, the nature of language, the process of assigning meaning to terms, the psychology and physiology of perception, and so on. An analyst such as Braithwaite (*1960*, 21) is thus simply concerned with:

the straight logical problems of the internal structure of scientific systems and of the roles played in such systems by the formal truths of logic and mathematics, and also the problems of inductive logic or epistemology concerned with the grounds for the reasonableness or otherwise of accepting well-established scientific systems.

The emphasis in much analytic work on the nature of explanation is thus upon its internal structure and its internal coherence. This approach, which is the basic line of approach in this book, poses some problems. In the first place, it tends to ignore explanation as an *activity*, as a *process*, as an *organised attempt at communicable understanding*. Indeed, it ignores the whole problem of the motives which we have in seeking for acceptable explanations. This problem is directly tackled by those writers, such as Kuhn (1962), Churchman (1961), and Kaplan (1964), who prefer to approach the problem of explanation from a behavioural rather than a formal standpoint. In the second place, the interpretation to be given to experience is itself ignored. Experience is thus regarded as some set of statements about reality which are commonly regarded or accepted as being 'factual' in a sense that remains undefined. Explanation is regarded as a formal connection between different factual statements, or between factual statements and more general 'theoretical' statements which can be brought to bear on such factual statements. Treating experience in this way is convenient from the point of view of the analyst concerned with the formal structure of explanation, but it is not always helpful to the geographer concerned with the problem of interpreting experience in a manner that is acceptable to other geographers.

In writing on explanation in geography, therefore, it is impossible to avoid some discussion of these issues. As geographers we are, after all, basically concerned with elucidating substantive geographical problems and not, as are many philosophers of science, with elucidating the form of explanation *per se*. At some stage or other explanatory form has to be brought into contact with experience. This interface between methodology and philosophy is, however, very difficult to write about without either writing a lengthy treatise upon the meaning of experience itself or making substantial presuppositions. Undoubtedly one of the most profound discussions of the nature of the 'geographical experience' is that provided in an essay by Lowenthal (1961) in which he examines the relationships that exist between our experience or perception of the world in which we live, our ability to make coherent statements about that experience, our ability to communicate that experience, and the role of the 'geographical imagination' in providing concepts and principles upon which to build a common geographical epistemology. Lowenthal's discussion is primarily concerned with philosophy rather than methodology (in the sense in which we are using these terms here) but it demonstrates very forcefully the significance of philosophical issues to our attempt to solve substantive geographical problems. In this chapter, therefore, we shall be concerned with the nature of explanation in all its aspects and attempt, in the process, to lay some broad general foundations for discussing some of the more formal aspects of explanation in later chapters.

I THE MEANING OF EXPLANATION

There has been considerable discussion and dispute regarding the meaning of explanation. In *Part II* of this book some of the major controversies regarding explanation in the natural sciences, in the social sciences, in history, and in geography, will be examined. For the moment, and for our immediate purposes, it is convenient to give a very broad interpretation of the term *explanation*. It will be regarded, therefore, as any satisfactory or reasonable answer to a 'Why' or 'How' question. This view requires some clarification. In particular, it is useful to ask how such questions might arise, how we might proceed to construct answers, and how we might judge whether or not the answers given are reasonable and satisfactory.

A The need for an explanation

Toulmin (*1960*A, 86) suggests that the desire for an explanation originates from a reaction of surprise to some experience. This surprise, he suggests, is generated by a conflict between our expectations in a given situation and our actual experience of it. We will, for the moment, ignore the problem of experience, and simply concentrate on the notion of conflict and its association with the desire for explanation. Toulmin's example is a simple one. A stick which everyone regards as being straight appears bent on being thrust into water. Some people might react to this experience by saying 'So what?', others might regard it as 'fun', others might regard it as rather strange and surprising. This last reaction leads to questions. Toulmin accounts for such questions in the following way:

> The thing which disconcerts you about the situation is the way in which the evidence of our senses, originally unequivocal and unanimous, has become ambiguous and conflicting. There are obvious conflicts of three kinds:
>
> (i) between the reports of the same observer about the same property at different times—first he said, 'It's straight'; now he says, 'It's bent';
> (ii) between the reports of different observers about the same property at the same time—some say, 'It's bent to the left'; others say, 'It's bent to the right'; others again, 'It's just foreshortened';
> (iii) between the evidence of different senses about the same property at the same time—looking at it, you would say it was bent but, to feel it, you would say it was straight.
>
> In consequence of these conflicts, you ask, 'What is *really* the case? Is it really bent or not? If so, to the left or to the right? And if not, why does it look as though it were?' And in asking these questions, you begin to demand an *explanation* of the phenomenon.

An explanation may thus be regarded as reducing an unexpected outcome, which is the source of conflict and surprise, to an expected outcome. We might explain the phenomenon of the stick in water by reference to Snell's law, and thus show that we should always expect such a phenomenon to occur in such circumstances. In the process of generating one explanation, however, we may find other surprises and conflicts which require explanation, and a process of question-answer interaction may get under way that eventually leads to an organised body of knowledge to which we can refer for satisfactory explanations of all kinds of phenomena.

This portrayal of the process of explanation is, of course, simplistic in the extreme. But it does suggest a basic psychological motive in seeking for explanation, namely, the reduction of conflict or, as the psychologists call it, the reduction of stress. For the geographer this account is meaningful. Early travellers found the conflict between expectation and experience quite extraordinary. The accounts of Hakluyt and Samuel Purchas, for example, contained much for people to marvel at, much to provoke questions that demanded explanation. These questions did not go unasked, and the subsequent search for an explanation of the variation in ways of living and in environment might well be regarded as the history of geography. At the present time the process continues. Consider plotting the distribution of city sizes against rank on log-log paper and finding the distribution very close to a straight line. How do we react to this? We know of no government order that decrees how many people live in each city and town. We know of no conscious human process of migration or population growth that is designed to achieve conformity of city sizes to Zipf's rank-size law. Such a phenomenon is certainly in conflict with most of our expectations. We therefore search for some adequate and satisfying explanation of it. We seek to show that the rank-size rule is in accordance with our expectations and not in conflict with them.

The example of the stick in the water, Marco Polo's experiences in Peking, Mungo Park's experiences in Africa, Brian Berry's puzzlement over the rank-size rule, all have something in common. This common feature is a reaction of surprise. Now it would be foolish to suppose that we automatically react with surprise to every experience that we have. For the most part we shrug them off, dismiss them as irrelevant, or simply regard them as pleasurable or painful and leave it at that. There is, therefore, some degree of preselection regarding the experiences we react to with surprise and those that we dismiss. In short, we filter out all kinds of experience and examine just a few that for some reason or other we find suprising. In organised disciplines of learning, the preselection is partly done for us. One generation of geographers thus tends to set up ready-made questions for the next. Part of the training of the geographer, for example, is a training designed to teach him how to set up questions, a training that provides him with a set of expectations by which to judge whether or not a particular experience, is surprising. Mungo Park if confronted by the rank-size rule, would probably have shrugged his shoulders and said 'So what?', or have just regarded it as an interesting phenomenon but one that did not really arouse his interest. The questions that we ask, therefore, are partly conditioned by our training. The explanations that we seek tend likewise to be conditioned. Occasionally, however, this tradition is broken. A particular generation may come to regard the dialogue of question and answer that has taken place in the past as leading up blind alleys. It is a charge that all disciplines at some time or other have been liable to, that they have become enamoured of questions that have no real interpretation in terms of concrete experience, or that they have simply set up unrealistic questions and a neat mechanism for providing seemingly satisfying, if equally unreal, answers. On other occasions it seems as if a discipline has worked out a particular vein of thought, and requires to shift the location of activity to some other plane. At these points in the history of a discipline we are likely to find an example of what Thomas Kuhn (1962) calls a scientific revolution, the shift from one paradigm to another (a notion we will consider later).

It is not, however, my concern to consider the history of geography in this light. But it is vital to remember that the notions of conflict, of surprise, of questioning, and of searching for some answer, demand some preselection on the part of the individual, and that this preselection or filtering is very much conditioned by his training. At any one time, therefore, a particular geographer is likely to be concerned with reducing a few seemingly unexpected results to expected results. It would be impossible for him to be concerned with all problems and all issues, even among those that come to his attention, while the number of potential conflicts and questions seems to be infinite.

B Constructing an explanation

The purpose of an explanation may be regarded as making an unexpected outcome an expected outcome, of making a curious event seem natural or normal. In general there are three ways in which we may do this. Some would argue that they are mutually exclusive, some that they are complementary, some that they really amount to the same thing in the long run. We shall, however, just briefly, examine their characteristics.

(i) Probably the most important approach to constructing explanation is that generally known as the deductive-predictive approach. It is developed in great detail by writers such as Braithwaite (1960), Nagel (1961), and, above all, by Hempel (1965), who has sought to extend an approach based on classical physics to explanation in all fields of enquiry. We will consider this mode of explanation in some detail in later chapters. For the moment it will be sufficient to note its main characteristics. The objective is to establish statements or 'laws' and to show, empirically, that these laws govern the behaviour of various types of event. The law is then assumed to be a *universally true statement* (that is, it applies independently of space or time). The law functions in an explanation as follows: a set of initial conditions, which the law covers, is first stated and it is then shown that these taken in conjunction with the law must necessarily result in the event to be explained. It may be noted here that prediction and explanation are symmetric in this case and that there is no essential difference between them. The basic problem which such a view of explanation poses is, of course, that of providing adequate support for the supposedly universally true statements.

(ii) An alternative view has been developed by a number of writers, such as Hanson (*1965*), Toulmin (*1960*B), Wisdom (1952), Bambrough (1964), and Ryle (1949). This view has been termed the 'relational' view (Workman, 1964). In the relational view explanation is regarded as a matter of relating the event to be explained to other events which we have experienced, and which, either through familiarity or analysis, we no longer find surprising. The behaviour of planets, for example, may be related to the familiar performance of apples falling from trees. The essence of explanation, then, lies in providing a network of connections between events. According to this view, laws are not necessarily universally true statements, but simply a convenient device for bringing information derived from particular instances to bear on happenings in other instances (Workman, 1964; Bambrough, 1964). The deductive form of explanation, therefore, which seeks to explain the particular by reference to the general, may be regarded as a particular form of relational explanation. Relational explanation is, however, rather broader in its approach than deductive explanation by itself.

(iii) Workman (1964) proposes a rather different form of explanation, which it might be convenient to term explanation by way of analogy, or, as it is sometimes called, *model* explanation. Given the confusion that surrounds the term 'model', however, we will avoid it here. Workman begins by noting how an unexpected occurrence can be made less unexpected by developing some 'picture' of events in such a way that the unfamiliar becomes or seems more familiar. The explanation thus contains a description of something that is not observed but which may, for example, be obtained by analogy. This description provides adequate predictions and can be shown to be true in the sense that it does not lead to contradictions. In these circumstances we may use analogy to render something that is difficult or curious to comprehend into something with which we are reasonably familiar. We may thus represent atoms by billiard balls, complicated chemical structures by physical models, and in the process satisfy our demand for an explanation.

These three approaches to explanation are not exhaustive of all modes of explanation, and in many respects they are not mutually exclusive. Indeed it could be argued that the relaxation of many of the principles of deductive explanation in the work of Hempel (1965) has brought deductive and relational explanation very close together, and that the differences are of degree and not of kind. Similarly it is difficult to differentiate between explanations by analogy (or *via* models) and explanations that make reference to abstract theoretical concepts. We shall not, therefore, argue strongly in favour of any one explanation structure, but generally accept all of them as valid approaches to constructing an explanation in given circumstances. We shall tend in forthcoming chapters, however, to make somewhat greater use of the deductive form, mainly because of its simplicity.

C The criteria for judging whether an explanation is satisfactory and reasonable

We may set up a generalisation of the sort that 'God determines everything' and thence conclude that the rank-size rule is a manifestation of God's will. Some would be satisfied with such an explanation and find nothing inherently surprising save, perhaps, the mysterious way in which God moves 'His wonders to perform'. Others might deny that belief in God or any tenet of religious faith has anything to do with the process of providing a reasonable explanation of an event. The criteria set up to judge whether or not a particular explanation is reasonable and satisfying are highly subjective, and there can be no denial of this fact. Science itself has tended to get round this problem by setting up conventions—rules of behaviour for the scientist—to which he must conform if he is to be seen to be explaining in a reasonable manner. The scientific method is nothing more than the explicit development of these rules. By setting up such rules and conventions, a community of practising scientists develops a norm by which to judge whether or not a particular explanation is reasonable. Modern statistical decision-theory, for example, provides a set of objective decision-rules which, if followed properly, ensure that different scientists will arrive at the same decisions with respect to a given hypothesis and set of data.

But the conventions and rules developed by science vary from place to place and from time to time. One school of geographers may regard the explanations offered by another as unacceptable and unreasonable. A regression analysis (r = ·9) may satisfy some, may be totally unsatisfactory to others. What is acceptable at Cambridge University, might not be acceptable at Bristol, and what is acceptable at Berkeley might not be acceptable at Northwestern. The conventions of science change quite substantially over time also. What was a satisfying and acceptable answer to a question in the late nineteenth century (to, say, Ratzel and Semple) is no longer accepted as a satisfying answer in the mid-twentieth century. Within particular traditions or conventions, it is possible to judge objectively whether or not a particular explanation is satisfying or reasonable. The problem is that we must needs choose between different conventions and this choice is subjective. It amounts to choosing what T. Kuhn (1962) terms a 'paradigm' for the practice and conduct of science. It is worth examining this notion of scientific paradigms in somewhat greater detail.

D On paradigms and world images

T. Kuhn (1962, x) interprets the term 'paradigm' as:

... universally recognized scientific achievements that for a time provide model problems and solutions to a community of practitioners.

This concept of the paradigm is useful to us because it expresses something rather important about explanation as a *process* and as an *activity*. In particular it brings together two aspects of explanation which cannot be divorced from an understanding of the behaviour of the investigator, namely, the questions which he asks and the criteria he sets up to judge whether or not a given explanation is reasonable and satisfying. Kuhn (1962, 37) puts it this way:

one of the things a scientific community acquires with a paradigm is a criterion for choosing problems that, while the paradigm is taken for granted, can be assumed to have solutions. To a great extent these are the only problems that the community will admit as scientific or encourage its members to undertake. Other problems, including many that had previously been standard, are rejected as metaphysical, as the concern of another discipline, or sometimes as just too problematic to be worth the time. A paradigm can, for that matter, even insulate the community from those socially important problems that are not reducible to the puzzle form, because they cannot be stated in terms of the conceptual and instrumental tools the paradigm supplies. Such problems can be a distraction, a lesson brilliantly illustrated ... by some of the contemporary social sciences. One of the reasons why normal science seems to progress so rapidly is that its practitioners concentrate on problems that only their own lack of ingenuity should keep them from solving.

To Kuhn, most scientific activity consists in seeking for solutions within a generally accepted (but often unspecified) set of rules and conventions. This activity is called 'puzzle-solving' and it is characteristic of what Kuhn terms 'normal science'. This activity is occasionally interrupted by a 'scientific revolution'. This revolution is seen as the response to a crisis generated by the accumulation of more and more problems which cannot be solved by reference to a prevailing paradigm. Confronted with prolonged and severe anomalies between results and expectation, the scientist searches for a new kind of explanation. Speculation comes to the fore, methodological issues are thoroughly debated, philosophy is invoked, novel experiments are thought up, until some new paradigm emerges which has characteristics which help solve the anomalies that before existed. As Kuhn (1962, 77) points out, however,

> The decision to reject one paradigm is always simultaneously the decision to accept another, and the judgment leading to that decision involves the comparison of both paradigms with nature *and* with each other.

The substitution of one paradigm for another is not a matter that can be settled entirely by reference to logic or experiment. It is a matter of judgement, an act of subjective choice, an act of faith, which can, it is true, be backed up by substantive evidence from logic and experiment.

Kuhn's analysis is interesting from two points of view. In the first place it provides a conceptual framework for examining the history of scientific endeavour; and the history of geography could well be treated in such a manner. Such a history would help us to understand why we now tend to reject as intractable, many of the problems and questions posed by, say, Ratzel or Griffith Taylor. It seems that our conceptual apparatus, our current paradigm, cannot extend to the consideration of such questions. The shifting geographical focus of political power traced by Griffith Taylor and projected into the future, seems to be a question that is metaphysical rather than a question that can be given a rational answer. In the second place, Kuhn's account provides us with some profound insights into the behaviour of scientists, into science as an activity. It helps us to understand the nature of paradigm conflict, the problems we face in choosing one mode of puzzle solving rather than another. The qualitative-quantitative dichotomy which some see in geography at the present time might well represent a conflict in paradigms—a conflict that has nothing to do with whether or not one employs a chi-square test or calculates a regression line, but which is symptomatic of a conflict between two different views of what are tractable and interesting geographical questions, and what are satisfying and reasonable answers. Chorley and Haggett (1967) thus regard the so-called quantitative movement as but a symptom of the search of many geographers for some new paradigm—a search process which, it must be admitted, this book is designed to promote. In seeking for some new paradigm, however, it is perhaps useful to remember that we may be required to reject traditional and interesting problems simply because we have not the necessary apparatus to handle them. A paradigm may provide an extraordinarily efficient way of solving problems, but in general it buys concentrated effort at the cost of sacrificing comprehensive coverage.

In transforming allegiance from one paradigm to another, the scientist also transforms his own behaviour. Questions which at one time did not arise, now do so because his expectations have changed. Experiences that seemed irrelevant now seem surprising and demand explanation. This general change in expectations involves a shift in the scientist's perceptions of the world around him. It involves a change in what Kuhn terms the 'world view' of the scientist. His reactions to experience thus change and his conceptualisation of that experience also changes. This shift in what Boulding (1956) calls the image of the investigator with respect to the world of experience is, again, an aspect of behaviour in the scientific community which has tremendous implications for the prosecution of the scientist's endeavour. It must again be emphasised, therefore, that the objectivity of science and the judgement of what is a relevant question and what an acceptable answer can only be understood in the context of this prevailing image, in the context of the prevailing rules and conventions, in other words, in the context of diverse and often conflicting paradigms which themselves reflect and result from diverse behaviours, value systems, and individual philosophies.

II EXPERIENCE, LANGUAGE AND EXPLANATION

A particular paradigm involves a particular image of the world and a particular interpretation of perceptual experience. This, as we have already seen, is a complex philosophical issue that has been the subject of speculative metaphysical argument through the ages. I do not intend to become involved in a discussion of problems in the philosophy of perception and language, yet, as will become apparent subsequently, it is impossible for the geographer concerned with substantive issues to avoid such complex problems. It will be useful, therefore, to develop a simple conceptual framework for considering these issues which does not presuppose too much with respect to the philosophy of perception or the philosophy of language. Since this book is primarily concerned with rational explanation (with scientific explanation interpreted in its broadest sense), we shall develop this schema with respect to this particular style of explanation.

Caws (1965, 69) writes:

> Starting from perception, which is regarded as a gross smoothing over of the surface of reality (owing to the crudity of our sense organs), science proceeds to the construction of conceptual schemes whose order reflects the order of perception, and links these with specialised languages for the purpose of making predictions.

We may therefore think of a set of connections running from sense perceptions (*percepts*), through mental constructs and images (*concepts*), to linguistic representations (*terms*). 'Percepts', 'concepts', and 'terms', cannot be regarded as truly isomorphic. They possess, in some degree, an independent existence. This independence is difficult to acknowledge and is itself a major area of philosophical controversy. Cassirer (*1957*, 112), for example, writes:

> even for man it is evident that long after he has learned to live in images, long after he has completely implicated himself in his self-made image worlds of language, myth, and art, he must pass through a long development before he acquires the specific *consciousness* of the image. In the beginning he nowhere distinguishes between the pure image plane and the causal plane; over and over again, he imputes to the sign not a representative function but a definite causal function, a character not of signification but of efficacy.

The transformation from percepts to words requires, therefore, that we understand in some way the relationship between them, since

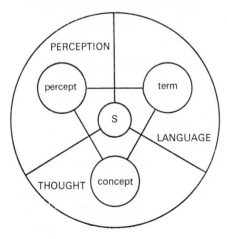

Fig. 2.1 A diagrammatic representation of the relationship between percepts, concepts and terms (*after Caws*, 1965).

only through such understanding are we able to discuss the relationship between man himself and the reality he is seeking to know. Cassirer (*1957*, 282) suggests that

the moment man ceases merely to live in and with reality and demands a knowledge of this reality, he moves into a new and fundamentally different relationship to it.

The relationships between percepts, concepts, and terms, have been schematically represented by Caws (1965; see Figure 2.1). We may represent the three domains of perception, thought, and language, by three segments of a circle. In the centre a small circle (S) represents what Caws calls 'the subjective pole of experience' which represents a vantage point from which perception, thought, and language, can be surveyed. Outside of the large circle lies everything unperceived, unheard of, and unthought of—'in other words, most

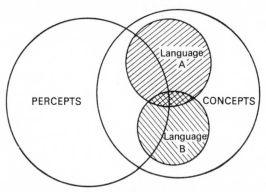

Fig. 2.2 A diagram to show how two rather different languages may be developed within the same context of perceptual experience and conceptual development. The languages have only a small area of overlap and hence only a few terms can be translated from one language to the other (*after Lenneberg*, 1962, 107).

of what there is.' Within each segment of the large circle, we may distinguish an area of percepts which can be related to terms, an area of concepts that may be related to terms, and so on. But there are large areas of conceptual thought that cannot be represented by terms, many terms that have no relation to percepts, and so on.

Any particular language will succeed in representing only a limited number of concepts and percepts and it is quite likely that different languages will represent rather different sets of concepts and percepts. This kind of situation is shown in Figure 2.2. Different languages therefore vary in their capacity to convey information: English is thus sometimes regarded as the language of empiricism,

French the language of rhetoric. Certainly different languages represent different concepts and reflect different perceptual experiences. Fortunately the vast rambling ambiguity of natural languages allows a tremendous flexibility in their use and an ability to represent a vast range of thought and experience. The vastness of this range should not, however, blind us to the fact that it is still only a restricted subset of actual thought and experience. To some extent also, there is a feedback effect; 'learning a particular language may induce a characteristic way of dividing up or "segmenting" what presents itself in perception' (Caws, 1965, 33). The relationship between the semantic structure of a language and our ability to perceive has been the subject of a continuing debate (Segal *et al.*, 1966) but it seems likely that to some extent (as yet undetermined) natural language systems censor sense-perception messages and affect our ability to perceive—perhaps, after all, the English are incapable of achieving the heights of grand passion because theirs is the language of empiricism! But even if perception is unaffected by its linguistic context, the ability to communicate our ideas drawn from perception depends upon the formation and acceptance of some common language. Science itself possesses many specialised languages—artificial languages, such as mathematics provides in abundance. These special languages are characteristic of what Boulding (1956, 15) calls the *subculture* of science. It is useful, therefore, to consider some of the general characteristics of such special artificial languages.

Any system of interpreted signs may be regarded as a language. Natural languages, such as English, French, and Japanese, contain plenty of abstract terms, terms which generalize, which subsume experience in classes and typologies, which represent abstract concepts of thought. Artificial languages further this process of abstraction and attempt to exert a firm control over the way in which the language itself operates. The sign, says Cassirer (*1957*, 333) 'tears itself free from the sphere of things, in order to become a purely relational and ordinal sign.' If the sign is to fulfil a scientific task, it must assume a fixed and determinate form unaffected by ambiguity of interpretation or function. It must, therefore, 'disengage itself from intuitive existence far more energetically than was the case in the sphere of language.' The result is the creation of an abstract system of signs and relata which have no empirical content or substantive meaning. As such, artificial languages 'signify a monstrous impoverishment.' But what such a language 'lacks in closeness to life and individual fullness it makes up for by its universal scope and validity. In this universality national as well as individual differences are annulled.'

The particular strength of such artificial languages is that the symbols they use are free from ambiguity and are precisely defined by their function within the language itself. Artificial languages are thus internally consistent and unambiguous although as we shall see in *Part IV* the empirical interpretation of these artificial languages is far from unambiguous or precise. It is resort to such artificial language systems which gives science much of its power of objectivity and universality. Mathematics is usually regarded as the language of science. In geography we make use very often of a rather special artificial symbolic system—namely the map—in order to convey and communicate information as unambiguously as possible. This is not to say that maps or mathematics are free from ambiguity in interpretation. They are (or at least should be) free of internal ambiguity. This power of artificial language system in the subculture of science is, however, bought at a price:

The price is a severe limitation of [the] field of enquiry and a value system which is as ruthless in its own way in the censoring of messages as the value system of primitive man. Messages that will not conform to the subculture are condemned as illusion. (*Boulding*, 1956, 71.)

Part and parcel of the paradigm which a scientist accepts, therefore, is a particular language which is restricted in the range of experience to which it can refer, but which can be used powerfully and unambiguously within that domain.

The relationships between percepts, concepts, and terms, are complex. Explanation is usually regarded as being communicable to others, and we may therefore think of it as taking place through the medium of language. Yet what we are trying to explain, in the empirical sciences at least, are events and experiences recorded *via* our perceptions. In examining the characteristics of explanation, therefore, it is occasionally useful to remind ourselves that we have already filtered out a vast amount of information by way of the translation from percepts and concepts into language itself. In discussing methodological issues in explanation we presuppose an adequate translation into language to have been already made. But in presupposing this, we are presupposing much of a philosophical nature. The power of methodology, however, is rather like the power of an artificial language system. It is possible to set forth rigorously the procedures to be followed in producing an explanation. The application of these procedures to actual substantive problems, in geography for example, requires that we take cognisance of the complexities of the percept-concept-term relation.

III EXPLANATION AS AN ACTIVITY

This chapter has been concerned with looking at explanation as an activity indulged in by real people who possess value systems and who do not shirk from referring to those value systems when making choices and decisions regarding the questions that generate explanations and when judging the worth of such explanations. In the chapters that follow we shall take a more formal approach to the problem of explanation and attempt to discuss the logical form of explanation, rather than the behavioural issues raised in this chapter. The formal analysis of explanation—such as that provided by Braithwaite, Hempel, Nagel, and many others—can provide us with deep insights into quite complex methodological problems. But in the application of these formal procedures to substantive geographical problems certain issues arise which cannot be solved without resorting to a much broader interpretation of the activity of explaining. We shall therefore have occasion to refer to the behavioural aspects of explaining in future chapters. The structure of this book, therefore, amounts to the development of a hard inner core of methodology—the analysis of explanation as a formal procedure—and a rather more general outer zone concerned with philosophy, speculation, perception, images, and the like. In this chapter we have examined some of the very general characteristics of this outer zone. The presentation is, in many respects, unsatisfactory, but at least it gives us something to hang on to as we progress towards the hard core of analytic methodology.

Basic Reading
Caws, P. (1965), chapter 10.
Boulding, K. E. (1956).
Kuhn, T. S. (1962).

Reading in Geography
Lowenthal, D. (1961).

PART TWO
The Methodological Background and Explanation in Geography

Chapter 3
Geography and Science—the Methodological Setting

The relationship between geography and other academic disciplines has never been easy to define. Geographers frequently form disparate allegiances. Some regard themselves as being in the natural science tradition, some involve themselves more in the work of the social sciences, while others associate with the humanities and in particular with history. Different national groups tend to develop rather different traditions in this respect. The French have traditionally maintained close contact with history, the British with geology, and so on. Within the various national schools of geography there are major differences. Thus in American geography the Berkeley school, under the inspiration of Carl Sauer (1963), look to the anthropologist, the quantifiers look to behavioural science, or to mathematics, while the geomorphologists look, not unnaturally, to geology and physics for support in their quest for explanation. At times geographers have appeared to reject *any* allegiance and have sought refuge entirely within the boundaries of their own discipline. At other times they have taken an extraordinarily broad view and come to regard themselves as the synthesisers of all systematic knowledge in terms of space.

Whatever the view, geographers have been open to influence by the methodological attitudes developed in neighbouring disciplines. At times the influence has been quite specific, at others it seems that faint echoes of the current *Zeitgeist* have been heard and responded to by geographers working in comparative academic isolation. Any methodological work, therefore, should take account of these influences and attempt to measure their impact and significance.

In general we may distinguish three sources of influence. One lies in the natural sciences where physics has forged a powerful paradigm of scientific explanation. Another comes from the social sciences, although the message from this source is less clear. Finally, history has provided a major influence upon geographical thinking.

It would be wrong to assume that within each of these groups of

subjects there is complete agreement as to the appropriate methodology. Indeed, there are substantial disagreements within disciplines.
It is thus rather unfortunate that when philosophers and logicians
discuss *scientific* explanation they tend to discuss explanation in
physics almost exclusively. This is mainly due to the impact of practising physicists, such as Heisenberg, Born, Frank, and Bridgman,
upon philosophers through their writings on their specific problems
of explanation. But many natural scientists have resisted the version
of scientific explanation derived from physics. Smart (1959), for
example, claims that biology cannot hope to subscribe to such a form
of explanation, simply because it is inappropriate for the kind of
subject-matter that the biologist is concerned with. He then suggests
that Woodger's (1937) attempt to provide an axiomatic foundation
for biological research is rather misguided. Geologists have similarly
demurred. Simpson (1963, 46) takes strong exception to the way in
which philosophers of science have concentrated on physics 'to give
a distorted, and in some instances quite false, idea of the philosophy
of science as a whole'. He thus regards geology as being very different
from physics and chooses to ally that discipline with what he calls
the *historical* sciences in general. 'Historical events,' Simpson (p. 29)
states, 'are unique, usually to a high degree, and hence cannot
embody laws defined as recurrent, repeatable relationships.' The
methodology of geology, if we accept this view, has more in common
with history than it has in common with physics. But the view has
not gone unchallenged. Watson (1966) has challenged Simpson's
view very strongly, objecting to what he terms the 'tangle of ambiguities' by which Simpson seeks to show that 'at least part of geology is logically "different in principle" as a science from such "non-
historical" sciences as chemistry.' In Watson's view Simpson only
succeeds in showing that 'due basically to empirical circumstances
in, e.g., geology, methodological techniques and formal results are
often different from (and weaker than) those in, e.g., chemistry.' This
argument, it turns out, is common to history, and various social
sciences, as well as geography.

It is not, however, the intention to enter into substantive controversy
at this juncture. The essential point is that there is no generally
accepted structural form of explanation that bestrides all the natural
sciences. There are even more serious and more widespread differences of opinion in the social sciences. There are those who search
for a fundamental understanding of human behaviour by *introspection*.
Sociologists and historians have frequently sought explanation
(or rather understanding) by empathy or *verstehen* as the method is
sometimes called. Even classical economics and Freudian psycho-

logy rely upon the intuitive reasonableness of their theories and not upon empirical verification. By contrast, there are those, such as the behavioural psychologists and the econometricians, who proceed by *observation* and *measurement* and who rely upon direct empirical evidence for the confirmation of hypotheses. This latter group probably have more in common with the physicist than they do with the historian as regards their methodological thinking.

Given these contrasts even within disciplines, a complete discussion of the methodological background to geographic thought becomes impossible. Yet there are a number of common themes in the arguments regarding an appropriate methodology. Smart's (1959) argument regarding biology, and Simpson's (1963) arguments regarding geology, have a very similar ring to those of the historian resisting the notion of laws in history. The idea of uniqueness is common to history, geology, geography, sociology, and so on. These arguments need to be examined somehow. We shall therefore begin by examining what might be called the *standard model* of scientific explanation. This is the methodological system which philosophers and logicians have derived from a study of explanation in the natural sciences—particularly physics. We shall ignore the deviant view in the natural sciences, but discuss the problems involved in the application of this *standard model* to explanation in the social sciences and history, for it is here that controversy has been most vigorous and the issues most clearly stated. By such means it is hoped that the central issues of methodological controversy in *all* sciences will become clearer. It is of course against this background that geographers have worked and developed their own methodological opinions. What will become clear, however, is that there is little, if anything, in the way of methodological controversy in geography that is not fully covered in other disciplines. We can expect, therefore, to derive a great deal from a look at methodological controversy elsewhere, particularly since the arguments are often far sharper and far clearer in other disciplines than they are in our own.

Chapter 4
Scientific Explanation—the Model of Natural Science

What might be termed the 'standard' model of scientific explanation has been derived from an analysis of explanation in the natural sciences in general and in physics in particular. This standard model has a number of important characteristics as well as a number of important limitations with respect to the questions that can be asked of it and its ability to provide answers. In spite of these limitations, it has been generally claimed that this standard model (or something that closely approximates to it) provides 'the only equipment so far invented for discovering empirically true statements about the world' (Caws, 1965, 68). Within its limits of operation the standard model of explanation has been extraordinarily effective with respect to the puzzles it has solved and the efficiency with which it has solved them (T. Kuhn, 1962, 165). It would be wrong to suppose that our understanding of the real world rests entirely on our ability to provide scientific explanations of events, but science nevertheless provides us with the most consistent, coherent, and empirically justified, body of information upon which to base such understanding. It is important, therefore, to evaluate the means by which science has managed to accumulate such a body of information.

It is useful to begin by considering the general aim and purpose of scientific enquiry. This aim is, according to Nagel (1961, 15),

to provide systematic and responsibly supported explanations . . . for individual occurrences, for recurring processes, or for invariable as well as statistical regularities.

Braithwaite (*1960*, 1) similarly states that the aim of scientific explanation

is to establish general laws covering the behaviour of the empirical events or objects with which the science in question is concerned, and thereby to enable us to connect together our knowledge of the separately known events, and to make reliable predictions of events as yet unknown.

The scientific method really consists of a set of prescribed rules of behaviour in the pursuit of such an aim. Logicians and philosophers of science have attempted to elucidate these rules, but it has proved difficult to model the scientific method in detail. This is in part because there are three distinct aspects to be considered. The first concerns what has been called 'the context of discovery'. Here we seem to be mainly in an area of activity governed by the intuition of the scientist, although recently there have been attempts to suggest some rough search procedures which appear to be efficient in generating hypotheses and new ideas. The second aspect concerns the way in which the scientist is required to call into play a number of different procedures in the process of providing responsible support for his conclusions. The third aspect is the way in which he states these conclusions and the way in which conclusions are linked together to form a coherent and consistent body of organised knowledge.

It is unfortunately dangerous to confuse these three aspects of scientific thinking and yet difficult to separate them. It would be erroneous to suppose, for example, that the procedures which a scientist quotes to support a conclusion were the same procedures which led him to that conclusion. Gilbert Ryle (1949, 275) has amusingly caricatured the 'tinge of unreality' about applying the rules of scientific investigation to the scientist's mind:

> How many cognitive acts did he perform before breakfast, and what did it feel like to do them? Were they tiring? Did he enjoy his passage from his premises to his conclusion, and did he make it cautiously or recklessly? Did the breakfast bell make him stop short halfway between his premises and his conclusion?

The actual processes going on in a scientist's mind are a matter for the psychologist to study. But science itself is essentially discussion in language not in thought. We therefore primarily discuss scientific method in terms of the linguistic procedures which we use to state conclusions and to support them. Analytic philosophy, which has been deeply concerned with linguistics in the last three decades or so, has done much to further our understanding of the language forms which the scientific method uses.

We may regard the statements which science makes about the real world as being ordered in a consistent hierarchy. The lowest-order statements we might call *factual* statements, the intermediate statements we might call *generalisations* or *empirical laws*, and the highest-order statements we might call *general* or *theoretical laws*. The achievement of such an inclusive system of explanation involves linking statements of extreme generality to statements of lesser generality and, ultimately, to sense-perception data (Nagel, 1961, chapter 3; Braithwaite, *1960*, 9-21). The kind of hierarchy of statements which has evolved in physical theory is demonstrated in Figure 4.1, taken from Kemeny (1959, 168).

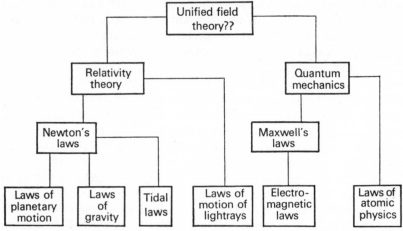

Fig. 4.1 A simplified hierarchical structure of scientific laws (*after Kemeny*, 1959, 168).

A The routes to scientific explanation

Although scientific explanation is essentially the result of what Koestler (1964) has called 'the act of creation', which in itself is difficult or perhaps impossible to explain, there are a number of ways in which we may argue in favour of accepting certain conclusions and giving them the status of scientific law. The subject of what forms an appropriate procedure is contentious and has undergone considerable revision as scientific explanation has become more precise and sought to cover itself more certainly against false inference. Generally speaking there are two alternative routes to be followed in establishing a scientific law. The first is by *induction*— proceeding from numerous particular instances to universal statements—and the second that of *deduction*—proceeding from some *a priori* universal premiss to statements about particular sets of events. The use of these terms—'induction' and 'deduction'—in this context invites confusion however, since they are also used in discussing the logical form of scientific statements. We shall therefore abandon them in the following exploration of the routes to scientific explanation.

(1) *Route 1*

Sense-perception data provide us with the lowest level information for fashioning scientific understanding. This information, when transformed into some language, forms a mass of poorly ordered statements which we sometimes refer to as 'factual'. It is partly ordered by the use of words and symbols to describe it. Then, by the processes of definition, measurement, and classification, we may place such partially ordered facts into groups and categories and therefore impose some degree of seemingly rational order upon the data. In the early stages of scientific development such ordering and classification of data may be the main activity of science, and the classifications so developed may have a weak explanatory function. Thus Braithwaite (*1960*, 1) states that

if the science is in an early stage of development . . . the laws may be merely the generalisations involved in classifying things into various classes . . . to classify a whale as a mammal is to assert the generalisation that all infant whales are provided with milk by their mothers, and this proposition is a general law, although of limited scope.

From further study of the interaction between classes and groups of phenomena, a number of regularities may be revealed. A regular association between two classes of event may suggest an empirical law. The whole collection of empirical laws so established may then constitute a body of knowledge which can be used to explain. The status of empirical laws established by such a route is a matter of some controversy. It should be noted that each step along this route so far involves inductive inference. Thus laws established by this route alone are sometimes called *inductive laws* (see Figure 4.2). Some maintain that *inductive laws* cannot be accorded the status of scientific law. This problem will be examined later (below, pp. 100–6), but for the moment we need merely note that one of the important criteria for establishing if a statement qualifies as a scientific law or not is the relationship of that statement to a surrounding structure of statements which constitute a coherent and consistent theory. If, therefore, the procedure of investigation is terminated at the point where inductive laws are established, it would seem that such laws at best can only weakly qualify as truly scientific laws. It is, however, unusual to terminate investigation at this point. Using the empirical regularities discovered, we may attempt to unite a number of them into some unified theoretical structure. Such a procedure involves transforming empirical statements into postulated universal laws, and attempting to link the laws governing different kinds of events by developing theoretical laws which succeed in predicting the empirical regularities which have been observed. In the process new empirical regularities may be predicted and these can be checked as further evidence for the effectiveness of the theoretical system proposed.

This route to scientific explanation (see Figure 4.2) coincides with the classic image of 'how a scientist proceeds', as suggested by

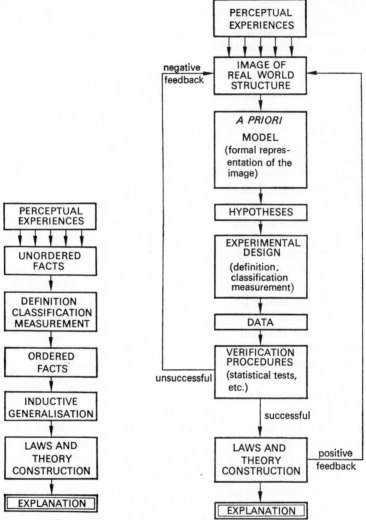

Fig. 4.2 The 'Baconian' Route to Scientific Explanation (*Route* 1).

Fig. 4.3 An Alternative Route to Scientific Explanation (*Route* 2).

Francis Bacon. It does not describe how the scientist in fact proceeds, but it does describe one of the ways in which a scientist might describe his actions so as to meet with the approval of other scientists. The defect of this particular structure, however, is the assumption that the processes of ordering and structuring data are somehow independent of the theory ultimately constructed. In his critique of this, the popular view of the scientific method, Churchman (1961,

71) has argued that 'facts, measurements and theories are methodologically the same'. Applying an *a priori* classification system to a set of data may thus be regarded as an activity similar in kind to postulating an *a priori* theory.

(2) *Route 2*

The second route (see Figure 4.3) whereby we may justify scientific conclusions clearly recognises the *a priori* nature of much of scientific knowledge. It firmly rests upon intuitive speculation regarding the nature of the reality we seek to know. At its simplest, this involves some kind of intuitive 'picturing' of how that reality is structured. Such *a priori* pictures we shall later identify as *a priori* models. With the aid of such pictures we may postulate a theory. That theory should have a logical structure which ensures consistency and a set of statements which connect the abstract notions contained in the theory to sense-perception data. The theory will enable us to deduce sets of hypotheses which, when given an empirical interpretation, may be tested against sense-perception data. The more hypotheses we can check in this fashion, the more confident we may feel in the validity of the theory provided, of course, that the tests prove positive. In the process of elaborating or seeking to test a theory we may resort to another kind of model—an *a posteriori* model—which expresses the notions contained in the theory in a different form—say in mathematical notation. In some circumstances model-building may here amount to developing an experimental-design procedure, and a primary function of this procedure is to lay down the rules whereby we may define, classify, and measure the variables which are relevant for testing the theory. By using such experimental designs we may amass evidence to confirm the hypotheses contained in the theory. But we can never prove an individual hypothesis in an absolute sense. All we can do is to establish a certain degree of confidence in the theory. Statements contained in a theory which commands considerable support we may call scientific laws. The difference between a hypothesis and a scientific law may be regarded, thus, as a matter of degree of confirmation or degree of confidence.

We may view scientific knowledge as a kind of controlled speculation. The control really amounts to ensuring that statements are logically consistent and insisting that at least some of the statements may be successfully related to sense-perception data. The procedures involved in this second route to scientific explanation are extremely complex. In later sections, therefore, we shall explicitly consider the nature of theories, laws, and models, and examine some of the procedures which a scientist calls into play in order to ensure the utility and validity of his theory. This presentation amounts, therefore, to a quick sketch of what is to come.

One major problem may be dealt with here. As we have already suggested, there are two types of logical inference involved in scientific methodology—deduction and induction. Both of these are important, but they function in rather different circumstances. We shall therefore examine their nature and function.

B Deductive and inductive forms of inference

The axioms, laws, and explanations offered by science require some logically sound method of inference to be effective. Most writers on the scientific method have thus stated that the appropriate logic is that of deduction. Thus the 'view that scientific explanations must always be rendered in the form of logical deduction has had wide acceptance' (Nagel, 1961, 29). Braithwaite (*1960*, 12) also characterises the systematic organisation of scientific knowledge as a *hypothetico-deductive* system:

> A scientific system consists of a set of hypotheses which form a deductive system; that is, which is arranged in such a way that from some of the hypotheses as premisses all the other hypotheses logically follow. The propositions in a deductive system may be considered as being arranged in an order of levels, the hypotheses at the highest level being those which occur as premisses in the system, and those at the lowest level being those which occur as conclusions of the system, and those at intermediate levels being those which occur as conclusions of deductions from higher level hypotheses and which serve as premisses for deductions to lower level hypotheses.

The advantage of deduction as a form of inference is that if the premisses are true then the conclusions are necessarily true. If, therefore, we have a certain degree of confidence in a set of premisses we may possess the same level of confidence with respect to any logically deduced consequences. This property has led to the use of deduction wherever possible. Theories are thus invariably stated as deductive systems of statements. Further, the application of such theories to the actual explanation of events is, wherever possible, rendered as a logical deduction. Thus Hempel (1965) has suggested that all explanation which aspires to be scientific should be rendered in the following manner:

$$
\begin{cases} C_1, C_2, \ldots, C_n & \text{(a set of initial conditions)} \\ L_1, L_2, \ldots, L_r & \text{(a set of appropriate laws)} \end{cases}
$$
$$\longrightarrow \text{therefore} \quad E \qquad \text{(the event to be explained)}$$

This form of explanation, which Hempel calls *deductive-nomological* involves stating a set of initial conditions and a set of laws which, when taken together, show that an event, E, must necessarily have occurred. In this form of explanation, prediction and explanation are symmetrical, and deduction ensures the logical certainty of the conclusion.

The difficulty with deductive systems of explanation is that deduction cannot, by itself, prove anything which we do not already know. There are, as Bambrough (1964) has pointed out, 'no propositions for which the ultimate reasons are deductive reasons.' Deduction has nothing to say about the truth or validity of the initial premises. Our degree of belief in the initial statements or, indeed, in the L-statements used in Hempel's model of an explanation can only be established inductively. As Carnap (1950, 2) has it,

The problem of induction in the widest sense—concerning a hypothesis of any, not necessarily universal form—is essentially the same as the logical relation between a hypothesis and some confirming evidence for it.

The application of the basic model of explanation proposed by Hempel may similarly be affected. Hempel thus proposes a model which involves *inductive systematisation*. This model involves the use of probability statements within a deductive system and we will examine its properties very shortly. When we speak of the scientific method as entailing deductive inference, we must necessarily exclude the problem of confirmation of theory as well as probability statements. In these two areas we are firmly in the field of induction.

The problem of induction has been a thorny one for many years. The essential weakness of this form of logical inference is that it is possible to draw false conclusions from correct premises. Hume made the point that because we conduct an experiment a thousand times and get the same result, we cannot infer with certainty that the next experiment conducted under the same conditions will necessarily yield the same result. For all inductive inferences, therefore, there is no logical justification for extending belief in the premises to belief in the conclusions. The failure of logicians and philosophers to find (or agree upon) such logical justification has led many to reject its use entirely in the presentation of scientific knowledge. Popper (*1965*, 29) rejects its use entirely on the grounds that it is 'superfluous and must lead to logical inconsistencies.' Attempts, such as those of Reichenbach (1949), to provide some logical foundation for induction have invariably met with strong criticism (e.g. Katz, 1962). Various apologists for the inductive method have sought to defend it on pragmatic grounds. Braithwaite (*1960*, 264) defends it because 'it yields hypotheses from which testable consequences can be deduced which are found to be true.' In the context of discovering theories there is no doubt that induction plays a key role. But Popper has argued that the peculiar hall-mark of scientific, as opposed to any other form of explanation, lies in the certainty with which the statements that make up a theory are linked together, and the certainty with which such statements may be applied to explain particular events. Any retreat from this insistence on certainty is a retreat from science. The psychological process of theory formation is not, in his view, distinctively associated with the scientific method. It is clear that the 'problem of induction' is no nearer solution today than it was two centuries ago.

It must be emphasised, however, that the methodological rejection of induction can only apply to certain aspects of the formulation of scientific knowledge. Science attempts to organise its propositions within a deductive frame of inference. In the early stages of the development of a science this aim may not be realisable, simply because we do not know enough or simply because our imagination cannot reach that far. In such a situation induction may prove important. The deductive form of scientific theories must be regarded as the end-product of scientific knowledge, rather than as the mould into which all scientific thought is cast from the very initiation of an investigation. But even assuming that a deductive theoretical structure has been successfully evolved, induction still plays an important function at certain stages in the articulation and verification of such a theoretical structure. Induction appears to retain importance in two particular respects.

(1) *The problem of verification*

The verification or confirmation of a theory *must* rely upon inductive inference. We may broadly distinguish between three schools of thought on this issue.

(a) Some philosophers of science, noting that no theory can ever be exposed to all possible relevant tests, have attempted to provide criteria for evaluating a theory in terms of the probability of that theory's being true in the light of the evidence. Such writers as Nagel (1939), Carnap (1950), and Hempel (1965) have examined this problem at great length. The aim of these writers has been to furnish a system of inductive logic which will enable the scientist to choose, as objectively as possible, between competing theories and alternate systems of explanation. Hempel (1965, 4) states:

While, in the practice of scientific research, judgements as to the confirming or disconfirming character of experiential data obtained in the test of a hypothesis are often made without hesitation and with a wide consensus of opinion, it can hardly be said that these judgements are based on an explicit theory providing general criteria of confirmation and disconfirmation.

The provision of such a 'logic of confirmation' faces two problems. The first is how to define the relevant tests for a given hypothesis, and the second is to establish the inductive rules by which we may assess the degree of confirmation which we may attribute to that hypothesis. Both problems have proved intractable and there certainly seems little hope of achieving a universally agreed-upon system of inductive logic for confirmation of hypotheses in the near future. But the progress which has been made towards solving this apparently intractable problem has yielded some important insights.

It is not difficult to regard the scientist as a decision-maker, and in fact the scientist's actions can, as Churchman (1961) has pointed out, be analysed in terms of decision theory. The first essential point is that we cannot understand the decisions of any particular scientist without understanding his value system, his goals, his objectives. In choosing one theory in preference to another, a scientist presumably tries to optimise his choice with respect to the value system which he possesses. Goals and objectives are subjectively determined. But within the context of such objectives it is possible to discuss optimising procedures. Thus the scientist cannot be understood independently of his own value system. Whether an inductive logic can be extracted from such a situation remains to be seen (see Churchman, 1961, chapter 14; Hempel, 1965, 73–96; Carnap, 1950). But it is clear that the confirmation of hypotheses has, in the modern view, a very close relationship with statistical decision theory. The details of statistical decision theory and its philosophical underpinnings will also be examined in chapter 15.

(b) Popper (*1965*) chooses to replace the notion of verification by that of falsification. In his view a theory is assumed to be true until it is shown to be false. The difficulty of this view has been pointed out by T. Kuhn (1962, 145–6):

if any and every failure to fit were ground for theory rejection, all theories ought to be rejected at all times. On the other hand, if only severe failure to fit justifies theory rejection, then the Popperians will require some criteria of 'improbability' or 'degree of falsification'.

(c) Kuhn himself proposes an even more behavioural view of verification than Churchman. He suggests that the acceptance or rejection of a hypothesis or theory is a matter of faith rather than logic. Verification and confirmation procedures are part and parcel of the rules which a scientific community accepts as part of a dominant paradigm. During periods of 'normal' scientific activity, the scientist tackles only those problems which are assumed to have a solution within the context of these rules. The anomalies that arise are the leaven for the rise of a new paradigm. But within the paradigm the rules for confirmation are firmly laid down even if they can be given no strict logical justification. Scientists could thus accept Boyle's law as proven even though there were only twenty-five observations, simply because the manner in which those observations were obtained conformed to the paradigm of scientific experiment. Certain heuristic rules govern the use of induction and these rules are established in accordance with the value system of a community of scientists. It is, according to Kuhn, a hopeless task to try to provide logical justification for such rules.

It is difficult to choose between these views on verification. The choice lies between accepting a prescribed inductive procedure for which there is no ultimate justification, or taking a behavioural view of the activity of scientists either as a group (by way of some paradigm) or as individuals attempting to maximise their decisions with respect to particular objectives. These views are not necessarily mutually exclusive. The view of the 'scientist as decision maker' entails the definition of some utility scale. Decisions cannot be made in a moral or ethical vacuum. This does not necessarily mean that the scientist 'makes his decisions for any other reason than the attainment of pure knowledge' (Churchman, 1961, 340). But it does raise the very important question of precisely what a scientist is trying to maximise when he accepts one theory in preference to another, or when he accepts a particular theory as being proven (Hempel, 1965, 73–79). There is not space to consider this behavioural element in a scientist's decision-making in any great detail, but it is nevertheless a major factor determining our view of the nature of scientific knowledge. In particular, it gives the lie to those who claim total objectivity for knowledge accumulated by way of the scientific model; this point is of vital significance in resolving some of the difficulties which face the application of the scientific model to social and historical circumstances.

(2) *The problem of inductive statements within deductive systems*

The problem of verification is of very general significance to the nature of scientific thought and understanding. A more restricted, but nevertheless difficult, problem is that of inductive statements contained within essentially deductive arguments. It is misleading to regard deduction and induction as mutually exclusive forms of inference. Although it is generally agreed that scientific knowledge should be organised as a hypothetico-deductive system and that the laws contained in that system can best be applied by a deductive explanatory procedure, there are many occasions when inductive steps may be used within these deductive frameworks. The most important cases are those where probability statements are invoked either in the hypothetico-deductive system or in the explanation itself. But there are other cases where the initial hypotheses or the initial conditions provide only partial support for the conclusions.

(a) *The use of probability statements—inductive systematisation.* Hempel (1965) has drawn attention to the difficult logical problem of explaining either individual events or finite numbers of events by way of probability laws which refer to infinite populations. Suppose the event to be explained is 'Joe migrates.' We might attempt to explain this in the following manner:

Initial Condition 1	Joe lives in town A and earns $50 per month.
Initial Condition 2	Jobs are available in town B at $100 per month.
Initial Condition 3	The distance between A and B is 50 miles.
Law Statement	The probability that a person migrates is proportional to the extra income to be gained by migrating and inversely proportional to the distance to be moved.
Event	Joe migrates.

The initial conditions and the law statement (which would presumably be more specific in a particular case) provide some support for the conclusion that Joe must necessarily move to town B. But the support is only partial. This kind of argument Hempel calls *inductive systematisation*. The argument is basically deductive in structure but the elements contained within it do not necessarily supply the conclusion and therefore belief in the premises does not necessarily imply belief in the conclusion. The use of probability statements in the course of an explanation usually involves inductive systematisation. Similarly a probability statement in a hypothetico-deductive system may also imply that the lower-level hypotheses in the system cannot necessarily be deduced from the initial hypotheses. If, however, the lower-level hypotheses refer to infinite populations, then probability statements can be deduced from other probability statements. Thus the whole of probability theory itself may be formalised as a deductive system (see chapter 15). But probability statements frequently imply difficulty in the application of explanatory forms to events and occasionally create difficulty in the attempt to unify scientific understanding in a hypothetico-deductive system of statements.

(b) *Incomplete theoretical systems.* The application of deductive inference requires completeness in the initial conditions or the initial premisses. From the point of view of explanation itself all relevant initial conditions need to be known so that the application of the law statements ensures that the conclusion is necessary. If all the relevant initial conditions cannot be specified, then the explanation is incomplete, and confidence in the known initial conditions and the known laws cannot be transferred to the conclusion. Similarly, in organising hypothetico-deductive systems, some of the initial premisses may not be known. Many theoretical statements, as we shall later see (below, pp. 96–9) are incomplete and therefore are quasi-deductive rather than truly deductive in structure. In most disciplines theoretical statements are of this form simply because not enough is yet known for complete formalisation of theory to be possible. Most hypothetico-deductive systems are, in the initial stages of a discipline's history, riddled with inductive steps. Thus Keynes (*1962*, 241) has observed:

> When our previous knowledge is considerable . . . the purely inductive part of the argument may take a very subsidiary place. But when our knowledge of the instances is slight, we may have to depend upon pure induction a good deal. In an advanced science it is a last resort—the least satisfactory of the methods. But sometimes it must be our first resort, the method upon which we must depend in the dawn of knowledge and in fundamental enquiries where we must presuppose nothing.

But at this point we come back to the important role of induction in the 'context of discovery'. We may conclude from this general survey, however, that scientists seek to organise knowledge by way of a hypothetico-deductive system and that they seek to apply that knowledge by way of deductive explanations. But there are many instances where the ideal deductive form of argument cannot be applied. Some instances arise because of lack of information, incomplete understanding, or because of the need to use probability statements. The most important case of all is in confirming and verifying scientific theories. In all these instances inductive inferences are important. This, of necessity, detracts from the certainty which science seeks in stating and applying scientific knowledge. In the case of confirmation theory it seems that some behaviouristic elements have to be included.

Given the aura of 'pure objectivity' which surrounds the discussion of scientific research and the 'sacred-cow' attitude taken by many to scientific understanding, it is important to note the many logical difficulties which face scientific investigation and in particular to note that scientific decisions cannot be made without behavioural constraints being introduced. The standard model developed by science has been extraordinarily efficient in organising and promoting our knowledge of the world around us. It remains to be seen if there are any other systems of explanation which provide a realistic alternative to the scientific model or if there are any insuperable barriers to the application of the scientific model to social and historical circumstances.

Basic Reading
Caws, P. (1959).

Further Reading
Braithwaite, R. B. (*1960*).
Nagel, E. (1961).

Chapter 5
Problems of Explanation in the Social Sciences and History

The relationship between explanation in the natural sciences and explanation in the social sciences and history has always been a controversial subject. In the mid-nineteenth century J. S. Mill put forward the view that all explanation possessed the same logical form. 'There can be,' he suggested, 'no fundamental logical difference between the principles according to which we explain natural changes and those according to which we explain social changes' (quoted in Winch, 1958, 71).

Mill's view has been attacked, defended, restated in more sophisticated form, and still attacked, over the last century or so. In some areas of social science, such as experimental psychology and econometrics, Mill's thesis has apparently been accepted, whereas in other areas, such as politics, the thesis is less generally accepted. Historians and philosophers of history have also debated this issue but few, until recently, have accepted it.

Perhaps the most extraordinary thing about this controversy is the way in which the debate has consistently confused a number of separate issues and then compounded the whole matter by failing to distinguish between the various views and activities that may be attached to the term 'scientific' explanation. The social scientist or historian bent on resisting the idea of 'scientific' explanation has usually pointed to just one of the varied set of opinions extant in the natural sciences. Often the model form rejected is that of Newtonian physics—social physics and mechanistic explanation—or of Darwinian biology—evolutionary explanation. There is a similar tendency to postulate a unity of view as to what constitutes an explanation in the social sciences—a unity of view which does not and need not necessarily exist (Kaplan, 1964, 4–5). But although individual disciplines may exercise autonomy as regards explanatory form, certain norms and conventions have developed within the social sciences and history governing explanatory form. These conventions have in the past led many to suggest that scientific explanation as developed in

the natural sciences in general has little or no relevance to the problems of explanation facing the researcher concerned with social
situations.

The sense in which explanatory form differs between the natural
sciences and the social sciences and history needs some clarification.
It is useful to attempt this by employing concepts developed by
philosophers of science to discuss the language of explanation. Carnap
(1958, 78–9) differentiates between three aspects of any linguistic
system:

> An investigation which refers explicitly to the speaker of the language—no
> matter whether other factors are drawn in or not—falls in the region of *pragmatics*.
> If the investigation ignores the speaker, but concentrates on the expression of the
> language and their designata, then the investigation belongs to the province of
> *semantics*. Finally, an investigation which makes no reference either to the speaker
> or to the designata of the expressions, but attends strictly to the expressions and
> their forms (the ways expressions are constructed out of signs in determinate
> order), is said to be a formal or syntactical investigation and is counted as belong-
> ing to the province of (logical) *syntax*.

Those who wish to show that explanation is necessarily different
between natural and social science must show that the language of
explanation is *necessarily* different. Further, they have to show in
which respect or respects—*pragmatics, semantics* or *syntax*—these languages are necessarily different. We may crudely translate these
linguistic terms and suggest that we may differentiate between the
techniques which a scientist is forced to adopt in the light of his
environmental situation (pragmatic considerations), the *conceptual
content* of an explanation (semantic considerations) and its *logical
structure* (the syntax developed). Mill's view does not necessarily
imply that techniques are necessarily the same, or that the conceptual content of explanations should be similar. But it does suggest
that the same logic of justification, the same syntax, may be employed in all disciplines irrespective of whether they are dealing with
natural or social phenomena (Kaplan, 1964). Confusion among these
separate issues lies at the root of much methodological controversy.
The main concern here is with the logical structure of explanation
but in order to isolate this aspect of explanation clearly, similarities
in techniques and conceptual content will first be discussed.

I TECHNIQUES OF INVESTIGATION

Generally speaking, any disciplines will develop a set of techniques
appropriate for solving the problems with which that discipline is
concerned. In some instances the problems may be defined with

reference to particular techniques of investigation. The interaction between techniques and problems is thus complex. But to say that techniques in natural and social science are radically different

is not to become aware of anything significant about the nature of social science. It is not even obvious that techniques in social and non-social sciences differ from each other more than the techniques of non-social sciences differ among themselves (Rudner, 1966, 5).

This is important because the argument for maintaining a distinction frequently amounts to drawing a line between sciences where experimental techniques can be employed and sciences where this technique is impossible. The experimental method has played, it is true, a far greater part in establishing laws and theory in the natural sciences than in the social sciences. But there are areas of study in the natural sciences which do not employ experimental techniques. In such situations 'vicarious' experimental techniques—such as simulation—may be employed and such techniques can also be applied in the social sciences and history. The practical difficulty of employing experimental techniques does, however, raise some difficult problems with respect to the verification of theory in the social sciences and history. This issue will be examined later. By and large, differentiation between explanation in natural and social science and history on the ground that different techniques are employed may be summarily dismissed.

II THE CONCEPTUAL CONTENT OF EXPLANATIONS

Methodological differentiation on the ground that the conceptual content of explanations is fundamentally different is a far more serious issue. In part it is serious because of the strong emotional reaction of some social scientists and historians to the suggestion that the concepts involved in the laws of physics could be applied with very little modification to the study of human behaviour. In particular it stems from the strong resentment which historians and social scientists felt towards the employment of 'mechanical' explanations in connection with human affairs. The form of such explanations—the theory of clock-works, as Boulding (1956) calls them—was well understood by the mid-nineteenth century. It was the thesis of the positivist philosophers, such as Comte and J. S. Mill, that the world of phenomena is a determined world and that all that was needed in the study of human affairs was sufficient persistence and insight to uncover the mechanical laws which governed human behaviour. Reducing explanation to some variation on the laws of mechanics provoked a vigorous reaction. Historians and sociologists such as Dilthey, Weber, and Pareto, reacted against the positivism of Comte, Spencer, Mill, and Marx, while Freud began to uncover evidence of mental processes that could not easily be brought within the framework of mechanistic explanation (Hughes, 1959). The controversy over mechanistic explanation at the beginning of the twentieth century, paralleled in geography by the determinist-possibilist argument, created an intellectual legacy of great significance.

The evidence now suggests that mechanistic explanations contained in Newtonian physics and, less appropriately, in Spencer's sociology, can be applied only in limited circumstances. In the mid-nineteenth century statistical explanations were already developing in physics, although it was not until Heisenberg enunciated the 'uncertainty principle' in quantum physics that chance variation became a central concept in physics. The overthrow of the Newtonian concept of a mechanistic universe in physics is but one example of a radical change in the concepts employed by a particular discipline as it grows and develops. Similarly, there are wide differences between disciplines as regards the concepts employed. The problems of biology are increasingly being approached by way of systems, and systems analysis has become a powerful tool in social science. Ackerman (1963) has suggested that geographers should also turn to a systems framework for formulating their problems. There are many different types of conceptual framework possible in rendering explanations and it may be questioned whether it is reasonable to presuppose that only one conceptual framework does or can dominate all science. Mechanistic, genetic, and systems explanations are examples of alternative conceptual frameworks.

Different disciplines adopt different conceptual frameworks. To some, mechanistic explanations may seem appropriate, to others the more complex stochastic form may seem more suitable. It is not possible to differentiate between natural and social science in terms of the concepts adopted since there is clearly as much variation within disciplines as there is between them. The demand for methodological differentiation between natural science and the social sciences and history on the grounds of major conceptual differences cannot be sustained.

This general conclusion that differentiation is not possible on the grounds of either techniques or conceptual development does not involve denying the difficulty of transferring techniques and concepts from one discipline to another. It is obvious that the experimental techniques of chemistry cannot be transferred to a discipline studying the behaviour of complex organisations. It is not so obvious when it comes to transferring concepts developed about matter to situations dealing with human behaviour. The development of social physics —particularly the gravity model in human geography—is a case in point. Such transference of concepts and principles can be stimulating and exciting. But it does involve dangers. It involves arguing by analogy or using analogue models as an aid to explanation. This procedure will be examined later in some detail (chapter 11), but provided the method is used circumspectly and with clear understanding of the respects in which the phenomena being studied and the model used are conformable, then the method appears justifiable. It can be argued that the employment of crude uncontrolled analogies from physics to social science or from evolutionary biology to social science (as in Spencer's sociology) is so damaging that all forms of such argument by analogy should be abandoned. The supposed methodological difference between the natural and social sciences must partly be attributed to the desire of responsible social scientists (such as Max Weber) to avoid spurious arguments by analogy. Whether or not there is *real* justification for methodological separatism on the grounds of fundamental differences in logic remains to be seen.

III THE LOGICAL STRUCTURE OF EXPLANATION

There are a number of complicated issues involved in any examination of the logical form (the syntax) of explanation in the social sciences and history. The argument in its most extreme form may be found among philosophers of history and for this reason it is easiest to concentrate on the argument as it has developed in that discipline. The argument centres on what Dray (1957) calls 'the covering law thesis'.

Dray, together with most practising historians, accepts that it is the function of the historian to provide responsibly supported explanations of historical events. There are some historians who would suggest that they are concerned with rational descriptions rather than explanations but this view is difficult to sustain in a modern context. The real controversy develops, however, around the rather different interpretations given to the terms *explain* and *responsible support* in the context of the historian's work. The latter turns out to be basically a problem of technique (pragmatics) although it has important metaphysical trappings associated with it. The former is essentially a syntactical problem and it will therefore be considered in detail here.

Popper (*1965*) and Hempel (1965) have argued strongly in favour of what has come to be known as the 'covering law' model of explanation in history. Hempel states:

> The explanation of the occurrence of an event of some specific kind, *E*, at a certain place and time consists . . . in indicating the causes or determining factors of *E*. Now the assertion that a set of events—say of the kinds C_1, C_2 . . ., C_n—have caused the event to be explained, amounts to the statements that, according to certain laws, a set of events of the kind mentioned is regularly accompanied by an event of the kind *E*.

Thus to Hempel all historical explanation has the fundamental form already identified of:

$$\begin{cases} C_1, C_2, \ldots, C_n & \text{(a set of initial conditions)} \\ L_1, L_2, \ldots, L_r & \text{(a set of appropriate laws)} \end{cases}$$
$$\longrightarrow \text{therefore} \quad E \qquad \text{(the event to be explained)}$$

There is therefore no logical difference between explanation in the natural sciences and history. The alternative form (inductive systematisation) that makes use of probability statements may also be used and Hempel suggests that this form may be particularly important in history where interactions are complex and experimentally uncontrollable.

Both Popper and Hempel take the view that historians *must* use this kind of rigorous form in offering explanations. The fact that most historians do not choose to make use of this form explicitly is irrelevant to the argument. Historians imply this form of argument but the non-rigorous way in which (in Hempel's view) they offer explanations means that historians offer what Hempel calls an *explanation sketch* rather than a rigorous explanation. Nevertheless historical explanations are syntactically the same as statements developed in the natural sciences.

The argument that has developed around the Popper-Hempel view is voluminous and complicated (e.g. Donagan, 1964; Dray, 1964; Mandelbaum, 1961). A major issue embedded in this argument concerns the nature of the covering laws implied by historians. It has been claimed by many that the law statements implied are radically different from the statements developed in the natural sciences, and that they perform a rather different function. Historians do not, by and large, concern themselves with establishing general laws. It is often claimed that historians are concerned to explain *unique* events. This *uniqueness thesis* has wide support among historians. Oakeshott

(1933, 154) took the view, for example, that 'the moment historical facts are regarded as instances of general laws, history is dismissed.' Dray (1957, 45) portrays the argument as follows:

> History is different in that it seeks to describe and explain what actually happened in all its concrete detail. It therefore follows *a priori* that since laws govern classes or types of things, and historical events are unique, it is not possible for the historian to explain his subject-matter by means of covering laws. If he is to understand at all, it will have to be by some kind of special insight into particular connections.

This declaration of 'methodological independence' on the part of history came in the latter part of the nineteenth century. Windelband, Dilthey, Rickert, and other historians, chose to differentiate between those subjects which they regarded as being susceptible to the *idiographic method* (the exploration of particular connections) and those which were concerned with establishing generalisations and were *nomothetic* in character. The fundamental concern with unique events was, to them, sufficient justification for separating history from many other sciences and for regarding explanatory form itself as being radically different. In the ensuing controversy over this proposed dichotomy different disciplines were forced to take sides. It is perhaps a tribute to the strength of the German school of historiography that the idiographic view has not only dominated the practice of history until the present day, but has overlapped into other areas of social and natural science. Thus the social scientist has similarly been enjoined to accept the uniqueness thesis. Max Weber (*1949*, 80), one of the father-figures of modern sociology, has written:

> Laws are important and valuable in the exact natural sciences, in the measure that those sciences are *universally valid*. For the knowledge of historical phenomena in their concreteness, the most general laws, because they are most devoid of content are also the least valuable. The more comprehensive the validity—or scope—of a term, the more it leads us away from the richness of reality since in order to include the common elements of the largest possible number of phenomena, it must necessarily be as abstract as possible and hence *devoid* of content. In the cultural sciences the knowledge of the universal or general is never valuable in itself.

Simpson (1963) has similarly described the methodological attitude of many geologists as conforming to the uniqueness thesis (above, p. 28). The argument has not been without force in geography either, and it is perhaps no surprise that German geographers, such as Hettner, claimed that geography was an idiographic rather than nomothetic science. Given the important influence of German methodological thinking in Hartshorne's work it is hardly surprising that the uniqueness thesis is still the subject of controversy in geography.

The dominance of the idiographic method in history, so powerfully presented by writers such as Collingwood (1946) and Oakeshott (1933), has recently been challenged. Some historians have never accepted it, and there have always been those—such as Spengler and Toynbee—who have sought for generalisations in their attempt to write 'grand history'. But lately the challenge has come on logical grounds such as those offered by Hempel and Popper or from practical considerations (Barraclough, 1955; Anderle, 1960). This challenge has had as one of its main objectives a clarification of the respect in which historians may regard events as being unique.

Joynt and Rescher (1961) have examined the uniqueness thesis in detail and summarised much of the argument. They conclude that the thesis cannot, without serious modification, be sustained either by an appeal to logic or by an appeal to the practice of historians. They argue:

(1) All events may be considered unique and, therefore, there is no logical difference between the situation in the natural and in the social and historical sciences.

(2) Events are rendered non-unique 'in thought only', by our choosing to use them as examples of a type or class.

(3) There can be no doubt that the use of categories, classes and generalisations is essential to the adequate discharge of the historian's task. Indeed, 'taken together, they constitute the framework and structure of history, the setting in which the recital of particulars unfolds'.

The only sense in which historians can be interested in unique events depends upon the definition of means and ends:

> It is clear that the historian in effect reverses the means-end relationship between fact and theory that we find in science. For the historian *is* interested in generalisations and *does* concern himself with them. But he does so not because generalisations constitute the aim and objective of his discipline, but because they help him illuminate the particular facts with which he deals. (Joynt and Rescher, 1961, 153.)

Historians are therefore consumers rather than producers of general laws. Only in this sense is it possible to insist that historians are concerned with unique events. Although the historian may regard the uniqueness of a particular event as being of prime interest, he cannot, apparently, accomplish this task without employing generalisations, principles, and perhaps laws. Whether this conclusion is sufficient to show that history is methodologically distinct from physics is to be doubted. Natural scientists are, after all, frequently concerned with apparently unique events (the origin of life and the origin of the universe are good examples). The application of scientific knowledge is often directed to the apparently unique case (in engineering each dam or bridge is 'unique in a sense). The means-end distinction becomes in the long run a little blurred although there can be no doubt that the activity of the historian looks very different from that of the engineer. In the long run it is worth noting Bambrough's (1964, 100) comment that:

All reasoning, including all mathematical, scientific, and moral reasoning, is ultimately concerned with particular cases, and laws, rules and principles are devices for bringing particular cases to bear on other particular cases.

This, however, is a very 'relational' view of explanation and it would perhaps seem that this broader approach is more appropriate to history and social science given the undoubted practical difficulties with which the application of the rather narrower deductive model is faced in these contexts.

If the historian merely applies the laws developed in the social sciences to particular cases, then his function is rather like that of the engineer in the natural sciences. Popper (1952) clearly regards most historians as being rather inefficient. In particular, the kinds of law to which the historian frequently appeals appear to be rather trivial generalisations about human behaviour—too trivial, Popper (1952; 1957) maintains, to pose any serious problems to the student of sociology or psychology. Implicit in the Popper-Hempel thesis is the view that explanation in history is self-evident commentary compared with the natural or even the social sciences. The failure to develop explanation in depth is partly due to the failure to use the power of deductive logic to make general statements which are consistent and powerful. Hempel (1965, 236) freely admits that the historian faces considerable difficulty in formulating such generalisations or laws 'with sufficient precision and at the same time in such a way that they are in agreement with all the relevant empirical evidence available.' Under such conditions it is hardly surprising that the historian is forced to offer 'explanatory sketches' rather than fully fledged explanations.

Such a weak explanatory form is not uncommon in disciplines in the early stages of development; for incomplete, partly formalised, and scarcely verified, theory is almost all that a discipline can expect in the initial stages. To resist the development of theory seems, however, to be self-defeating. Thus Anderle (1960, 40–1) has commented:

Though the canonisation of the idiographic method is supposed to establish the independence of the historical from the natural sciences, it really just commits the former to an earlier stage of the latter. Descriptive historiography is not a new science with an independent method but just an antiquated form of natural science.

Such assertions have not passed unchallenged. Part of the dispute concerns the nature of the 'covering laws' which historians are supposed to refer to in offering explanations. In the natural sciences laws are supposed to be unrestricted and universal statements; clearly historical generalisations are not valid for all times and all places and they are not therefore universal or unrestricted statements (Joynt and Rescher, 1961, 157). The generalisations which the historian refers to are therefore *mere* generalisations and cannot by the canons of scientific explanation be regarded as laws. This is a serious issue which will be considered in some detail later (below, pp. 100–6). The two replies to this claim are worth noting. First, the generalisations used by historians might be regarded as first-stage empirical generalisations which might, at some later date, be subsumed under more sophisticated laws of universal validity. Second, the problem of localised generalisation is common to many other areas of natural and social science, and there are signs that the notions of universality often developed by philosophers of science are too rigorous for most purposes and therefore require some modification.

Even in the light of these two replies the opponents of the covering-law thesis have some ground for complaint, for it appears rather as if scientific explanation were being forced on history. The adjustments proposed bear, according to Dray (1964, 7)

the marks of expediency rather than principle. A theory that begins by elaborating the essential meaning of explanation *a priori*, rather than trying to discover what the practitioners of the discipline concerned themselves call explanation, is surely on weak ground when it relaxes its requirements in the face of difficulties of explanation.

Dray (1957, 39) has also argued that covering-law statements in history perform a different function in history to law statements in natural science. Such statements merely establish

the *principle of the historian's inference*, when he says that from the set of factors specified, a result of this kind could reasonably be predicted. The historian's inference may be said to be *in accordance with* this principle. But it is quite another matter to say that his explanation entails a corresponding *empirical law*.

Dray therefore maintains that the historian is fundamentally concerned with providing a rational explanation for a particular historical event and this involves establishing what was the rationalisation behind a particular decision or set of decisions. Explanation does not involve an appeal to any set of laws, but involves showing that a particular person or group of people had a rational reason for acting in a particular way at a particular time. Once these rational reasons have been discovered the historian has completed his task. Further, it is quite sufficient for the historian to use 'common sense' explanations of events in terms of the logic of the situation. Such common-sense explanations are perfectly adequate, according to Dray, and although there is no reason to deny the possibility of developing laws of any significance in history, there is no need to invoke the existence of such laws to provide perfectly adequate explanations of historical events. This view of Dray's has provoked a strong counter-argument from Hempel (1965, 469–87), but at this point the controversy becomes extraordinarily complicated.

Through this tangled web of argument and controversy in the philosophy of history a number of conclusions may dimly be perceived. These conclusions extend to those disciplines which have a strong relationship to history and look to the idiographic method:

(i) The *uniqueness thesis* cannot be sustained without substantial modification.
(ii) There can be no explanation or rational description without the employment of generalisations, classes, concepts, and principles.

Beyond this it is difficult to find any general agreement and it is only possible to summarise some of the major areas of controversy:

(iii) Whether the generalisations which historians undoubtedly employ in providing explanations qualify as laws depends upon the view taken of the function of the generalisation in historical explanation and the criteria set up to determine whether a particular statement may be regarded as a law.
(iv) Although there is no doubt that explanation in history could conform to the norms derived from natural science, there is considerable disagreement as to whether such a procedure would in any way ease the difficulty of providing explanations of complex historical events.

IV VERIFICATION—THE PROBLEM OF PROVIDING RESPONSIBLE SUPPORT FOR STATEMENTS IN SOCIAL SCIENCE AND HISTORY

The search for an independent methodology in social science and history has, in the examination so far, been based mainly on *a priori* arguments regarding the appropriateness of the scientific model of explanation for understanding social and historical events. There are, however, other arguments which appeal to the practical problems involved in verifying general statements in the context of human activity. Here essentially pragmatic problems appear to have profound philosophical repercussions. The argument here centres around two questions:

(i)   Can the study of social phenomena be objective?
(ii) If not, then what kind of justification can be found for the statements of social science and history?

There are many writers who have persistently argued that the social sciences and history cannot employ the same objective techniques as do the natural sciences. Thus Max Weber (*1949*, 80) writes:

> An 'objective' analysis of cultural events . . . is meaningless. It is not meaningless, as is often maintained, because cultural or psychic events for instance are 'objectively' less governed by laws. It is meaningless for a number of other reasons. Firstly, because the knowledge of social laws is not knowledge of social reality but is rather one of the various aids used by our minds for attaining this end; secondly, because knowledge of *cultural* events is inconceivable except on a basis of the *significance* which the concrete constellations of reality have for us in certain *individual* concrete situations.

Weber regarded all investigation of social phenomena as being dependent upon the value-orientation of the investigator and the investigated. The meaning or significance of an action could not be determined independently of the cultural background. This means that the verification of any hypothesis cannot be, as it supposedly is in natural science, objective and independent of the value-system of the investigator. A sophisticated argument along these lines has been presented by Winch (1958). He suggests that any social act is governed by rules. These rules are not independent of the concepts which we use to describe them. There is a strong interaction between the development of concepts and the development of the rules which govern behaviour in society. Given this subtle interaction it is impossible to develop any ethically neutral language with which to describe social behaviour. Further, it is impossible for an investigator to understand the rules that govern behaviour without learning the rules himself. And once the rule is learned, then the investigator's value-system is altered. Understanding in the social sciences and history, therefore, depends not on scientific investigation but on learning the meaning of social events by understanding the rules and concepts which govern these events. To Winch, therefore, objectivity is impossible and the scientific method meaningless when applied to social events.

The Weber-Winch thesis denies that the social sciences and history can achieve a sufficient degree of objectivity to allow the independent verification of hypotheses. This automatically restricts the development of theory in the usual scientific sense and restricts the kinds of question that can be asked regarding social phenomena.

This view then poses the problem of finding some alternative mode of validating statements. To Weber and Winch explanation in the social sciences involves understanding a particular event by empathy with the individual or individuals involved in the event. Putting oneself in another person's shoes—an operation sometimes called *verstehen*—provides us with the only kind of understanding which is possible regarding human behaviour. This view is common in social science and history. Collingwood (1946, 283) writes:

> For science, the event is discovered by perceiving it, and the further search for its cause is conducted by assigning it to its class and determining the relation between that class and others. For history, the object to be discovered is not the mere event, but the thought expressed in it. To discover that thought is already to understand it.

The strong relationship which has developed between this view of explanation and the uniqueness thesis forms a major prop for those who maintain that social and historical enquiry are methodologically distinct. Dray's view of 'rational explanation given the logic of the situation' has a similar ring to it.

This view restricts explanation to a statement of the intentions, reasons, motives, and dispositions, involved in a given act and such a restriction also limits the kind of question that the social scientist and historian can ask. At this level of analysis it is difficult (especially in historical situations) to envisage any other form of validation apart from *verstehen*, and given that *verstehen* forms an exclusive and exhaustive form of validation, it is difficult to envisage explanations ranging beyond a statement of intentions, dispositions, motives, and reasons. This vicious circle is difficult to break out of. But there are ways out, indicated by writers such as Brown (1963) and Rudner (1966), and it is worth while examining these.

There is no doubt that the social sciences and history are faced with very real problems associated with the variability of human behaviour and observer-involvement in the observation process. It is also true that social events may be explained in some sense by reference to motives, intentions, and the like. It is interesting to note that J. S. Mill thought that the 'laws of the mind' were high-level causal generalisations in themselves and could therefore be used to explain events in a reasonably objective manner. Winch (1958, 83) suggests that such a procedure is bound to be rule-bound for

> learning what a motive is belongs to learning the standards governing life in the society in which one lives; and that again belongs to the process of learning to live as a social being.

Brown (1963, 98) regards this as self-evidently true:

> To say, then, that social scientists depend heavily upon the notion of rule-conforming behaviour is to utter a platitude. Reference to rule-conforming behaviour is not less an explanation for that reason, but . . . a social scientist has little need to make explicit reference to a convention as a means of explaining a particular act or kind of act . . . In exactly the same way that the investigator cannot solve the problems that confront him by means of intention-explanations alone, so he cannot rest content with those referring to norms. In each case he is impelled to ask 'Why this intention . . .?' or 'Why this rule . . .?' and the chain of answers will soon enough depend on referring to neither goals nor rules for its explanatory power.

In Brown's view explanation in terms of intentions, motives, and the like, is inevitable to a certain extent, but these items enter into explanations as factual statements. They form the initial conditions which enter into an explanation, and it is the function of social science to try to develop adequate generalisations and a conceptual framework which, in conjunction with these initial conditions, can provide explanation in depth. Winch would doubtless counter this view by pointing out that the very language which we use to formulate concepts and generalisations affects and is affected by rule-bound social behaviour.

Rudner (1966, 78–80) tackles this problem by separating two senses in which we can attach meaning or significance to an event or concept. In the first sense, as in a completely abstract calculus, a term derives its meaning from its position and function within the set of propositions developed in the calculus. In chess, for example, we could specify what is meant by a knight or a rook by specifying the possible field of play and the rules that govern the movement of each piece. It is possible, therefore, to examine the meaning of a term by reference to its *syntactical* function. In the second sense there is a value judgement involved as to the meaning and importance of a term, for we are giving it an interpretation in relation to experience —i.e. the term is being evaluated *semantically*. This latter kind of evaluation is undoubtedly culture-bound but, as Brown and numerous other writers (e.g. Churchman, 1961; Fishburn, 1964) have pointed out, the evaluation itself is susceptible of analysis. With respect to syntactical function, it is possible to assign an objective,

unambiguous meaning to a term, but in offering a semantic interpretation of the term a subjective judgement is involved. Thus
Winch maintains that a neutral language for discussing social problems cannot be devised. This view applies as much to the natural
sciences as it does to history and social science however. As Kuhn
(1962, 125-6) points out, no ethically neutral 'pure' observation
language has yet been designed for the discussion of empirical
problems and 'philosophical investigation has not yet provided even
a hint of what a language able to do that would be like.' To concede
Winch's point, therefore, does not amount to conceding a basic
methodological difference between natural and social science. The
behavioural attitude taken by Kuhn (1962) and Churchman (1961)
to the verification problem in natural science does not seem radically different from the view taken by Winch (1958, 84) when he
writes:

> To understand the activities of an individual scientific investigator we must take
> account of two sets of relations: first his relation to the phenomena which he
> investigates; second his relation to his fellow-scientists. Both of these are essential
> to the sense of saying that he is 'detecting regularities' or 'discovering uniformities';
> but writers on scientific 'methodology' too often concentrate on the first and
> overlook the importance of the second.

Any scientific community develops a language which is used to
communicate ideas within that community. The meaning of their
investigations—be it of natural or social phenomena—is partly
defined by that language. Verification procedures are also developed
in the context of that language and are not independent of its form.
It makes sense, therefore, to speak of objectivity only within the
context of some accepted language or paradigm (see above, pp. 19–
21). It follows, finally, that objectivity is a relative standard rather
than an absolute measure and that this situation is ultimately true of
both natural and social science.

The standard which Winch would wish us to accept is that the
truth of a particular statement can be ascertained, and indeed the
meaning of a statement can be understood, only if the observer has
experience of the situation. This, Rudner (1966, 83) regards as
being a subtle form of the 'reproductive fallacy'. Thus:

> The claim that the only understanding appropriate to social science is one that
> consists of a reproduction of the conditions or states of affairs being studied, is
> logically the same as the claim that the only understanding appropriate to the
> investigation of tornados is that gained in the direct experience of tornados.

Rudner goes on to suggest that direct experience is not irrelevant to
understanding in social situations but that it is far from being the

only mode of approach open to the social scientist or a substitute for scientific explanation in a social context. He concludes that

Neither Weber's arguments nor the . . . arguments of Winch are decisive, then, in compelling the conclusion either that social science must fail of achieving the methodological objectivity of the rest of science or that social science must employ a radically distinct methodology.

The employment of *verstehen* therefore may be fundamental to hypothesis formation, for it may lead us to the imaginative creation of hypotheses (Abel, 1948), but it does not add to our store of knowledge because it amounts to applying knowledge already validated by personal experience. Nor 'does it serve as a means of verification. The probability of a connection can be ascertained only by means of objective, experimental, and statistical tests.'

It will probably never be proved conclusively that history and social science *should* or *should not* adopt the norms of explanation set up in the basic model of scientific explanation. If we wish to employ the scientific model we may undoubtedly do so. The objections to its employment on the grounds of lack of objectivity refer as much to natural science as to social science and history.

V EXPLANATION IN THE SOCIAL SCIENCES AND HISTORY—A CONCLUSION

This chapter began by examining Mill's contention that all explanation possessed the same logical structure and closed by concluding that the claim can probably never fully be proved or disproved. The claim that all explanation is *necessarily* the same is just as unprovable as the claim that it is not. Under such circumstances there can be no barrier placed to prevent the employment of the scientific model to discuss and analyse human affairs.

The reasonableness of employing the scientific model must ultimately be judged by its use and effectiveness. If, as Caws (above, pp. 30-1) claims, the scientific model provides us with the only equipment for a rational understanding of empirical phenomena, then it would be foolish to deny the application of the model on essentially pragmatic grounds. It is, of course, extraordinarily difficult to assess the effectiveness of the model in the social-science context. The early employment of the scientific method in a human context was not very productive of results which could be regarded as reasonably objective. At the present time it is possible to refer to a very wide range of studies in econometrics, sociology, decision theory, operations research, quality control, experimental psychology, and so on, which have achieved a reasonable status by most standards of judgement as to what is or is not scientific investigation. Most disciplines are, however, forced to find some kind of compromise between the methodological norms set up by science as a whole and the peculiar problems of investigation which individual practitioners of that discipline face either on account of the difficulties of controlling the phenomena being examined or on account of poor theoretical development. Nagel (1961, 503) has pointed out

None of the methodological difficulties often alleged to confront the search for systematic explanations of social phenomena is unique to the social sciences or is inherently insuperable. On the other hand, problems are not resolved merely by showing that they are not necessarily insoluble; and the present state of social inquiry clearly indicates that some of the difficulties . . . are indeed serious.

It is hardly surprising that in the face of these difficulties social scientists have either rejected the model form of scientific explanation or have not been particularly rigorous in their application of it. Each discipline tends to be autonomous with respect to the criteria it employs in deciding whether or not statements are justified (Kaplan, 1964). Zetterberg (*1965*, 151) has an interesting comment on the situation in sociology:

How stiff should our criteria be? The ideal of science prescribes standards that few, if any, concrete research projects ever meet. The surest way of damning any research report is to compare it with the ideals of science. The best way of evaluating a research report is to compare it with other research reports, the most reputable ones in our field. Looking at the most reputable specimens in sociological research, we find, not unexpectedly, that standards vary from place to place, from time to time, from topic to topic.

There are variable criteria of acceptability within and among different disciplines. The point of setting up a model form of scientific explanation is not to provide a weapon to damn creative and formative research. It functions rather as an ultimate objective, an ultimate goal, at which we may aim in our pursuit of powerful, consistent, and reasonable descriptions and explanations. It is salutary on occasion to remind ourselves of the nature of this norm. It is also salutary to remind ourselves that scientific research is, above all else, creative imaginative effort. There is no substitute for good judgement, sympathetic understanding, and creative insight in evaluating propositions in the natural sciences, the social sciences and in history. The real choice that faces the researcher on human activity appears to be between trying to cut the thick and tangled Gordian knot that faces the employment of the normative model of a scientific explanation in the context of human affairs, and relying, as in the past, upon intuition and judgement alone. Clearly the ideal is creative imaginative effort backed by the control which scientific method gives us over the reasonableness and consistency of statements which we make about reality. Such an ideal is surely not so unattainable?

Basic Reading

Dray, W. (1964).
Rudner, R. S. (1966).

Subsidiary References

Brown, R. (1963).
Hempel, C. G. (1965).
Kaplan, A. (1964).
Nagel, E. (1961), chapters 13, 14, and 15.

Chapter 6
Explanation in Geography—Some General Problems

The discussion of the previous two chapters has centred on the nature of scientific method and on the difficulty of extending that method to the social and historical sciences. The views discussed were those of philosophers of science and the methodologists of the various disciplines. The relationship between these views and the conduct of empirical investigation by the practitioners of some discipline were only incidentally considered. In examining explanation in geography this relationship becomes of great importance. Methodological investigation in geography should not be regarded as an end in itself. It is, rather, concerned with elucidating the conduct of empirical enquiry.

It has been suggested on occasion, however, that the work of philosophers and logicians bears little relationship to the conduct of empirical work. It is sometimes claimed that much of the philosophy of science is too normative to be useful and that the standards set up are far too rigorous for the practical purposes of most disciplines (*cf.* Churchman, 1961, 341–3; Wilson, E. B., 1955). Zetterberg's comment, quoted at the end of the last chapter, is indicative of such an attitude towards the rigorous employment of scientific method in the social sciences. But philosophy of science has changed. The practical problems of extending the scientific method to the social and historical sciences (examined in some depth by Kaplan (1963), Nagel (1961), and Hempel (1965), for example) and the growing recognition of the behavioural aspects of scientific decision-making (identified by Churchman (1961), and Kuhn (1962) for example) have provoked a continuing revolution in the attitudes of philosophers of science. Explanation in physics is no longer their sole concern. It is still relevant therefore to examine the extent to which philosophers of science have made general statements which are of use to geographers in the conduct of empirical work.

The methodology of a particular academic discipline is not determined by philosophers of science. It is partly evolved in 'rule of thumb' fashion by the practitioners of that discipline and partly developed by methodologists trying to formalise procedure within that discipline. To some extent both practitioners and methodologists may be influenced by philosophy of science. But it is possible for, say, the methodologist of geography to make statements which are at variance with those of philosophers of science or to make statements which are at variance with actual practice in most geographic research. The purpose of this section, therefore, is to try to identify such contradictions where they exist, attempt to resolve certain methodological arguments in geography, and finally to attempt to provide a broad general statement of the explanatory frameworks appropriate for geographic investigation. In this, a number of important general questions need to be considered:

(i) What is the relationship between the methodological argument as developed in geography and the methodological argument regarding knowledge in general? Put another way, how far do the views of the methodologists of geography tally with the views of philosophers of science and, if there are differences, what rational basis can be provided for such differences?
(ii) What is the relationship between the statements made by the methodologists of geography and the practice of geographers as revealed in their empirical work?
(iii) What is the relationship between the explanatory form accepted by practising geographers and the explanatory form accepted by the practitioners of other disciplines?

Most attention will be paid to the first question, mainly because the relationships are easier to determine. Some comments will be made on questions (ii) and (iii) here, but question (iii) will be examined in some detail in later sections (see below, pp. 107–29).

I PHILOSOPHY OF SCIENCE, GEOGRAPHIC METHODOLOGY, AND EXPLANATION IN GEOGRAPHY—SOME BASIC RELATIONSHIPS

The methodology of a discipline is not determined by the practitioners of that discipline in isolation. Philosophers of science have frequently been involved in direct debate with the practitioners of some discipline as to the nature and form of the explanations they pursue. Some disciplines have tended to avoid such direct debate and a gap has thus developed between the methodologists of that discipline and the philosophers of science. If prizes were to be awarded for the discipline where this gap was most marked then geography would come close to taking first prize. For the most part geographers have not argued about explanatory form but have argued about objectives. Where explanatory form has been considered there has usually been scant reference to the enormous literature on explanation in general.

Hartshorne's two studies, for example, scarcely mention the work of modern analytic philosophy even in those sections where the concern is with explanatory form rather than objectives. The references in *The Nature of Geography* (1939, 8, 374–8) amount to ten items, only two of which may be considered 'analytic' in their approach, while the *Perspective* (1959) makes no reference to works of this type. It was not, of course, Hartshorne's stated intention to examine current philosophical concepts and geographical explanation. Indeed, he is not concerned with 'explanation' directly, but with the 'nature, scope, and purpose' of geography. This, Hartshorne (1959, 7) maintains, is 'primarily a problem in empirical research'. There is nevertheless a good deal in *The Nature* which has to do with explanation simply because a particular set of objectives sometimes entails a particular framework for explanation (below, p. 67). Hartshorne's attempt to distil the essence of geography out of extraordinarily diverse approaches leads to difficulty. It ignores, for example, the strong stimuli which practising geographers may have received from philosophical writing and it fails to take into account the way in which both objectives and styles of explanation change over time. It is clearly important to examine the relationship between the current of philosophical opinion in a particular society at a particular time and the conduct of empirical work in geography.

The historians of geographic thought provide a few (but by no means enough) examples. One of the best is documented in Lukermann's (1965) study of the influence of French mathematical and philosophical work on the thinking of Vidal de la Blache. This study does much to elucidate the nature of French methodological thinking in a formative period in the evolution of geographic thought. It is interesting to examine Hartshorne's methodological position in the same way. Thus Hartshorne relies heavily on Hettner's works and Hettner was apparently influenced by the German school of historiographers. Thus, of the few external influences which Hartshorne specifically mentions, that of Dilthey and Rickert may be highly significant. The views of the German historiographers which centred around the uniqueness thesis and the idiographic method have already been examined (above, pp. 49–53). The criticisms since developed by historians and philosophers of science have provided a powerful challenge to these views, yet the Hettner-Hartshorne view of geography expounds these views without rebutting these challenges. We may find, in pursuing our understanding of geography in isolation, that we are in the unedifying position of expounding without reasonable foundation views which have generally been discredited in all other disciplines as well as in the philosophy of science. The philosophical underpinnings of Hartshorne's methodology thus appear to relate to the philosophy of history in the latter half of the nineteenth century rather than to the philosophy of science in the mid-twentieth. It is possible, of course, that the uniqueness thesis and the idiographic method can be given a modern defence, but such a defence needs to parry the challenge posed to the uniqueness thesis in modern philosophy of science. Certainly geographic methodologists cannot realistically proceed in isolation, although many have appeared to attempt the feat.

The independent attempt to state a methodology for geography has provoked a reaction. Ackerman (1963, 431–2) writes:

Our search for a professional identity led to an intellectual independence and eventually to a degree of isolation against which a number of the rising younger generation of geographers have now reacted . . . In our desire to make our declaration of independence viable, we neglected to maintain a view of the advancing front of science as a whole. We acted as though we did not believe in anything more than the broadest generalities about the universality of the scientific method. In effect we neglected to appraise continuously the most profound current of change in our time. We neglected an axiom: the course of science as a whole determines the progress of its parts, in their greater or lesser degrees.

Yet we still find geographers, with some exceptions, very remote from the 'ferment of ideas', as Chapman (1966, 133) calls it, characteristic of the social and natural sciences at the present time. The temptation is still, according to Chapman, 'to preserve the *status quo* by bathing in a euphoria of inertia'.

A failure to admit the current *Zeitgeist* by the front door has led to its creeping entry round the back. The climatologist thus draws much in the way of basic methodology from the closely related disciplines of atmospheric physics and physics. The biogeographer draws much from the soil scientist, the biologist, the chemist, and so on. The historical geographer tends to look to history, the economic geographer may look to economics, and so on. Thus methodological separatism has grown in geography as each subdivision of the subject has matured and as the number of specialised subdivisions has increased. There seems to be little in common in the actual methodological approach of the historical geographer and the climatologist. Wrigley (1965, 17) has recently written:

Geographical writing and research work has in recent years lacked any generally accepted, over-all view of the subject even though techniques have proliferated.

This, where it has been recognised, has widely been regarded as a bad thing. A unifying vision is a comforting thing, but one may perhaps question whether it is as vital a thing as sometimes supposed. Without it there is always the danger of the drifting apart of the congeries of interests which together make up the subject. With it, on the other hand, there is also danger from rigidity and from the creation of orthodoxy.

Hardly surprisingly, the failure to indulge in methodological debate which specifically considers the relationship between the methodology of geography and scientific epistemology in general, has led to the development of multiple methodological frameworks within which specialist geographers pursue their own interests in relative isolation. In such a situation it becomes particularly difficult to establish the nature of geographic explanation by a study of what geographers do. It would also be unrealistic to expect some unified view of geographic methodology to emerge from such a study. One particular view may be postulated *a priori*, but it makes little sense to suppose, without further argument, that the geomorphologist must *necessarily* be concerned with unique cases or that the historical geographer must *necessarily* be concerned with the research for the general laws which govern spatial evolution. The multiple methodological frameworks which now co-exist, rather uneasily, in geography can be identified by reference to several recent symposia. Considerable variation of viewpoint is evident, for example, in *American Geography—Inventory and Prospect* (James and Jones, 1954), but more vivid differences of opinion are portrayed in a recent report on *The Science of Geography* (N.A.S., 1965). In this publication four areas of geographic research were examined in some detail. The differences between physical geography, cultural geography, political geography, and location theory, in terms of the kind of problem studied and the kind of explanatory framework envisaged, are extremely marked. In Britain, *Frontiers in Geographical Teaching* (Chorley and Haggett, 1965A) and *Models in Geography* (Chorley and Haggett, 1967) also exhibit marked differences in methodological attitude among the various contributors (the former much more so than the latter).

A disturbing conclusion emerges from this brief discussion: namely, that the methodological assumptions made by the practitioner of geography bear as little relationship to the views of geographic methodologists as the latter do to the vast researches into the nature of empirical knowledge conducted by writers such as Carnap, Braithwaite, Hempel, Nagel, Popper, Reichenbach, and many others. There are of course significant exceptions. But such exceptions are far too few for our own methodological comfort.

II SOME METHODOLOGICAL CONTROVERSIES IN GEOGRAPHY

It is perhaps significant that most of the British and American literature in geographic thought has been concerned with defining the objectives, scope, and nature, of geography. Most controversy has thus centred on philosophical rather than methodological issues. But by adopting a particular stance with respect to the objectives of geography, geographers have sometimes been forced to adopt some stance with respect to explanatory form. In some cases this has resulted from a largely spurious association between a particular view of the objectives of geography and a particular style of explanation. In other cases a particular objective does indeed imply a certain commitment as regards explanation. But the relationships are complex. In most disciplines the interaction between the questions which a discipline asks and explanatory form is extremely important. In geography this usually fruitful interaction has been reduced to a rather more sterile one way dependence. In particular, the tendency to be critical of the questions being asked on *a priori* or metaphysical grounds, without examining the explanatory form involved has led to a good deal of unnecessary, and often unedifying, argument. If, for example, a particular type of question entails an explanatory form which is generally regarded as lacking in power—a good example here would be the kind of question that entails answers in terms of 'intentions' (see above, p. 57)—then it may be inferred that we are either asking the wrong kind of question or, at the very least, asking it in the wrong kind of way. This simply amounts to saying that we frequently phrase a question in such a way as to presuppose the explanatory form of the answer. Under these conditions it is possible to dismiss questions on two grounds. First, the objective may appear irrelevant to the needs of modern geography or modern society in general. Second, the explanatory framework entailed in asking that kind of question may appear very weak. Our task, therefore, is part metaphysical since there is a need to identify questions which appear relevant to the needs of society, and part logical, since there is a need to phrase those questions in such a way that they can be given powerful rather than weakly grounded answers. In geography these two tasks have usually been confused. To demonstrate this a number of controversies will be considered. Since it is the explanatory form that is so frequently neglected we shall examine these from the point of view of the explanatory form involved rather than from the point of view of objectives.

Most geographers, curiously enough, regard their discipline as some kind of science, but also submit that the questions which geographers ask cannot be answered by way of the rigorous employment of scientific method. Hartshorne (1939, 375) states:

> Geography attempts to acquire knowledge of the world in which we live, both facts and relationships, which shall be as objective and accurate as possible. It seeks to present that knowledge in the form of concepts, relationships, and principles that shall, as far as possible, apply to all parts of the world. Finally, it seeks to organise the dependable knowledge so obtained in logical systems, reduced by mutual connections into as small a number of independent systems as possible.

The overall aim of geography is not, therefore, at variance with scientific enquiry in general. It is, however, frequently held that the questions which geographers ask lie partly beyond science (Hartshorne, 1959, 167). There are, it is claimed, limits to the application of the scientific method and these limits make geography a rather special kind of science. This view ignores the accepted limitations of scientific enquiry in *any* context and thus assumes a far rosier view of the achievements of the scientific method than exists in the minds of all but the most partisan of the analytic philosophers. There are, of course, problems in the application of scientific method in geography, but these problems are only different in degree, not in kind, from those experienced in any empirical enquiry. Certainly geography has no greater difficulty than, say, biology, zoology, economics, anthropology, and psychology. Nevertheless there is an important gap between the normative view of scientific method and geographic methodology. This gap is most marked in discussion of laws in geography.

Hartshorne envisages the development of generalisations and principles in geography, but his survey of the geographic literature before 1939 led him to doubt if scientific laws of similar status to those employed in the physical sciences could be developed in a geographic context. A distinction is sometimes drawn, however, between the physical and human branches of the subject. Thus Wooldridge and East (1951, 30) state:

> It is futile to assert that 'human' or 'social' geography can be seen in terms of formal categories and universal principles and processes as can physical geography. This imputes to it no inferiority; it is rather to admit that it is infinitely more complex, subtler, more flexible, manifold.

Wrigley (1965, 5) has recently commented also on the methodological difficulty of 'running in harness, as it were, physical geography and social geography.' By accepting the view that explanation in the social sciences is fundamentally different from explanation in the physical sciences, Wrigley implies the existence of two radically different frameworks for explanatory thinking in geography. In physical geography law-statements are of importance, therefore, but in human geography such statements are irrelevant. This geographic manifestation of the Weber-Winch thesis regarding laws in the social sciences need not be accepted, however, and there are strong grounds for rejecting such a view (above, chapter 5).

It may thus be claimed that laws can be established in both human and physical geography. Some writers dissent in general from this view and claim that laws cannot be established because of the multivariate nature of the subject-matter, because the number of cases about which one may generalise is often small, and because the occasional exceptional circumstance may have far-reaching consequences (Hartshorne, 1959, 148–53). This view may be challenged on a number of grounds:

(i)   In order to show that laws cannot be established we need some clear criteria by which to judge whether or not a particular statement qualifies as a law.

(ii) Given such criteria it has to be shown that there is no way in which such statements can be developed and used in a geographic context.

(iii) It also has to be shown that there is some real alternative to the use of law-statements which produces satisfactory and reasonable explanations. In this the argument over covering laws, explanatory sketches, and so on, is exceedingly relevant to geography.

The criteria for establishing the 'lawfulness' of a particular statement will be examined later (below, pp. 100–6). For the moment it suffices to remark that the criteria available are far from clear and appear to have changed significantly in the last few decades. Given certain criteria it may be argued that laws in a strict sense cannot be developed in any empirical context except, perhaps, in physics. Given other criteria it can be shown that laws can be developed in geography. In either case the argument for geography's being different, from, say, biology and economics, must be dismissed. Accepting the less rigid criteria, we may assume that laws can be developed. Whether or not such laws are useful or non-trivial is another matter. Most writers in the 1920s and 1930s (apart from the rather discredited determinists) abandoned the attempt to formulate such laws and contented themselves with generalisations and principles which were directed towards the study of areas on the earth's surface which were regarded as unique. This procedure partly reflected the needs of geographers working at that time, but it was also partly the result of what Ackerman (1963, 430) has called a too close association between geography and the disciplines of geology and history—disciplines which, as we have already seen (above, pp. 28-9, 49-54) were dominated by the notion of uniqueness and the idiographic method. The net result for geography was adherence to a particular view of the nature of geography coupled to a particular view of explanation in geography. This association between a particular set of objectives (description and interpretation of unique areas) and a particular explanatory form (the idiographic method) formed a powerful orthodoxy from which geographers found great difficulty in breaking free.

This view is expressed in its most rigorous form in the argument over *exceptionalism* in geography. This argument centres on a statement about the place of geography in the system of knowledge which was made by Kant and which has subsequently become a basic tenet of orthodox geography as expounded by Hettner and Hartshorne. There is a very considerable literature on exceptionalism in geography (Hartshorne, 1939; 1955; 1958; 1959; Schaefer, 1953; Bunge, *1966*, chapter 1; Blaut, 1962; Haggett, 1965A, 2-4; Lewis, 1965), and it will, therefore, be examined here.

Kant apparently characterised the position of both geography and history in relation to the other sciences as follows:

> We may classify our empirical knowledge in either of two ways; either according to conceptions or according to time and space in which they are actually found. . . . Through the former we obtain a system of nature, such as that of Linnaeus; through the latter, a geographical description of nature . . . Geography and history fill up the entire circumference of our perceptions: geography that of space, history that of time. (Hartshorne, 1939, 134-5.)

The inferences usually drawn from this statement may be summarised as follows:

(i) If geography is to treat of the sum of our perceptions in space then there can be no limit placed on the class of objects which geography studies (Hartshorne, 1939, 371-4; 1959, 34-5).

(ii) If there is no limit to the factual content of geography then the discipline must be defined by its distinctive method of approach rather than in terms of its subject-matter (Hartshorne, 1939, 374). Thus geography is frequently characterised as a 'point of view' rather than a subject which deals with a characteristic subject-matter.

(iii) If we are concerned with the sum total of reality in all its aspects as we perceive it in terms of spatial location, then it follows that we are essentially concerned with unique collections of events or objects, rather than with developing generalisations about classes of events. Locations, it was argued, are unique.

(iv) If locations are unique, then the description and interpretation of what existed at those unique locations could not be accomplished by referring to general laws. It required, rather, understanding in the sense of empathy or *verstehen*, i.e. the employment of the idiographic method.

The logic of this argument needs some examination, for not all of these conclusions can be drawn from the premisses as stated by Kant and Hartshorne. The conclusions in fact involve a number of hidden premisses and a number of logical difficulties. But the acceptance of an argument often has little to do with the logic of its statement. Before discussing the underlying assumptions and inherent logic of this argument, it is worth while examining some of the reasons for the evidently deep intuitive appeal of the Kantian thesis for geographers.

The Kantian thesis was apparently used by Hettner to establish that geography, along with history and certain other disciplines, was an idiographic rather than a nomothetic science (Hartshorne, 1958). How far this position was directly influenced by the work of the German historiographers requires investigation—but the influence was certainly not negligible (Hartshorne, 1959, 149). There is sometimes cause to doubt whether a particular methodological credo is as influential on the course of empirical research as the historians of a discipline are wont to maintain. The Kantian thesis appears, however, to have been particularly important because it seemed to fit, in a general kind of way, much of the professional activity of geographers during the 1920s and 1930s. At this time there was a strong reaction against the so-called determinist school and a consequent rejection of the crude laws put forward as aids to explanation by writers such as Semple, Huntington, and Griffith Taylor. Research thus tended to focus on small areas. It is hardly surprising, therefore, that a methodological credo which embraced the uniqueness of areas and the idiographic method as its major tool should elicit a good deal of support. At the same time it was the subject of some concern to geographers, that their discipline was beginning to spread its field of enquiry into all kinds of subject-matter which were the basic concern of other disciplines, both physical and socio-economic. Such a wide range of subject-matter could happily be justified under the umbrella of the Kantian thesis. Geographers even dared to hope that these diverse systematic studies were but the prelude to some final synthesis of all knowledge in terms of a spatial structure of unique geographic areas. The aim of regional synthesis emerged as the teleology of geography. But as each systematic aspect of geography develops and matures, this particular millennium appears to fade more and more into the distance. Latterly, the Kantian thesis has been used more specifically to buttress a particular research tradition (i.e. the idiographic method) against the challenge of a younger generation whose work is more nomothetic in style (Blaut, 1962, 5). Here, however, it is tempting to be cynical and suggest that this merely amounts to invoking an eminent philosopher's name to support the *status quo* without really considering whether the statement made by Kant is reasonable from either a geographic or philosophic point of view. Kant was, after all, a prolific lecturer and writer and many aspects of his philosophy—such as the notion of *synthetic a priori* knowledge which was closely bound up with his views of space—have been either profoundly modified or dismissed for a hundred years or more.

The Kantian thesis also assumes that space can be examined and spatial concepts developed independently of matter. This assumption has not been clearly stated in the past. It amounts to postulating an absolute space. Given the assumption of an absolute, rather than a relative, space, it is possible to derive some of the statements usually made about the place of geography among the sciences. But the absolute philosophy of space has not been current in scientific thinking in general since the early nineteenth century. It appears, therefore, as if geographers have accepted a particular view of space which is at variance with that of philosophers of science. This is not necessarily a bad thing. What is regrettable, however, is that the assumption of an absolute space has not been explicitly discussed nor has it been recognised as one of the basic assumptions of the Kantian thesis. The background argument about the appropriate philosophy of space is therefore discussed in chapter 14.

The assumption of absolute space has great relevance to the issue of uniqueness in geography. In so far as geographers are concerned with objects and events, all the arguments which have been examined previously (above, pp. 49–54) may be brought to bear. There can be no doubt that either a major modification or an outright rejection of the notion of uniqueness is called for. But the argument in geography is different in that it is claimed that geographers are concerned with locations rather than with objects and events. There is, as we shall see later (below, pp. 212–17), some justification for regarding the distinction between location (by way of spatial languages) and properties (by way of substance languages) as of great significance in geographic methodology. In the geographic context, therefore, the epithet 'unique' is applied to locations rather than to properties. This raises the basic question of the uniqueness of locations. This issue has been discussed by a number of writers (Hartshorne, 1955; 1958; Bunge, *1966*; Schaefer, 1953; Grigg, 1965; 1967;

Haggett, 1965A) and it involves the problem of the 'geographical individual'. The problem here is that no matter whether one is arguing in favour of the uniqueness thesis or whether one is arguing for regionalisation by way of classification and grouping procedures, it is necessary to identify an *individual* or some basic unit of space to facilitate discussion. The short answer to this problem is that two types of individual may be identified—the first by way of its space-time co-ordinates and the second by way of its properties. Geographic work has frequently confused the two and this has led to considerable confusion in the exposition of geographic problems and in methodological discussion in geography. For the moment it makes sense to assume that geographers are essentially concerned with individuals identified by their space-time co-ordinates (since there are ready-made arguments for dismissing the notion of uniqueness with respect to individuals defined by way of properties).

It follows from assuming an absolute space that locations are unique. The exponents of the Kantian thesis have never made this statement directly, but they have tended to assume *a priori* a set of regional entities which exist and hence constitute geographic individuals. Much of the search for regional divisions may thus be regarded as an attempt to identify geographic individuals. In other instances it has been assumed that space possesses an atomistic structure which can somehow be aggregated into distinct regions. Given a relative view of space, however, the idea of uniqueness of locations has to be profoundly modified. Within any co-ordinate system locations may be uniquely determined, but the relative view of space postulates an infinite number of possible co-ordinate systems. Thus the distance between two points in space will vary according to the co-ordinate system selected. Here the concept of a transformation becomes extremely important and the relationship between geography and geometry becomes of like importance. There are, however, transformations which are not unique, and it is therefore technically possible to transform one map into another in such a way that locations projected are not unique. Given a relative view of space, therefore, locations are either not unique or, at best, unique only within a selected co-ordinate system. This issue will be considered further in chapter 14.

Given a relative view of space the problem is to identify the co-ordinate system which is most appropriate for a given geographic purpose. It is usually held by philosophers of science that this is an empirical problem and that its solution depends on the kind of activity being studied. Activity involves discussing properties and therefore the choice of co-ordinate system is dependent upon the phenomena being studied. Thus the view that geography is *not* concerned with any specific types of activity needs to be re-evaluated. The same kind of problem emerges even under the assumption of an absolute space. In this context it is sometimes called the 'problem of significance'. In practice geographers do not study *everything* in spatial context, but limit consideration to a selection of phenomena. The question arises as to the grounds for this selection. Hartshorne (1959, chapter 5) examines this bothersome problem but the only criterion of significance he could establish was that the phenomena should be 'significant to man'. This criterion can be applied to all knowledge, however, and without further refinement it remains empty of any meaning. In practical terms the same problem arises in the context of regional division. Thus although it is claimed that geographers are solely concerned with locations, the criteria for judging whether or not a particular regional division is reasonable or not is derived from the properties of that region. Again, it becomes clear that objects and events have some place in geographic thinking, for without reference to particular types of phenomena it is impossible to determine an appropriate co-ordinate system, to judge whether or not a system of regional division is appropriate or not, or even to judge whether the objects and events examined in terms of spatial location are reasonably selected or not. There are, it is clear, limits to the philosophy of 'geography as a point of view' in the way usually stated. Most disciplines on examination turn out to be partly determined by the subject-matter they choose to study and partly by the point of view they cultivate with respect to that subject-matter.

In spite of the considerable literature on the objectives of geography, geographers have rarely tackled this problem directly. The basic question really amounts to asking how the 'point of view' of the geographer can be distinguished. It is characterised, as Blaut (1962) and Berry (1964A) have suggested, by the system of interlocking concepts and theories which geographers have developed about their subject-matter. In most cases the nature of a discipline may be identified by the explicit theory developed in that discipline. Theory thus defines the point of view with precision. In some cases this definition remains implicit because theory has not been explicitly developed. The point of view contained in the development of the Kantian thesis regarding the nature of geography thus contains an implicit theory about the absolute structure of space—it replaces notions about objects and events by notions about locations. Until recently geographers had remained content with an implicit definition of their point of view and tended to avoid specific theory. Where theory has been developed it has been purely speculative and non-scientific.

Theory thus forms the hall-mark of a discipline. It invests objects and events with significance, it defines the framework (e.g. the co-ordinate system) into which events and objects may be fitted, and it provides systematic general statements which may be employed in explaining, understanding, describing, and interpreting, events. Scientific theory, as opposed to purely speculative statement, ensures the consistency, power, and reasonableness, of a statement by sub-jecting it to a number of independent tests. It is therefore of vital importance to understand the nature of scientific theory and this issue will be discussed later (below, pp. 87–99). Theory in geography is, however, not well developed. It thus becomes difficult to identify with precision the 'point of view' which characterises geography and difficult to state the criteria of significance which that point of view defines. The problem of significance as defined by Hartshorne has no solution independent of geographic theory. Whatever the conclusion may be—and some tentative suggestions are made in a later section (below, pp. 113–40) it is clear that geographers are concerned with both properties and locations (i.e. two different types of individual) and that neither of these individuals can be satisfactorily discussed given the idea of uniqueness as it is sometimes stated by the pro-ponents of the Kantian thesis.

One of the major arguments against the uniqueness thesis, of course, was that it was difficult to provide a realistic framework for explaining and describing without doing violation to the notion of uniqueness. Clearly, the Popper-Hempel thesis regarding explana-tion in history and the social sciences has relevance here and it can be transferred into a geographic context intact. The fact that it has not been discussed in geography is but one further indication of the methodological isolationism to which geography has recently been prone. The only discussion remotely related to geography is that by Sprout and Sprout (1965).

The Popper-Hempel thesis may be formulated in geographic con-text as follows. Explanations of any intrinsic worth ought to be rendered as the necessary conclusion to some deductive argument. It has to be shown that an event had to occur, given the circum-stances. Such an argument requires the use of law-statements or something equivalent. In many cases we may not be able to specify any law with precision and we may choose to leave the law-statement unspecified. But a law-statement is nevertheless implied. These covering laws are of interest because they are vital to the whole pro-cedure of explaining. Geographers have, for the most part, remained content with implying rather than specifying laws and have been content with rough explanation sketches rather than with more rigorous explanations.

The general principles on which geographers rely function as law-statements but these are, for the most part, rather weakly supported inductive statements. This may be inevitable in the early stages of a discipline's development, and if geography were a new-born discipline it would not be so surprising to find that 'geography until the 1940s did not have balance between the empirical-inductive and the theoretical-deductive approaches, but leaned heavily towards the former' (N.A.S., 1965, 12). But given the length of the tradition in geography it is surprising that more effort has not been made to explore 'the dialogue between the empirical-inductive and the theoretical-deductive methods of thought and investigation' (N.A.S., 1965, 12).

The geographer's fear of explicit theory has not been entirely irrational. The practical problems of extending the scientific method into the social and historical sciences are considerable. Similar problems arise in geography. The complicated multivariate system which geographers were trying to analyse (without the advantage of experimental method) is difficult to handle. Theory ultimately requires the use of mathematical languages, for only by using such languages can the complexities of interaction be handled consistently. Data-analysis requires the high speed computer and adequate statistical methods, and hypothesis-testing also requires such methods. To some extent the geographer's reluctance in developing theory reflects the slow growth of appropriate mathematical methods for handling geographical problems. Without such methods the geographer's problems must have appeared analytically intractable. Crude systematic attempts at explanation were made by the determinists but by the 1920s these were in disgrace. Yet there seemed nothing capable of taking their place. At the same time philosophers of science appeared to be insisting on a particularly rigorous framework for scientific explanation—one which geographers could not hope to conform to. Thus, according to Blaut (1962, 5):

> Lacking the qualification of a proper systematic science, we turned to philosophy for a set of special credentials—which, of course, were freely supplied in the form of metaphysical concepts of object, relationship, and space. It was a question of metaphysicalising our methodology or letting the reductionism of science dissolve us into parts.

The special credentials were, of course, those provided by the Kantian thesis. But the situation has changed. Blaut continues:

> Philosophy of science has lately matured; has, in fact, caught up with sciences like geography, which study soft systems rather than hard objects.

There is, therefore, less excuse at the present time for avoiding the construction of geographic theory. It might be claimed, however, that geography, like history, is a consumer rather than a producer of theories and laws. The covering laws which enter into geographic explanation may thus be regarded as derivative of some other discipline. If theory in geography is derivative then the geographer should be aware of the wide range of theoretical constructs which are available to him. Derivative theory, as we shall later see (below, pp. 117–24), plays an important role in geography but, until recently, there has been a tendency to avoid the full responsibilities inherent in deriving theory from other disciplines. For the most part geographers have tended to imply either rather trivial observations about human behaviour as covering laws in human geography, or have implied crude deterministic (physical environmental) laws — laws which most geographers have claimed to reject. An examination of almost any regional textbook will serve to demonstrate how strong the environmentalist tradition still is. This is not to say, *a priori*, that there is anything inherently wrong with the environmentalist thesis. But there is something wrong with pretending to an objective regional synthesis which implies a set of environmentalist laws which have previously been disclaimed.

Derivative theory is undoubtedly more strongly developed in geography than indigenous theory. Much of the theorising now current in geography is of this type. The question therefore arises as to whether *indigenous* theory can be developed in geography as opposed to *derivative* theory and, if so, what the relationships are between them. This issue will be examined later, but for the moment it is worth noting a preliminary conclusion; that when geographers develop their analytic thinking in terms of space-time languages (the spatial framework for thought as proposed by Kant or later relativistic spatial frameworks), then indigenous theory may be developed, but when geographers resort to property languages the resultant theory is clearly either actually or potentially derivative of some other discipline. This important statement will not be justified here for it is the subject of analysis in chapter 9 (see also Harvey, 1967A; 1967B).

More generally we may conclude that there is no logical reason for supposing that theory cannot be developed in geography or that the whole battery of methods employed in scientific explanation cannot be brought to bear on geographic problems. There are, it must be admitted, some serious practical problems involved. But these practical difficulties certainly cannot be invoked to prove that geographic thought is essentially different with respect to explanatory form from all other disciplines except history and, perhaps, geology.

The Kantian thesis is not totally irrelevant to geographic thinking—indeed it contains some important insights into the structure of geographic thinking. But it contains some premisses which cannot be accepted at the present time (e.g. the premiss of an absolute space) and it contains some conclusions which are not rationally derivable from the premisses. The Kantian thesis, it must be concluded, needs profound modification if it is to be satisfactory for the current needs of geography as an independent discipline embedded in an over-all structure of knowledge.

III EXPLANATION IN GEOGRAPHY

From the preceding section it may be concluded that geography is 'short on theory and long on facts' (Ballabon, 1957, 218). Yet the development of theory appears vital both to satisfactory explanation and to the identification of geography as an independent field of study. Theory 'provides the sieve through which myriads of facts are sorted, and without it the facts remain a meaningless jumble' (Burton, 1963, 156).

Scientific theory may be evolved by two different routes (above, pp. 32–6). The theoretical-deductive route is probably now most favoured since it clearly recognises the hypothetical nature of much of scientific thought. In general this style of thinking has not been prevalent in geography although there has been plenty of a priori thinking. Writers such as Griffith Taylor and Carl Sauer developed theory in a sense: but theories only achieve a scientific status if they yield hypotheses which can somehow be tested and, for the most part, the theories developed by such writers, although stimulating and exciting, are scarcely capable of verification. This was partly because these theories were stated in such a way that they were incapable of deductive elaboration, and partly because the links in scientific methodology which run from the statement of theory through hypothesis-formation, model-building, experimental design, and verification procedures, have only been very weakly and very recently forged.

The route implicit in Hartshornian orthodoxy is different. It appears to run from the study of unordered observations (the facts), through classification and generalisation, to the formation of principles which may then be used to assist in the explanatory description of areas. The strength of such a route depends entirely on the power of inductive logic and it appears, therefore, to be a rather weak route to the formation of valid general statements which may function as covering-laws. It also tacitly assumes the ability to identify 'the facts' independent of some theory—an assumption that many will not be prepared to grant. Most research in geography, until recently, has tended to be concerned with the collection, ordering and classification of data and in this respect has conformed to Hartshornian orthodoxy without necessarily accepting the Kantian thesis. This approach to description and explanation appears to be inferior even to the Baconian approach (above, pp. 32–5) in that it avoids attempted unification of the general principles into some unified theoretical structure.

The failure to achieve a hypothetico-deductive unification of geographic principles—or to postulate such a structure—has serious implications. It has not only relegated most geographic thinking and activity simply to the task of ordering and classifying data, but it has restricted our ability to order and classify in any meaningful way. Where explanations have been attempted, they have tended to be *ad hoc* and unsystematic in form. Nevertheless it is possible to identify a number of explanatory forms partly by reference to methodological statements and partly by reference to empirical work. These explanatory forms will be examined in greater detail in subsequent chapters. For the moment it is sufficient to identify them briefly.

(a) *Cognitive description.* Under this heading are included the collection, ordering, and classification of data. No theory may be explicitly involved in such procedures, but it is important to note that a theory of some kind is implied. Thus classification involves some *a priori* notions about structure and these notions really amount to a primitive theory. In the early stages of a discipline's development such theoretical assumptions may be amorphous and ill-defined. In the later stages classification procedures tend to become a part of experimental design and, hence, determined by the particular theory being examined, or the measurement and classification of data may be derived directly from theory. Cognitive descriptions may thus range in quality from simple primary observations through to sophisticated descriptive statements.

(b) *Morphometric analysis.* In some ways morphometric analysis may be regarded as a particular type of cognitive description; one that involves a space-time language rather than a property language. Morphometric analysis thus provides a framework within which the geographer examines shapes and forms in space. In general the assumptions are geometric ones and this amounts to identifying a co-ordinate system suitable for discussing the particular problem in hand. In particular this allows the discussion of shape and pattern of town locations, the structure of networks, and so on. The analysis is explanatory in the sense that given two sides and one angle of a triangle in Euclidean space it is possible to predict the length of the third side and the other two angles. In geographic context we might, thus, predict the occurrence of settlements given a number (say two) of initial settlements and the geometric laws of central-place theory (Dacey, 1965A). Geometric predictions of this type are becoming of considerable importance in geography at the present time.

(c) *Cause-and-effect analysis.* The main contribution of Ritter and Humboldt to explanation in geography was their insistence that cause-and-effect laws could be established to explain the occurrence of geographic distributions. Cause-and-effect thus became one of the dominant forms of explanation in geography during the nineteenth century. Its unfortunate association with mechanistic and deterministic metaphysical concepts has led to some reaction against its use in the twentieth. But as later analysts have pointed out (Jones, 1956; Blalock, *1964*) there is no need to regard cause-and-effect analysis as necessarily implying causal deterministic explanation. The mix-up between determinism and cause-and-effect led to the far more muted (and occasionally disguised) use of this explanatory form in geography. The search for the 'factors' that govern geographic distributions is a good example of the restrained use of causal analysis at the present time.

(d) *Temporal modes of explanation.* It is but a short step from causal explanations to causal-chain explanations which stretch back over a long period of time. The general mode of explanation which follows this tack will be termed *temporal.* The assumption is that a particular set of circumstances may be explained by examining the origin and subsequent development of phenomena by the operation of process laws. Thùs Darby (1953) once commented that 'the foundations of geographic study lie in geomorphology and historical geography—' both of which were strongly dominated by temporal modes of explanation. Like causal analysis the various temporal modes of analysis have become associated with a number of metaphysical assumptions regarding real world processes. Historicism, for example, postulated that the nature of anything could only be comprehended in terms of its development (this view is sometimes termed with genetic fallacy), and historicism was, in turn, associated with deterministic assumptions about inexorable historical laws which moulded the evolution of both cultural and physical forms over time. Temporal modes of explanation need not be regarded as the only approach to explanation, nor need they necessarily be associated with deterministic or historicist philosophies. They provide us, quite simply, with one dimension through which we may try to comprehend geographic distributions—a dimension which, by its insistence on the study of temporal change, inculcates a deep awareness of the nature of temporal processes.

(e) *Functional and ecological analysis.* The deliberate attempt to avoid causal and causal-chain explanations because of the metaphysical trappings associated with them, led to the development of alternative explanatory frameworks. In social anthropology, for example, functionalism became a dominant framework for analysis largely through the efforts of Malinowski. Functional analysis attempts to analyse phenomena in terms of the role they play within a particular organisation. Towns may be analysed in terms of the function they perform within an economy (thus functional classifications of towns are developed), rivers may be analysed in terms of their role in denudation, and so on. Ecological and functional thinking have been important in geography. Wrigley (1965) has pointed out, for example, how the approach of Vidal de la Blache and Brunhes comes very close to functionalism in its form, while Harlan Barrows' paper in 1923 exerted some influence. Indeed Hartshorne's definition of geography as the study of interrelationships within areas has a distinctively functional-ecological ring to it. At the present time there are numerous geographers who regard ecological concepts as providing an important basis for geographic explanation (Stoddart, 1965; 1967A; Brookfield, 1964).

(f) *Systems analysis.* It is a short step from examining the function of a particular phenomenon within an organisational framework to studying the structure of that organisation as a system of interlocking parts and processes. There is a direct path from functional analysis, through ecology, to systems analysis which provides a framework for examining

interpenetrating part processes . . . discrete and bounded only to the extent that boundary processes are unimportant in relation to internal processes . . . [this] forms a part of a larger, including system, and its parts are themselves smaller, included systems. (Blaut, 1962, 2.)

Systems analysis provides a framework for describing the whole complex and structure of activity. It is peculiarly suited to geographic analysis since geography characteristically deals with complex multivariate situations. Thus Berry (1964B) and Chorley (1964; 1962) have suggested that systems analysis and general systems theory have an important role to play in facilitating geographic understanding.

These six headings cover much of the thinking of geographers as regards explanatory form. The categories are not mutually exclusive and there are many instances of overlap. It is possible to develop, for example, genetic-systems approaches, genetic-morphological approaches, genetic-classificatory approaches, and so on. The choice of explanatory form depends, of course, upon the kind of question asked. Here it is worth noting specifically the relationship between the kind of question and the explanatory form entailed. The explanatory forms briefly outlined are thus connected with the following questions:

(i) How may the phenomena being studied be ordered and grouped?
(ii) How are the phenomena organised in terms of their spatial structure and form?
(iii) How were phenomena caused?
(iv) How did the phenomena originate and develop?
(v) How do particular phenomena relate to and interact with phenomena in general?
(vi) How are phenomena organised as a coherent system?

It is also evident that the type of theory which will emerge is partly conditional upon the nature of the question asked and the nature of the explanatory framework chosen. At this point objectives and logical form converge to determine the nature of geographic explanation.

We may regard the six explanatory frameworks as outlined above as *model forms* for explanation. We shall examine each of these forms in some detail in later chapters. Given that geographers tend to treat problems in one or other of these frameworks it becomes important to understand the strengths and weaknesses, the pitfalls and the positive properties, of each framework for explanation. Only given such understanding will it be possible to evaluate the reasonableness of the theoretical structures which may emerge. If there were a thoroughly developed theoretical structure to our discipline there might be little need to examine these model forms of explanation in detail. But the conscious use of such model frameworks for explanation will only yield appropriate geographic theory if we clearly understand the assumptions implicit in each model. Otherwise there is always the danger that we might adopt the bath water as well as the baby. Such methodological discussion as there has been in geography over the use of models has frequently shown a dangerous unawareness of the precise relationship between models and theories. Similarly there is a great deal of misunderstanding as to the meaning of theories and laws in science and as to the meaning of these terms in a geographic context. These issues themselves need considerable debate.

This chapter has drawn a great deal on former chapters in that it has attempted to sketch in the position of explanation in geography with respect to explanation in the physical, social, and historical sciences. But it has also referred to future chapters. The rest of this book is thus concerned with elucidating some of the basic problems opened up for discussion in this chapter. In *Part III* the meaning of terms such as theory, hypothesis, and law, together with the nature of theory in geography will be examined. In *Part IV* the meaning of the *language* of explanation will be considered, with some general remarks on mathematical languages, followed by a detailed study of the nature of spatial languages and probability languages, both of which appear to be of great importance to geographic research. Here the nature of spatial co-ordinate languages as opposed to property languages will be examined and the Kantian dichotomy between space and substance be given a more modern interpretation. In *Part V* and *Part VI* the model frameworks for explanation outlined above will be examined in detail and the properties and utility of each will be considered in geographic context. This chapter has thus served merely to identify the problems. The subsequent chapters attempt to provide the foundations for a solution of these problems.

Basic Reading

Ackerman, A. E. (1963).
Berry, B. J. L. (1964A).
Blaut, J. M. (1962).
Chorley, R. J., and Haggett, P., (eds.) (1967), chapter 1.
Hartshorne, R. (1939).
Hartshorne, R. (1955).
Hartshorne, R. (1958).
Lewis, P. W. (1965).
National Academy of Sciences (N.A.S.) (1965).
Schaefer, F. K. (1953).
Wrigley, E. A. (1965).

PART THREE

The Role of Theories, Laws and Models in Explanation in Geography

. . . nothing seems to me clearer than that geography has already suffered too long from the disuse of imagination, invention, deduction, and the various other mental faculties that contribute towards the attainment of a well-tested explanation. It is like walking on one foot, or looking with one eye, to exclude from geography the 'theoretical' half of the brain power . . . Indeed, it is only as a result of a misunderstanding that an antipathy is implied between theory and practice; for in geography, as in all sound scientific work, the two advance most amiably and effectively together.

W. M. DAVIS (1899)

Chapter 7
Theories

'The quest for an explanation,' writes Zetterberg (*1965*, 11), 'is a quest for theory.' The development of theory is at the heart of all explanation, and most writers doubt if observation or description can be theory-free. Thus 'our ordinary language is full of theories', and 'observation is always observation in the light of theories' (Popper, *1965*, 59). 'I doubt,' writes Kemeny (1959, 89), 'that we can state a fact entirely divorced from theoretical interpretations.' Indeed some writers, Berry (1964A) and Blaut (1962) among them, regard the evolution of a distinctive theoretical structure for explaining certain sets of phenomena as being the main justification for regarding geography as a distinctive and independent discipline within the empirical sciences. If this be true, then a clarification of the 'nature' of geography depends upon the prior clarification of the nature, form, and function, of theory in geography.

Theories are, in Einstein's phrase, 'free creations of the human mind'. Any speculative fantasy may thus be regarded as a theory of some sort. The connection between 'images', 'perceptions', 'imaginative re-creations', 'empathic understandings', and theory construction is strong. Such psychological features are vital to theory-formation. Philosophical beliefs and the 'geographical imagination' provide the motive power for the construction of speculative theory. In this, philosophy and metaphysics assert their primacy over methodology. Metaphysical speculation has been a source of stimulating ideas in all areas of scientific research. Such speculations have, to use Körner's (1955) terms, acted as 'directives' or 'regulative principles' in the search for scientific theory. In geography we possess a multitude of such directives within our traditional literature.

A speculative theory does not necessarily possess the status of a scientific theory. It is unfortunately open to anyone to fabricate what James Hutton long ago called 'a system of apparent wisdom in the folly of hypothetical delusion' (Chorley *et al.*, 1964, 37). The success of scientific explanation lies mainly in the way it has taken such speculations and transformed them from badly understood and uncomfortable intrusions upon our powers of 'pure' objective description, into highly articulate systems of statements of enormous explanatory power.

A theory is, thus, a system of statements. At its simplest a theory may be regarded 'as a language for discussing the facts the theory is said to explain' (Ramsey, *1960*, 212). The traditional view is, thus, that a scientific theory is an 'interpreted calculus or is something that can in principle be cast as such a calculus' (Bromberger, 1963, 79). In practice most scientific theories, even in the natural sciences, are not stated in terms of a calculus and it is not altogether necessary that such a statement should be made (although there are, as we shall see, distinctive advantages to such a statement). In geography it is relatively rare for explicit development of such calculi, but it is nevertheless important to elucidate the structure of a scientific theory for it is against such a standard structure that we may measure the nature and effectiveness of geographic theories.

A The structure of scientific theories

A scientific theory 'may be considered as a set of sentences expressed in terms of a specific vocabulary' (Hempel, 1965, 182). The nature of this vocabulary has been discussed by logicians. The full details of the logical structure of theories need not be examined here (Rudner, 1966; and Braithwaite, *1960*, provide adequate accounts). The vocabulary may contain *primitive* terms which cannot be defined and *defined* terms which may be formed from the primitive terms. The sentences may similarly be divided into primitive sentences— *axiomatic statements*—and derivative sentences—*theorems*. In the classic case of Euclidean geometry, terms such as 'point', 'line', 'plane', form the primitive terms collected together in an initial set of axiomatic statements from which the whole structure of Euclidean geometry is derived.

In addition to the primitive terms and the axiomatic statements scientific theories also possess certain rules which govern the formation of the derivative sentences. In general these rules are those of deduction. These primitive terms, axiomatic statements, and formation rules, make up a *calculus*.

But a theory is useful in empirical science only if it is given some interpretation with reference to empirical phenomena. Thus in Euclidean geometry primitive terms such as 'point' and 'line' may be interpreted by 'dots' and 'pencil lines'. By elaborating a formal structure we ensure the logical truth of the propositions contained in the theory. These propositions are linked to empirical phenomena by a set of interpretative sentences—sometimes called a *text* or a set of *correspondence rules* (Brown, 1963, 147–8; Nagel, 1961, 90–105). The *text* for the theory performs two important functions. First it provides a translation from the completely abstract theoretical language to the language of empirical observation. Without such a translation there is no possibility of empirical support for the theory. We can demonstrate the kind of procedure involved with reference to von Neumann's and Morgenstern's (*1964*, 73–5) formal statement of game-theory where the basic postulates are stated in both a theoretical and an observation language. The first three postulates of the theory were:

Theoretical statement	*Observation statement*
(i) A number v	(i) v is the *length* of the game G.
(ii) A finite set θ	(ii) θ is the *set of all plays* of G.
(iii) For every $k = 1, \ldots n$: A function $F_k = F_k(p)$, p in θ etc.	(iii) $F_k(p)$ is the *outcome* of the play p for the player k.

The *text* of a theory not only identifies the empirical subject-matter which the theory refers to. It also identifies the *domain* of the theory. The *domain* of a theory may be regarded as the section or sections of reality which the theory adequately covers. The theory itself is simply an abstract set of *relata*, the *text* states how and under what circumstances such an abstract system may be applied to actual events. The extent of the *domain* of the theory varies according to the number of terms within it that have to be given a specific translation in terms of a specific subject-matter. On the one hand there is the calculus of probability which may be referred to all manner of phenomena (see chapter 15) and on the other we may have a theory of economic equilibrium which may contain terms such as 'perfect competition' which automatically restrict the domain of the theory to economic phenomena. One of the main trends in the history of science is, of course, the development of theories of greater and greater generality which can be used to derive theories of lower order generality.

A scientific theory may be regarded as exhibiting this standard formal structure. Axiomatic statements and primitive terms are set up from which, given the rules of deduction, a large number of theorems may be derived which, when linked to an empirical subject-matter by way of an appropriate *text*, form the empirical laws governing the behaviour of that empirical subject-matter. Not all scientific theories exhibit such a formal structure, but the important point here is that a scientific theory, if it is truly scientific, should in principle be representable in terms of such a structure. Only by creating such a structure can we ensure that the law statements we use in explanation are consistent with respect to each other. In fact axiomatic statement in the sciences tends to be rare, although mathematical systems (Stoll, 1961; Cohen and Nagel, 1934) are frequently represented in such a fashion and increasingly in the natural sciences such as biology (Woodger, 1937; Gregg and Harris, 1964), and in the social sciences such as psychology (von Neumann and Morgenstern, *1964*), theory is being given formal axiomatic treatment.

The development of formal theory is not necessarily beneficial, and Rudner (1966, 52) suggests that premature formalisation may in some instances prove stultifying. There are, however, benefits to be had from the formal statement of a theory:

(i)   The formal statement of a theory requires the elimination of inexactness and, as a consequence of this, ensures complete certainty as to the *logical* validity of the conclusions. The elimination of inexactness involves paying a price. In particular it imposes 'a gap between the idealised subject-matter and the subject-matter being idealised' (Körner, 1966, 91). In other words an exact class of objects is idealised out of an inexact empirical world. This generalising procedure is carried to its limits in formal theory construction. We simply use terms such as v, θ, p, etc., to use the example of game-theory, in a manner that assumes homogeneity within the class of events. The empirical success of a theory relies entirely upon the success of the *text* in linking the abstract symbols of the theory to real world events. As Kemeny (1959, 89) has pointed out, 'establishing a connection between these two worlds is one of the most difficult tasks a scientist must face'. The formal statement of a theory may increase this difficulty while ensuring at the same time that the logic of the theory is completely error-free.

(ii) The elaboration of formal theory, provided the basic postulates are good ones, can help suggest new ideas, prove unsuspected conclusions, and indicate new empirical laws. If it achieves nothing else, it may simply unify the theoretical structure relevant to some field of study in a complete and rigorous manner.

(iii) The formal statement of a theory requires turning a spatial or temporal sequence into a completely non-spatial and non-temporal set of *relata*. Even in theories which explicitly include time or space as variables the treatment is abstracted. This complete abstraction from location in either space or time implies that we must necessarily treat events *as if* they were universally true. It is therefore characteristic of formal theory to state all propositions—whether primitive or derived—*as if* they were *universal* propositions. Again, it is the *text* which has to perform the difficult task of linking these universal propositions to empirical events which have a location in space and time. In some cases, as with the laws of physics, such a process does not involve great difficulty, but in the social sciences it has become clear that considerable difficulties need to be overcome if formal theory is to perform a useful task.

This brief discussion regarding the advantages of formal-theory construction has indicated that the key problem in the *application* of formal theory in the empirical sciences is the provision of an adequate *text*. It seems necessary, therefore, to consider this problem in greater detail.

B The text of scientific theories

The *text* associated with a formal theory performs two essential tasks. First, it identifies an abstract symbol with a particular class of real-world phenomena—a dot with the Euclidean concept of a point for example. Second, it may place the abstract symbols within a particular context which may include specific mention of location in space and time. We shall consider both of these functions separately.

Relating a set of abstract symbols to real-world sets of events may appear to involve simply the development of an appropriate classification system for real-world phenomena. When the relationship is direct this is, by and large, what is involved. But unfortunately a direct relationship between the abstract symbol and real-world phenomena rarely exists, and it is characteristic of more sophisticated theoretical structures to involve many indirect steps in the generalising process. The steps involved may be characterised as follows:

real-world phenomena $\rightarrow$ definition and classification $\rightarrow$ idealisation and concept formation $\rightarrow$ abstract symbols

The symbols of a theoretical system may represent idealisations, abstract concepts, theoretical entities, and so on, which, in themselves, have no empirical content (see above, pp. 18–22). Many of the theorems deduced from the axioms may thus not contain any empirical references even when the *text* is provided. In this case we may regard these statements as the *theoretical laws* in the system. Only deduced statements which contain on translation empirically identifiable subject-matter may be regarded as *empirical laws*.

Considerable discussion has centred upon the nature of theoretical statements (Nagel, 1961, chapter 6; Braithwaite, *1960*, chapter 3; Hempel, 1965, chapter 8). On the one hand some writers have accorded theoretical statements the status of propositions whose truth or falsity is capable of being asserted. Others have argued that since the terms used—such as 'atom', 'electron', 'neutron', etc., have no direct empirical status (since they cannot be observed) it is illogical to attempt to assert the truth or falsity of statements which contain imaginary terms. Under these conditions we may only assert that a theory is useful or not useful, helpful or not helpful, in attempting to understand the complex phenomena of the real world. This controversy regarding the cognitive status of theories need not concern us in detail here since, as Nagel (1961, 152) points out, the conflict is largely a matter of 'conflict over preferred modes of speech'.

Science does not demand that *all* deduced propositions be capable of direct empirical testing. It does demand, however, that *some* of the deduced propositions should be so stated that empirical testing becomes possible. The *text* associated with a formal theory should not only link abstract symbols with abstract concepts. It should also specify how abstract concepts may be reduced to factual statements. This therefore raises the whole problem of the nature of concept formation in the sciences and the operational problem of giving idealisations and theoretical concepts some adequate definition.

There are innumerable idealisations to be found in both the natural and social sciences. Indeed explanation would be impracticable without such idealisations. Concepts such as 'frictionless engine', 'ideal gas', 'graded profile', 'perfect competition', 'point', 'culture', and so on, are all idealisations of some kind. Such terms are 'a convenient sort of shorthand employed to represent, and thus avoid using, sets of relatively complex statements' (Rudner, 1966, 57). We may conveniently divide such idealisations according to whether they are intuitively or theoretically defined.

An 'ideal type' or 'theoretical concept' may be defined in such a way as to make the term itself redundant. We may replace 'perfect competition', for example, by a number of defining terms which tell us the conditions under which perfect competition occurs—we might use defining terms like 'perfect information', 'economically rational individual', and so on. These defining terms are themselves capable of further definition, but unless we are to become involved in an infinite regression of definition we need to establish certain basic statements to which all definitions may be ultimately referred. These basic statements are, of course, the *axioms* of the theoretical system as a whole. Theoretically defined statements make reference to these axioms. But there are many statements which we cannot define in this way, and here we rely purely upon the intuitive appeal of the 'ideal type'.

Theoretically defined statements are characteristic of the natural sciences where physics, in particular, has achieved a high degree of unification in its theoretical structure. A concept such as an 'ideal gas' can be defined in terms of physical theory, while the extent to which an actual gas deviates from the ideal can be measured. This ability enables a very precise *text* to be provided for Boyle's law, since the exact conditions under which the law operates can be defined and the nature of an ideal gas be given precise meaning. Idealisations such as this may be obtained within 'the framework of a given theory, as special cases of more inclusive principles' (Hempel, 1965, 169). Ultimately the body of theory on which the idealisations are based will refer to idealisations or theoretical concepts which appear to be undefinable. These are the *primitive* terms of the theory. But as science progresses, so axiomatic statements and primitive terms developed for the explanation of a relatively restricted set of phenomena may later be subsumed under a more embracing theoretical system. Thus 'the most comprehensive theory of today may be but a systematic idealisation within the broader theoretical framework of tomorrow' (Hempel, *loc. cit.*).

Many idealisations in both the natural and social sciences cannot be referred to any well-established theoretical structure, either because the idealisation is itself inappropriate, or because the requisite general theory has yet to be established. The concept of 'grade' in river erosion has an interesting history in this respect (Dury, 1966). Originally an intuitive concept it has frequently been re-defined and elaborated since the late nineteenth century. In spite of repeated failure to provide a theoretically based definition (or even to agree on an intuitive one) the concept has played a very important role in explaining landscape evolution. More recently, however, it has been suggested that the concept could be defined, given the fundamental laws of physics and more special hydrological laws. To provide an analytic definition has proved too difficult, but Leopold and Langbein (1962) managed to provide a very general definition using a probabilistic formulation. This work suggests that the concept of grade, which holds such a central place in the Davisian system, is nothing more than the probabilistic tendency for the long profile of a river to assume an exponential form. This leads Dury (1966, 231) to conclude that the concept is 'unserviceable both in the study of actual terrains and in the theoretical analysis of landforms generally.' In this case the conclusion may well be that the concept itself is inappropriate.

In other cases idealisations cannot be referred to any theoretical structure and there is, therefore, no way of judging their appropriateness except by appeal to the intuitive reasonableness of the idealisation. This is typical of many of the social sciences and led Weber (*1949*, 89-112) to insist that the idealisations of social science were fundamentally different from the theoretical concepts of the natural sciences. This position has been examined in detail by a number of writers (Watkins, 1952; Hempel, 1965, chapter 7; Rudner, 1966). Hempel's comments on the idealisations of economics are worth noting. He suggests that they are deficient in two respects. Firstly, the idealisations cannot be deduced as special cases from a broader theory 'which covers also the non-rational and non-economic factors affecting human conduct.' Secondly, no clear specification of the areas of real-world behaviour which are supposed to be described by economic theory can be given. In other words the necessary *text* is deficient, and the *domain* of economic theory cannot be defined. In general 'the failure to achieve significant explanatory power in the social sciences is the result of the paucity in such disciplines of the requisite general theory' (Rudner, 1966, 62-3).

These deficiencies in the idealisations of social science (and human geography is no exception) would not be so serious were it not for the fact that they seriously inhibit the development of adequate verification procedures. It has been characteristic of economics, for example, to rely upon the inherent intuitive appeal of its basic idealisations as 'worldly' support for the validity of its theory, rather than upon close empirical testing of deduced consequences in the market place. This difficulty can be overcome to some degree by ensuring that *some* of the propositions deduced contain empirical subject-matter. Clarkson (1963) has thus shown how the theory of demand can be stated in such a way as to conform to the natural scientific model of explanation and, hence, be subject to empirical verification. In particular, Clarkson searches a hitherto 'uninterpreted' theoretical structure which made reference to only theoretical concepts, and provides some kind of *text* for the theory partly by reconstructing the theory and partly by reducing theoretical idealisations to propositions which are mainly empirical in content.

But the problem of finding adequate empirical definition of theoretical concepts can be solved only by the provision of an adequate general theory. The development of powerful basic axiomatic statements will make possible precise definition of the idealisations on which current theory rests. This procedure may lead to the reduction of the large number of idealisations and concepts in social science to special cases of more general axiomatic statements. In the natural sciences physics and to a lesser extent chemistry have spread their influences wide. Many of the concepts and idealisations used in the natural sciences may be ultimately defined by reference to the basic concepts of physics. The unification of disparate theoretical structures into one system of statements involves the *reduction* of disparate idealisations to special cases of a few basic postulates. This phenomenon of *reduction* (Nagel, 1949) may also be found in the social sciences, and the development of general theory in the social sciences may well depend on such reduction. The postulates of economics may be reducible to a particular subset of postulates in psychology. Clarkson (1963) reduced the theory of demand to a theory of individual consumer behaviour, while much of the work directed to defining and measuring abstract economic concepts such as 'utility', 'value', and so on, point to the general interest of economic research in seeking adequate definition by reference to psychological postulates (Fishburn, 1964; Shelly and Bryan, 1964). Some writers, such as Carnap (1956) have further suggested that the basic postulates of psychology might be reduced to the basic postulates of physics. Anthropologists, such as D. Freeman (1966), have also suggested that the route to a 'value-free' and truly scientific anthropology lies through studying the neurophysiological determinants of human behaviour. The degree to which such reduction can take place, however, is a controversial issue, and even if it is conceded that total reduction is ultimately possible, this is so far from being practicable at the present time that it seems irrelevant to the current problems of empirical enquiry. On the other hand it cannot be denied that there is considerable benefit to be had from the integration of diverse concepts and statements into some more general theoretical framework.

We may conclude, therefore, that the development of general theory in the social sciences—and the reduction of some concepts which this implies—may enable more precise definition of certain idealisations and hence facilitate the statement of an appropriate *text* for some of the theories developed in the social sciences. Without such *texts* the domain of these theories cannot be defined precisely, and hence verification relies more upon the intuitive appeal of the theory than upon empirical testing. To this extent Weber's point that theories in the social sciences are different from those in the natural sciences does highlight a significant practical difference. But, as Hempel has pointed out, this practical difference does not necessarily mean a basic methodological difference. If this is so, then there need be no fear of 'letting the reductionism of science dissolve us into parts' (Blaut, 1962, 5).

The *text* for a theory does not simply link the abstract symbols of the calculus with theoretical concepts and empirical classes of phenomena. It may also identify the external conditions which may modify the operation of a theory. In this the natural sciences again possess the advantage of an elaborate theoretical structure since, for example, an 'ideal gas' may not only be defined with reference to physical theory, but the factors that may interfere with the empirical identification of an 'ideal gas' can also be defined. Therefore the factors that need to be held experimentally constant (or adjusted for) can be fully identified. Again, in many of the social sciences these interfering conditions cannot be identified since the appropriate general theory does not exist. If it did, one of the most important variables to be considered is absolute location in time and space. Theories of diffusion or migration, for example, may be applicable only at certain times and in certain societies. This problem is undoubtedly one of the most important methodological problems facing disciplines such as history and geography where the application of theories and their empirical verification must take place against a background of shifting patterns of human behaviour over space and over time. In physics—the discipline which has provided the model for 'scientific' explanation—this problem does not arise. The universality of the theoretical statements may be transferred to any situation in space and time since it appears that the statements are universal *in fact*. In the social sciences this is clearly not so, and therefore one of the important functions of the *text* of a theory is to identify the domain of objects and events to which such theories can be applied—this domain *may* simply be defined by a set of spatial and temporal co-ordinates.

A theory without a *text* and a well-defined domain is useless for prediction. To a greater or lesser degree theories are provided with appropriate *texts*. No *text* can be absolutely perfect, but undoubtedly the greater success of the physical sciences, relative to the social sciences, in providing a *text* for theoretical structures accounts for the greater predictive success of the natural, as opposed to the social, sciences.

C Incomplete theories—the problem of partial formalisation

We have already suggested that it is comparatively rare for theories in either the natural or social sciences to be stated in a completely formal manner. In some cases this may simply be because sufficient information is not yet available for such a formal statement to be made. This raises, therefore, the problem of how theories are *in fact* stated, how far such theories can be partially formalised, and what criteria we need to employ in distinguishing speculative fantasies from 'scientific theory'. Here we seem to be trying to distinguish natural breaks in a continuum of theoretical formulations, at one end of which lies the pure formal theory and at the other end lies the purely verbal speculative statement. The problems associated with 'partial' or 'incomplete' theoretical formulations have not been given a great deal of attention until relatively recently and therefore the guide-lines available from the philosophers and logicians are not as numerous as one would like. There are fortunately some exceptions (e.g. Hempel, 1965; Rudner, 1966).

We shall attempt a brief classification of theoretical structures according to their degree of formalisation. There are four main types.

Type 1: Deductively complete theories possess a completely formal structure with the axioms fully specified and all steps in the deductive elaboration fully stated. A textbook in Euclidean geometry, for example, exhibits this kind of structure.

Type 2: Systematic presupposition (Rudner, 1966, 48) in theories involves reference to another set of theories. There are two sub-types which may be distinguished:

(i) Elliptic formulations (Hempel, 1965, 415) presuppose a body of theory which is deductively complete. We might, for instance, refer to one of the theorems of geometry without citing the whole proof. Such ellipt'cally formulated explanations are incomplete but as Hempel points out, this is in 'rather a harmless sense'.

(ii) In other cases the body of theory referred to may itself be incomplete or even non-existent. 'Common-sense' presuppositions are frequently referred to in incomplete theories of this kind. As Rudner (*loc. cit.*) points out, the presuppositions of a theory often remain inexplicit either because of the technical difficulty of making a full statement of them or because of 'the theoretician's ignorance of what his theory does presuppose'. Inexplicit reference to the supposed conclusions of some other area of scientific enquiry probably is 'one of the most frequently encountered ways in which current theories fall short of full formalisation.'

Type 3: Quasi-deductive theories may be regarded as incomplete because the primitive terms of the theory or the deductive elaboration of it do not conform to the standards of formal theory. There are three sub-types:

(i) Inductive systematisations may be regarded as a form of quasi-deduction since the conclusions follow only probabilistically upon the premisses.

(ii) Incomplete deductive elaboration may occur in the 'harmless' sense that although the steps could be shown they are left out for the sake of brevity of exposition. But in other cases quasi-deduction involves making more serious assumptions. The steps in the argument may be too complicated or technically too difficult for explicit deductive procedures to be employed. At best, this may mean that, for example, a system of differential equations cannot easily be solved analytically and, therefore, simulation procedures are employed to find an approximate solution. At worst, this may mean a purely intuitive leap over difficult deductive steps. Such an intuitive leap may or may not turn out to be justified in the long run. But it surely means that our confidence in the logical validity of the theory must be substantially reduced.

(iii) Theories making use of *relative primitives* (Rudner, 1966, 50–1) are automatically quasi-deductive in structure since the primitive terms and concepts are only partly established. In the initial stages of theory-formation it may be difficult (indeed impossible) to establish which of the indigenous terms in the theory should be regarded as primitive and which should be regarded as derivative. Indeed, all of the primitive terms may not yet be developed. Thus, apart from situations where reference may be made to external primitive terms, as in Type 3 situations, the incompleteness of a theoretical formulation may be due entirely to failures in concept-formation and identification. Without precise concept-formation the primitive terms of the theory may remain fuzzy and obscure. In the initial stages of theory-construction concepts may be developed as 'temporary' primitives from which deductions are made. One of the main aims of theory development is to replace such temporary concepts by more permanent concepts which may act as the primitive terms of the theory with precision. In the long run, as Hempel has pointed out, the primitive terms of one discipline may become deducible from the theoretical structure of another. But the failure to identify the *indigenous* primitives of a theoretical structure forms one of the major ways in which actual formulations depart from full formalisation.

Type 4: Non-formal theories may be regarded as statements made with theoretical intention, but for which no theoretical language has been developed. Theories stated in everyday language may range in sophistication from carefully thought-out systems of linked statements to the kind of 'explanation sketch' frequently used by historians. We may distinguish two sub-types:

(i) Verbal explanations which can, without any substantial modification of concepts or manipulation, be rendered at least partially into formal structure. An extremely good example of this kind of statement is the theory of interaction in social groups proposed by Homans (1950). This sophisticated verbal statement could be represented by Simon (1957, chapter 6) by a formal structure without any substantial changes in the basic concepts or the relationships established.

(ii) Verbal explanations which cannot be even partially formalised without a substantial modification of the concepts used and clarification of the deductive relationships proposed. Such theories may be regarded in their initial state as 'pseudo-theories' since they purport to be theories appropriate for explanation without conforming in any way to the basic model of scientific explanation. It is, of course, often extremely difficult to determine whether verbally stated theories lie in this category or in category 4(i). Such an investigation may well prove to be one of the major areas of research effort in disciplines where theoretical development is poor.

We may summarise these views regarding the structure of scientific theory as follows:

(i) It is possible to construct a standard normative view of what is meant by a *scientific* theory as opposed to any hopeful explanation. In practice theories diverge from this absolute standard because:

(ii) Some theories cannot be expressed in formal terms simply because the primitive terms cannot be specified and because of technical difficulty in the deductive process. We may attempt to classify theories in terms of the way in which they diverge from the standard structure. This classification is concerned wholly with the *logical* status of the theories. But theories may also prove deficient because:

(iii) The empirical status of the theory cannot be assessed because the appropriate *text* for the theory has yet to be established. We may thus have theories which are quite powerfully formalised but which have poor empirical status—this is true of, for example, game-theory and much of classic economic theory. On the other hand we may possess poorly formalised theories which are, nevertheless, strongly supported by empirical evidence. The best examples of this may, perhaps, be found in the history of physics where, for example, Boyle's law was initially poorly formulated but given quite strong empirical status.

Chapter 8
Hypotheses and Laws

Most logicians regard a hypothesis as being a proposition whose truth or falsity is capable of being asserted.[1] Once the truth or falsity of the proposition has been determined the proposition becomes a true or false statement. One might, therefore, put forward the hypothesis that Cairo is south of Madrid and then determine whether this is a true or false statement by referring to an atlas. This example demonstates a general problem about such a procedure. The truth or falsity of the hypothesis can be determined only given a prior definition of 'north' and some measure of the 'degree of northness'. Such definitions, when provided, turn out to be very closely tied to some theory. Thus the truth or falsity of a hypothesis can be determined, in most cases, only with respect to the domain of some theory.

In scientific investigation, however, the term 'hypothesis' is often given a somewhat more restricted meaning. Thus Braithwaite (*1960*, 2) asserts:

> A scientific hypothesis is a general proposition about all the things of a certain sort. It is an empirical proposition in the sense that it is testable by experience; experience is relevant to the question as to whether or not the hypothesis is true, i.e. as to whether or not it is a scientific law.

In Braithwaite's view a scientific hypothesis is a particular kind of proposition which, if true, will be accorded the status of a scientific law. The testability of a hypothesis is crucial but, as we have seen, there are many hypotheses within a theoretical system which cannot be directly tested against sense-perception data. Thus:

> The empirical testing of the deductive system is effected by testing the lowest-level hypotheses in the system. The confirmation or refutation of these is the criterion by which the truth of all the hypotheses in the system is tested. (Braithwaite, *1960*, 13.)

Whether or not the 'truth' or 'falsity' of the empirical hypotheses

---

[1] It is worth while noting the difference between 'hypothesis' and 'hypothetical' which, apart from grammatical function, appear to have rather different connotations. A theory may, thus, sometimes be spoken of as a set of hypothetical propositions. Since a theory cannot be shown to be true or false, this meaning turns out to be rather different from the definition of 'hypothesis' given here.

can be transferred to the theoretical propositions from which they are derived depends upon the view taken regarding the cognitive status of theories in general.

Given Braithwaite's use of the term 'hypothesis', it would seem that the difference between this term and a scientific law is simply a matter of confirmation. But between the more general meaning of 'hypothesis' and scientific law there lies a further difference. Any true proposition is not, thus, accorded the status of scientific law. But the precise nature of the criteria to be employed in judging whether or not a statement qualifies as a scientific law is difficult to determine. Braithwaite (*1960*, 10) comments that 'while there is general agreement that a scientific law includes a generalisation, there is no agreement as to whether or not it includes anything else'.

A generalisation asserts a constant conjunction between two events. It is of the form 'every A is also B' or 'every A stands in a certain relationship, R, to B'. To most writers generalisations assert constant conjunctions but nothing more, whereas laws imply some necessary connection over and above that of constant conjunction (Nagel, 1961, chapter 4; Toulmin, *1960B*, chapter 3). Thus Nagel draws a *prima-facie* difference between *accidental* and what he terms *nomic* universality. The latter contains some 'explanation' or at least asserts some form of necessary conjunction other than that occurring by accident or by definition. Nagel goes on to point out, however, that it is impossible to sustain logically the *prima-facie* difference asserted. Braithwaite (*1960*, 10) similarly notes the inability to differentiate between generalisations and laws on logical grounds, and he therefore commences his analysis with the assumption that laws 'assert no more (and no less) than the *de facto* generalisations which they include'.

The numerous analyses by logicians and philosophers suggest that two criteria are of major significance in identifying laws. The first is the *universality* of the statement. The second is the relationship of a statement to surrounding statements and in particular the way in which *one* statement fits into a whole collection of statements which themselves form a scientific theory. These two criteria will, therefore, be examined in the following two sections.

A The universality of law-statements

A law, it is frequently stated, should be unrestricted in its application over space and time. It is thus a *universal statement* of unrestricted range (Braithwaite, *1960*, 12; Nagel, 1961, 57–9; Popper,

1965, chapter 3). This suggests at least one important criterion for distinguishing a law. Unfortunately the interpretation to be put upon universality is not entirely clear. We shall begin by taking the notion of universality in its most rigid form.

The universality criterion requires that laws should not make specific or tacit reference to proper names. Consider the proposition that 'towns of similar size and function are found at similar distances apart.' (This proposition may be taken from Thomas's, 1962, work.) The term 'town' can be defined only with reference to human social organisation and it carries with it, therefore, an implicit reference to the proper name 'earth'. Within such a context the statement may be true, but the universality criterion has undoubtedly been offended. To get round this difficulty we may attempt to define 'town' in terms of a set of properties which we claim are possessed by towns and only towns. In an infinite universe, however, there may well be some phenomenon which possesses all the properties listed without being a town. Again we are not justified in regarding the statement as being a proper law.

J. J. Smart (1959) used the argument of the above paragraph to show that laws could not be developed in biology. According to Smart the only strict laws to be found in the whole of science are those to be found in physics and, perhaps, chemistry. These, he suggested, were truly universal in nature. This view automatically excludes the development of laws in biology, zoology, geology, physical geography, etc., except in so far as such disciplines can reduce their statements to those of physics. The social sciences and human geography are even more seriously affected.

But there are powerful arguments against interpreting universality in such a strict manner. There are two ways in which we may justify some relaxation of it. With 'purely' empirical propositions it may prove useful to draw a distinction between *philosophical* and *methodological* universality. Philosophical universality involves the belief that universally true statements can be made. Such a belief may be supported by reference to some set of metaphysical propositions—such as the Platonic doctrine of universal essences—or else it depends upon showing that a statement is in fact universally true. The latter course is essentially an inductive step and, therefore, a degree of uncertainty is involved. A proposition can never be shown empirically to be universally true. This applies as much to the 'strict' laws of physics as it does to the 'mere generalisations' of biology and economics. Philosophical universality implies methodological universality, but the reverse relationship does not hold. We may regard statements *as if* they were universally true without necessarily believing that they are or even assuming that they will ultimately be shown to be so. This position we shall term methodological universality. In such a case it becomes a matter of deciding whether it is useful and reasonable to regard a statement *as if* it were universally true, and, hence, *law-like*.

It is interesting to note that Smart suggests that laws in the strict sense cannot be developed to account for the behaviour of complex systems. This seems to reflect the declining utility and reasonableness of regarding extremely complex organisations as being subject to laws. But even such apparently intractable phenomena as wars can be treated *as if* they were subject to laws, with intriguing results. Of course the laws which writers such as Richardson (1939) and Rapoport (1960) have discussed cannot easily be regarded as 'universally valid' and certainly neither of those writers would claim that status for their results. But then many of the so-called strict laws of physics and chemistry are also questionable in this respect. This brings us to the second point of interpretation of the universality criterion which may lead to some relaxation in its application.

When considering the nature of scientific theories it was noted that a theory requires the development of an abstract calculus. One of the functions of this abstract system was to turn empirical relationships located in space and time into abstract sets of *relata*. It follows from this that all deduced consequences of such a theoretical structure must be stated *as if* they were universally true propositions. If we are dealing with fully formalised systems the statements are purely analytic and have no empirical status. They are given empirical status by means of the appropriate *text*. Invariably that *text* makes use of proper names and this in itself makes a fiction of the supposedly real universality of the statement. This applies as much to the propositions of natural science as to those of social science. As Brown (1963, 147–8) has pointed out:

> The qualification laid upon a scientific hypothesis by its 'text' includes the definition of its terms and an accurate statement of the limits within which it is supposed to hold. Neither of these need appear in the generalisation itself, and however obvious each is, critics of the social sciences often speak as though . . . hypotheses are deficient in not containing such qualifications on their face. It is worth time, therefore, to remind ourselves that the generalisations of the natural sciences do not differ on these points from those of the social sciences.

In the sense that all formal statements of scientific theory involve abstract analytic statements, all theoretical statements are universally true. But only when deduced theorems have been translated into empirical statements which can be verified, can we raise the issue of the universality of the empirical statement. If, as it seems, the very process of translation may involve mention of proper names or even specific locations or time periods, then any attempted rigid application of the criterion of universality becomes meaningless. In other words the universality criterion can be applied only by referring to an already-established domain of objects and events. In most cases, however, we cannot define that domain with precision and it may therefore be argued that universality has no absolute meaning. It has relative meaning in terms of the conceptual domains, of objects and events whose definition often requires the use of proper names and whose boundaries are subject to considerable change as our understanding increases. The universality criterion seems more applicable as a directive for formulating propositions than it does to determining the inherent empirical lawfulness of the proposition itself. In the latter case the problem is one of drawing a line along a continuum of statements of ever increasing generality; statements which appear to approach universality asymptotically! As Ackoff (1962, 1) has it:

> The less general a statement the more *fact-like* it is; the more general a statement, the more *law-like* it is. Hence, facts and laws represent ranges along the scale of generality. There is no well-defined point of separation between these ranges.

Somewhere along this continuum a division may be drawn but there are evidently no precise criteria available and necessarily scientists may vary in their judgement. Zetterberg (*1965*, 14) comments, for example, on the situation in sociology where the Berelson-Steiner thousand or so propositions contain 'anywhere from five to fifty' laws 'depending on how strict we make our criteria'.

B The relationship between laws and theories

A substantial part of Braithwaite's analysis of scientific explanation is concerned with establishing how laws are related to a surrounding structure of theory. In other words it is impossible to determine whether a statement is or is not a law simply by reference to the truth or falsity of the generalisation it contains. Consider again the statement that 'towns of similar size and function are found at similar distances apart.' Although we may have reservations about regarding this as a law (the notions of similarity and function are too inexplicit for example) we certainly regard it as being more law-like than a statement that 'all towns contain collections of buildings.' The significant difference between these statements is not basically a function of either empirical status or universality. It is simply that the first statement has some place within the structure of central-place theory which, however crudely it has been formulated, has a putative deductive structure in the formulations of Christaller and Lösch, whereas the second statement has no theoretical structure of explanation surrounding it and is in any case trivially true. Accordingly, a major criterion in determining whether a statement is or is not a law is the relationship of that statement to the system of statements that constitutes a theory.

If this criterion is accepted, then we also need to adjust our ideas regarding the verification procedures necessary to transform a scientific hypothesis into a scientific law. A generalisation may be established as true or false simply by direct reference to empirical subject-matter. The truth of an empirical law has to be established by this method too, but in addition it requires support from other established laws, theoretical laws (which cannot be given any direct test), and also, perhaps, from other lower-level empirical laws which it helps to predict. This double appeal to 'fact' and 'theory' may involve conflict. Statements derived from theory may turn out to be empirically unsupportable, while empirical statements which intuitively appear to be of great significance sometimes cannot be linked to any existing theoretical structure. The first case leads to doubt regarding the degree of confidence which we can attribute to a theory, the second case suggests that new theory should be formulated. This conflict is characteristic of all areas where scientific knowledge is advancing and it forms part of the stimulus to the development of newer and more powerful theoretical structures.

The criterion of theoretical supportability is again a rather imprecise one when we try to determine whether a particular statement is or is not a scientific law. The degree of theoretical support required, the degree of empirical support, the degree of confidence necessary in the theoretical structure of explanation as a whole, and so on, may vary significantly from one person to another. Yet it is certain that somewhere along this continuum we begin increasingly to speak of statements as scientific laws. The precise nature of the criteria may be obscure, but this does not mean that they are useless and insignificant.

A scientific law may be interpreted most rigidly as a generalisation which is empirically universally true, and one which is also an integral part of a theoretical system in which we have supreme confidence. Such a rigid interpretation would probably mean that scientific laws would be non-existent in all of the sciences. Scientists therefore relax these criteria to some degree in their practical application of the term. The precise degree of relaxation remains very much a matter for individual judgement, although it is worth noting that a particular group of practising scientists will frequently maintain broadly similar standards of judgement on this issue which yet again demonstrates that scientific judgements are frequently better understood as behavioural conventions rather than as matters of irrefutable logic (above, pp. 16–23).

Basic Reading on Theories, Hypotheses and Laws

Braithwaite, R. B. (*1960*).
Brown, R. (1963).
Hempel, C. G. (1965).
Nagel, E. (1961).
Rudner, R. S. (1966).

(*N.B*: the above are a selection of standard texts on the philosophy of science, and many of them cover similar ground. They all contain good statements regarding the nature of theories, laws and hypotheses, and a selection of two of the above titles will provide the necessary background.)

Chapter 9
Laws and Theories in Geography

Given the exposition of the previous two sections we may now consider the nature of the statements we make in geographic research and the form we use to link those statements together into a coherent explanatory structure. The aim of this section, therefore, is to examine the degree to which geographers have resorted to scientific explanations, and to investigate the potential of a more conscious use of 'the scientific method' in geographic explanation. This general form of explanation—and we are here referring to the very broad model of scientific explanation—has been extremely efficient in both the natural and social sciences. This is not to claim that all our outstanding substantive problems will be solved merely by the touch of the glittering wand of scientific explanation. Far from it. Scientific explanation, as all mature users of the method fully recognise, has its limits. The efficiency of the method is, as Boulding (1956, 73) has pointed out, bought at the price of 'a severe limitation of its field of enquiry and a value system which is as ruthless in its own way in the censoring of messages as the value system of primitive man.' But we shall become conscious of these limits only when we have explored fully what can be done within their confines. This, by and large, we have failed to do.

I LAWS IN GEOGRAPHY

Do we and can we employ laws to explain geographical events? Given the very imprecise criteria available for distinguishing scientific laws from other kinds of statement, such a question may appear rather senseless. But since the argument regarding laws in geography is so central to the 'methodological image' which geographers have of themselves, and since laws perform a vital function in scientific explanation, we are forced to attempt a seemingly impossible task. The attempt will, perforce, prove rather inconclusive, but the methodological by-products of it will prove very useful.

If we employ very rigid criteria for distinguishing scientific laws, then we can scarcely expect geographical statements to achieve such a status. If we follow Smart's (1959) analysis, even physical geography, which Wooldridge and East (1951, 30) suggest is susceptible to treatment by universal principles, can only hope to employ laws which are derivative of physics or chemistry. The laws which enter into physico-geographical explanations will simply be the 'fundamental' laws of physics and chemistry applied to geographical circumstances. Physical geography may thus 'consume' laws but specifically geographical laws cannot be developed.

Such an analysis leaves human geography with two alternatives. The first would be to relate all human behaviour to the 'strict' laws of physics and chemistry. Without denying the importance of work in neuro-physiology and the reduction of statements which this may ultimately accomplish (Carnap, 1956), it seems that this solution is, by and large, rejected by most social scientists as a practical impossibility in the immediate future. The second possibility is to establish 'strict laws' of behaviour which would provide for the needs of the social sciences in much the same way that physics provides for the natural sciences. Sociologists have not been averse to such an attempt. Dodd (1962) has suggested the reduction of all sociological theory to an examination of 'tell-hear acts'. This fundamental unit of communication can be examined independently of space, time, and value system, and can therefore be used to build up a system of truly universal sociological laws. Such laws, if developed, could be used and consumed in human geography.

The adoption of such a rigid interpretation of the term 'law' means that geography certainly cannot hope to produce laws, and their use in geography will be possible only to the degree that the 'strict laws' of science can be 'consumed' in the process of geographic explanation. For the rest, we are doomed to discuss mere generalisations. It should be noted that this argument does not place geography in particular into an invidious position. It merely serves to rank physical geography along with biology, zoology, botany, and so on, and human geography along with all the social sciences.

Using less rigid criteria, the identification of laws in geography becomes partly a matter of identifying the relevant theory, and partly a matter of our own willingness to regard geographical phenomena *as if* they were subject to universal laws, even when they patently are not so governed. We shall first consider the idea of methodological universality in relation to explanation in geography.

Ever since the victory of the 'uniformitarian' views of Hutton, geographers have assumed that the phenomena they study are subject to universal laws. Initially this assumption was an indigenous discovery, although now it can be firmly embedded in physical theory (Chorley *et al.*, 1964). To an increasing degree the laws which enter into the explanation of events in physical geography may simply be regarded as special instances of physical laws or derivative therefrom. Barry (1967) has thus suggested that meteorological and climatological analysis rests on six basic laws, two of which are the first and second laws of thermodynamics, while the remainder are more specific meteorological laws which rest firmly upon Newton's laws of motion. Physical geography thus consumes a great deal from physics and chemistry. But there are still specific types of relationship which are, as yet, indigenous laws accepted as such. Horton's law of stream numbers, Krumbein's laws of beach-profile development, Penck's laws of slope retreat, and the more complex laws of landscape evolution concealed beneath the superficial descriptive gloss of the Davisian system, are examples of the use of law-like statements in physical geography. Almost all studies in physical geography assume the processes they are examining to be universal (this was Hutton's major contribution) and the assumption seems perfectly reasonable to us. Thus even though the ultimate aim of geomorphological enquiry may have been (as it was for the most part until the 1940s) to elucidate the genesis of landscape (a unique history as it were), the processes themselves were certainly regarded as manifestations of scientific law.

Human geographers have frequently resisted the idea of treating individual events *as if* they were similarly subject to scientific law (*cf.* Wooldridge and East, quoted above p. 68). More recently opinion has swung in the other direction, and a growing number of geographers are willing to examine the phenomena of human geography *as if* they could be understood in terms of universal laws. The recent presentations by Bunge (*1966*) and Haggett (1965A) are clearly indicative of this trend, and many more examples could be cited. The principle of 'hidden order within chaos' appears in such works as a basic assumption. If we can believe Hempel's covering-law model of explanation, however, it becomes clear that there is only a difference of emphasis between those, such as Bunge and Haggett, who seek explicitly for statements of a 'universal' kind and those such as Sauer who do not. The modern movement towards methodological universality may be regarded as an attempt to be more explicit about the law-like statements often contained in the 'explanatory sketches' that geographers usually offer.

It cannot be denied, however, that geographers frequently require statements of very low-order generality for 'unlike other sciences geography can find great use for principles that are applicable to relatively small portions of the earth's surface' (McCarty, 1954). This does not preclude the development of more general statements—indeed it may make them even more desirable, since it will be only through such high-order generalisations that comparative studies will become possible. But the traditional preoccupation with 'localised' generalisations does make the development of some system of higher-order generalisations that much more difficult. Thus the statement that 'towns of similar size and function are similar distances apart' may not be applicable to all societies at all times. Berry (1967A) has thus attempted to survey the relevance of central place concepts in different cultural contexts. There is reason to doubt whether such a statement is universally true, but even if it is not, this does not diminish the utility of the statement in those situations where it has been found to be reasonable. Indeed, there are enough studies under sufficiently diverse conditions to suggest that we might regard the statement *as if* it was universally true—in other words a law of restricted but unspecified domain. But science does not remain content with a statement of this type. It lacks sharpness and precision. It demands an operational definition of 'similarity' of function and population which is difficult to provide. Attempts to sharpen statements of this type may lead to the complete abandonment of the initial statement in favour of something more general and more satisfying. Consider the following example.

In the study of many aspects of human behaviour, such as migration, journey to work, retail-centre patronage, etc., it has been shown that volume of movement over distance declines as some function of distance. Thus Stewart (1948) and Zipf (1949) postulated as 'inverse distance law' which, when applied to migration for example, stated that migration between two centres was proportional to the populations of the centres and inversely proportional to the square of the distance between them. This law has been the subject of protracted study and discussion by economists, sociologists, and geographers (see the review by Olsson, 1965A). There are a number of problems in developing and applying this 'inverse-distance law'. It has been shown, for example, that a large number of deterministic and probabilistic functions can fit the data equally well. We could formulate the law in a multitude of ways and there are no apparent criteria for choosing between alternative formulations. Even if we confine attention to just one formulation—say the gravity formulation originally given by Zipf and Stewart—problems arise. Although the formulation provides a good fit in a large number of circumstances, the parameters fluctuate wildly. Thus the *rate* of change in interaction over distance varies over time, according to place, and according to social characteristics in the population (Hagerstrand, 1957, 112–20; Lowry, 1964). The good fit of the inverse-distance law in a wide variety of circumstances is of interest, but it is rather like commenting on the inevitability of water's flowing downhill. A really satisfactory scientific law tells us the rate, yet little attention has been paid to the problem of predicting the parameters of the inverse-distance law. There have been attempts by Schneider (1959), Harris (1964) and Lowry (1964, 26) 'to derive the parameters of a gravity model from theoretical probability distributions'. But a good deal more investigation is needed before this issue is even partly solved. Our present situation is rather like that of a hydrologist who can predict that water will flow down hill but who cannot predict the rate because he is unaware of the laws of gravity or the degree to which slope, volume of water, channel shape, and so on, modify this fundamental force. What we need in these circumstances is (i) a clear analytic derivation of the law itself (see Wilson, 1967) and (ii) a clear statement of its domain. Given these specifications there is no methodological reason why the laws of spatial interaction which seem on the surface to be so radically different from the laws of the physical sciences, should not achieve the same status. In practice, however, it may be politic to recognise that 'the laws which control . . . social actions and interactions may themselves be subject to rapid change' (M. G. Kendall, 1961, 12). Under such conditions the assumption of methodological universality appears essential because laws of some kind are necessary to explanation yet we are forced to remain sceptical regarding their absolute universal applicability within a defined domain.

Methodological universality may also be an important assumption in relation to 'cultural relativism' in human geography. This problem may be regarded as a specific form of the argument regarding variable value-systems which has led many social scientists to reject the idea of an 'objective' social science. Cultures exhibit radically different value-systems. Does this irrefutable fact debar a scientific study of societies of different type? Some anthropologists seem to suggest that it does. Given the strong influences that have entered cultural geography from anthropology, it is hardly surprising that Sauer (*1963*) and the Berkeley school (see the review by Brookfield, 1964) have assumed a degree of this kind of 'cultural relativism' in their writings. In its most rigid form this view simply assumes the world to be divided into a mosaic of landscape types, each regarded as a unique expression of culture type and each only describable in terms of internal cultural coherence. Cultural geography thus becomes a matter of examining the specific interactions between culture and environment within a particular region. Clearly universal laws have no place in such studies, and there is no point in regarding geographical phenomena *as if* they were governed by universal processes.

But it is rare for cultural geography to be formulated in such a rigid form. In practice two kinds of reaction can be distinguished. The first is to develop some kind of scientific method that transcends cultural differences. D. Freeman's (1966) proposal for studying the physiological determinants of human behaviour, Dodd's (1962), to develop 'momental laws' or the attempt by Segal *et al.* (1966) to examine cross-cultural variation in visual perception by a standard psychological test procedure are examples of this kind of reaction. This reaction corresponds to Brookfield's (1964) plea that geographers should not fail to push back their 'explanations' to take account of work in the various social sciences on the processes governing the evolution of a distinctive 'human ecosystem'. Comparative studies are impossible unless we possess concepts which are capable of transcending individual cultures in some well-defined respect.

The second reaction is to adopt a position of *methodological* as opposed to *philosophical* relativism. Thus Obeyesekere (1966) has proposed that we examine culture systems *as if* they were the expression of the same universal human processes. This corresponds to Brookfield's (1964) second plea that cultural geographers should be more concerned with developing the comparative method whereby very broad generalisations regarding the interaction between cultural form and environment may be set up.

The first reaction is in the direction of *reductionism*, whereas the second clearly involves the employment of *as if* thinking with respect to universal statements. Both reactions involve the search for geographical laws, and if we reject the very rigid criteria of what constitutes a law, there seems no reason why both attempts should not be blessed with a certain degree of success.

There can be no doubt, therefore, that geographers have in the past frequently resorted to the idea of methodological universality in the explanations they have offered. To the degree that they have resorted to the 'comparative method' such an assumption has been inevitable, but even in more localised studies they have made similar assumptions. There have been occasions when the employment of such an assumption has been rejected and the postulate of 'uniqueness' adopted in its place. This latter postulate has the disadvantage that it leads ultimately to the rejection of *any* kind of explanation. The current tendency, therefore, is to seek for some partial reduction of particular statements to universal laws, and meantime to use 'mere generalisations' or speculative statements *as if* they were statements of scientific law appropriate for the explanation of certain types of events. This latter procedure poses methodological difficulty, since it often means that we need to assume statements to be laws even though we have very little supporting evidence. In itself this is not necessarily wrong. The only real problem arises if we forget that we have made such an assumption or if we neglect to temper our inferences by referring to the quite substantial assumptions that we made at the very start of the process of offering an explanation. There can be no doubt that the natural sciences (and physical geography) are in an advantageous position relative to the social sciences (and human geography), since the laws they possess are, by and large, better substantiated (and consequently require less in the way of assumptions in order to be employed) than are the laws in social science (and human geography). But as we pointed out in chapter 5, it cannot be inferred from this that powerful laws cannot ever be developed in social science and human geography. There is every reason to expect scientific laws to be formulated in all areas of geographic research, and there is absolutely no justification for the view that laws cannot be developed in human geography because of the complexity and waywardness of the subject-matter.

II THEORIES IN GEOGRAPHY

Laws are generally accepted as being something more than 'conveniently telescoped summaries of finite sets of data concerning particular instances' (Hempel, 1965, 377). The difference between *accidental* and *nomic* universality is not easily defined, but the main criterion here appears to be the relationship of a particular statement to an established theoretical structure. The question of the actual or potential use of laws in geographic explanation can thus be resolved into the far broader question of the nature and status of geographical theories.

A full account of 'theory in geography' would demand a book in itself. A distinction may be drawn, however, between the ideas and concepts being theorised about and the degree to which theoretical statements achieve a 'scientific' status as measured against the standard provided by philosophers and logicians. The essential concern of the methodologist is with the second question, but as we have already indicated, it is impossible to discuss the second question without at least some reference to the first.

A On themes and theories

Metaphysical speculation regarding the 'nature' of geography provides 'directives' and 'regulative principles' in the search for geographic theory. These directives are embedded in our traditional
literature. Theory-formation in geography has been a conscious and
positive reaction to these directives. It is important, therefore, that
we establish in a very general way the nature of these directives while
fully recognising that 'the act of conceiving or inventing a theory,
seems . . . neither to call for logical analysis nor to be susceptible of
it' (Popper, *1965*, 31).

The richness and inventiveness of 'the geographical imagination'
has provided us with a wealth of ideas to theorise about. The literature upon which we can draw is immense and growing—though
apparently at a somewhat slower rate than in many other disciplines
(Stoddart, 1967B). Even Hartshorne's (1939) massive survey of the
literature before 1939 refers mainly to development in German and
Anglo-American geography (the French geographers, as Hartshorne
(1959) later admitted, are given scant attention), while even within
these national groups the treatment was necessarily selective—Sauer
(*1963*, 352) regarded the treatment of historical geography as being
far too brief. Nevertheless, Hartshorne's work forms our most comprehensive source-book in English regarding the views which geographers have held regarding the nature of their subject (also Taylor,
1951). The problem, then, is how to characterise the vast reservoir
of ideas and concepts in our traditional literature? Haggett (1965A,
9–17) has suggested that geographers tend to organise their thoughts
around five major themes. It will be useful to examine the nature of
these themes. The presentation depends largely on Haggett and must
perforce be brief.

(a) *The areal differentiation theme.* Hartshorne's researches into the
history of geographic thought led him to the conclusion that the
fundamental aim of geographic research was the study of the areal
differentiation of the earth's surface. He also concluded that this
study ought to be made by synthesising our systematic understanding
within the context of 'the region'. There can be no doubt that this
has been, and still is, one of the major themes of geographic research,
although it may be doubted if it is *the* overriding aim to which all the
other 'themes' of geography must necessarily remain subservient.

(b) *The landscape theme.* The concept of 'landscape' as the central
focus for geographic research came largely from Germany, and

Although Hartshorne's review of an admittedly confusing concept led him to conclude that this was not a particularly rewarding focus for geographic research, it has remained of significant importance until the present day. Ever since Sauer (*1963*, chapter 16) first firmly placed the concept into American geography in 1925, it has functioned as a major research theme—particularly among the cultural geographers of the 'Berkeley school'. The latter group evolved a distinctive method of investigation (Brookfield, 1964, 288–91) which involved distinguishing between the physical and cultural landscape and examining the interaction between them. To a certain degree physical geographers also directed their attention to the physical landscape and it is interesting to note that the Davisian method and the method of the Berkeley school are essentially genetic.

c) *The man-environment theme.* A common theme within the context of areal differentiation and landscape geography has been the idea of the man-environment relationship. This theme formed a major focus for research among the 'determinists' who regarded the physical environment as the moving cause and neglected interaction or feedback effects. The possibilists, on the other hand, postulated essentially the same man-environment relationship, with man as the moving cause. A more balanced view is now provided by those who regard the essential focus of geography as being the 'human ecosystem'. The view of geography as human ecology has quite a long history, although its firmest statement has really come in the last few years (Brookfield, 1964; Eyre and Jones, 1966; Stoddart, 1967A). The theme of the man-environment relation has never been far from the heart of geographic research, and for many it has functioned as *the* overriding theme.

(d) *The spatial distribution theme.* The view has frequently been put forward that the overriding aim of geography is to describe and explain the distribution of phenomena over the earth's surface. Hartshorne regarded such a study as being an essential preliminary to the study of areal differentiation, but to others this aim became in itself a sufficient focus for geographic research. To a degree opinion on this issue splits according to the regional-systematic dichotomy, and we therefore find many of the systematic aspects of geographic research (as in climatology and economic geography) developing around this fundamental theme as their focal point of interest. Locational analysis, at present an active area in geographic research, may conveniently be regarded as a manifestation of interest in the theme of spatial distribution.

(e) *The geometric theme.* The 'geometric' tradition in geography is an extremely old but relatively neglecte done (van Paassen, 1957; Haggett, 1965A, 15–16). But since 1950 interest in this tradition has increased markedly and with the statements by Bunge (*1966*) and Haggett (1965A) we must regard this theme as of major significance in geographic research (see chapter 14).

The five major themes presented here are neither mutually exclusive nor entirely inclusive of all geographic work. But each in its way is capable of providing an operational definition of the 'nature' of geography and it is within the context of such operational definitions that geographers have begun to formulate concepts and theories. In some instances, as with locational analysis and geometric analysis, theory has become more explicit and a formal approach has begun to be adopted. In other instances the degree of formal theory construction has been negligible and the main aim descriptive. But

> The moment that a geographer begins to describe an area . . . he becomes selective (for it is not possible to describe everything), and in the very act of selection demonstrates a conscious or unconscious theory or hypothesis concerning what is significant. (Burton, 1963, 156.)

Geographers are clearly just as prone to theory-laden fact-gathering as are other academics. A 'theme' acts as a directive by indicating the sort of facts the geographer ought to collect and by suggesting a mode of organisation for those facts. Any definition of the 'nature' of geography also provides a crude definition of the domain of geographic investigation and hence an *a priori* domain for geographic theory. Each theme acts as a set of instructions as to how the geographer ought to proceed. Each gives rise to a limited set of questions—how are the facts of geography integrated within an areal context? how did the landscape originate? and so on. And in answer to these questions geographers have developed principles and concepts which, with refinement and testing, may become the postulates upon which explicit theory is based.

Themes give rise to theories. The degree to which this has been an explicit process needs now to be considered. We can best tackle this by examining three aspects of scientific theory—the nature of the basic postulates, the degree of formal representation, and empirical status.

B The basic postulates

An explicitly developed scientific theory requires a number of axiomatic statements from which the theorems can be derived. To achieve an empirical status these axiomatic statements (containing primitive terms) require translation to either observable classes of events or to theoretical concepts from which the behaviour of observable classes of events can be derived. The concepts which correspond to the axioms of the theory we will call the *basic postulates*.

Geographers have in the past developed 'concepts' and 'principles' to facilitate explanation and to justify the way in which they organise a particular set of facts. Thus Hartshorne (1959, 160-1), following Hettner and Ackerman, comments that

scientific advance in geography depends on the development of generic concepts and the establishment and application of principles of generic relationships. . . . It is in the search for universals, generic concepts as well as generic principles which may be constructed from them, that we are pursuing fundamental research.

Such concepts may ultimately act as the basic postulates for theory. In most cases explicit theory has not been developed from them. They thus function as either *relative primitives* or else as the far vaguer postulates for *Type 4* theories. In some instances particularly in location theory, geometric analysis, and in analysis in physical geography—the basic postulates have been more explicitly developed and function more in accordance with standard scientific theory. But in general we cannot identify the basic postulates upon which geographic thinking is based nor does it seem desirable, given the dangers attendant upon premature formalisation, that we should attempt to cast geographic thought into such a formal framework. But it is useful to set up a hypothetical argument regarding the basic postulates and concepts for an adequate geographic theory, for this argument can help us clarify certain important philosophical problems. We shall begin by classifying the concepts used by geographers into those which are *indigenous* and those that are *derivative* (Harvey, 1967B).

(1) *Derivative concepts*

Derivative concepts are of two types. On the one hand there are those which were developed indigenously within geography but which have since been successfully reduced to the concepts of some other discipline; on the other hand there are those concepts which have been taken directly from some other discipline since they appear to be of utility in explaining geographical phenomena. In many cases, however, it is difficult to tell which is which. An extremely vague concept developed indigenously may be sharpened by reference to a related concept from some other discipline. We shall not, therefore, concern ourselves with this distinction any further.

The use of derivative concepts may involve the 'consumption' of some theoretical structure from another discipline. Since geographers have traditionally been concerned with some aspect of spatial organisation, this means the elaboration of the theory of some other discipline in a spatial or areal context. Such an elaboration is by no means easy and, quite frequently, leads to intractable problems. We may demonstrate this process with respect to a number of theories derived from other disciplines.

(a) *Economic concepts* have frequently been used as the foundation for geographic theory. Economics has, perhaps, been the most successful of the social sciences in developing formal theory (even if the empirical status of that theory is open to doubt). Interestingly enough much of that theory has been derived from a simple definition of the 'nature' of economics. According to Lionel Robbins (1932, 75)

> Economics is concerned with the disposal of scarce goods with alternative uses. That is our fundamental conception. And from this conception we are enabled to derive the whole complicated structure of modern Price Theory. . . . On the analytic side Economics proves to be a series of deductions from the fundamental concept of the scarcity of time and materials.

The adequacy of this definition and the elaboration of economic theory need not concern us. But many of the postulates and theorems of economics have been absorbed into geographic theory. In particular the whole of location theory, which has been 'especially concerned with the development of the theoretical-deductive method in geography' (N.A.S., 1965, 4), can be related to economic postulates. The numerous cases need not be cited here. But it is interesting to examine just one case in detail.

Central-place theory has frequently been described as the 'one relatively well-developed branch of theoretical economic geography' (Burton, 1963, 159). The foundations of the theory were laid by Christaller (*1966*) in 1933 and it is instructive to quote his basic assumptions at length.

> We believe that the geography of settlements is a discipline of the social sciences. It is quite obvious that for the creation, development, and decline of towns to occur, a demand must exist for the things which the town can offer. Thus, economic factors are decisive for the existence of towns. . . . Therefore, the geography of settlements is a part of economic geography. Like economic geography, it must draw upon economic theory if it is to explain the character of towns. If there are now laws of economic theory, then there must be also laws of the geography of settlements, economic laws of a special character, which we shall call *special economic-geographical laws*.

In the first part of his book Christaller uses elementary demand analysis to define a fundamental spatial concept—*the range of a good* —and this, in conjunction with other economic arguments, led Christaller to define an 'optimal' spatial organisation of a hierarchy

of settlements. Christaller did not resort to formal deductive procedures nor did he attempt to develop a formal theory of settlement
location. Lösch (*1954*) treated the location of settlements as part of
the general location problem and, grounding his analysis firmly in
Chamberlinian economic theory, gave a far more powerful theoretical foundation for the settlement theories of Christaller. Lösch's
treatment was partly formal and partly intuitive. Later elaborations
of the theory by Isard (1956) and many other economists and geographers (see the bibliography by Berry and Pred, 1961) have served
to tighten up parts of the theoretical argument and to indicate the
empirical status of the theory. The question has also arisen of the
precise geometric form which a logically developed Löschian central-
place system should assume. Dacey (1965A; 1966A) has provided a
geometric version of a probabilistic central-place system. We can
thus trace the articulation of a theoretical structure from basic
economic postulates through the spatial structure of settlement systems to the geometric form which that spatial system assumes. Not
all links in this argument are complete, and certainly the formal
articulation of the theory leaves much to be desired. But the linea-
ments of a theoretical structure are there.

Central-place theory provides just one example out of many to
demonstrate how geographical theory may be derived from the basic
postulates of economics. The existence of such postulates was undoubtedly an important necessary condition for the emergence of
a theoretical human geography. Burton (1963, 159) has thus suggested that 'one role of an economic geographer is to refine and adapt
available economic theory'. Such a derivative position carries with
it certain penalties. In particular it means that the solution to a
geographical problem may well have to wait upon the development
of the relevant branches of economics. This does not seem a particularly serious issue with respect to economics but it becomes of great
significance when we attempt to derive geographical theory from
some of the other social sciences.

(b) *Psychological and sociological postulates* have also been introduced in
the construction of geographical theory. Human geographers have
long recognised that geographical patterns are the end-product 'of
a large number of individual decisions made at different times for
often very different reasons' (Harvey, 1966A, 370), and that it was
necessary to employ some psychological notions in explaining those
patterns. Brunhes (*1920*, 605) wrote in 1912 that

Geographical criticism is not content with observing facts in themselves. It
must distinguish the natural and general psychological effect that these facts

produce upon men, upon men obeying certain instinctive or traditional suggestions, seeking the satisfaction of certain needs. . . . It must never forget . . . *the psychological influence of geographical causes upon the human being, in proportion to his own appetites, needs, or whims*—this is the subtle and complex factor that must prevail in the study of human geography.

In 1940 Sauer (*1963*, 360) wrote:

> The culture area, as a community with a way of living, is therefore a growth on a particular 'soil' or home, an historical and geographical expression. Its mode of living . . . is its way of maximising the satisfactions it seeks and of minimising the efforts it expends. That is perhaps what adaptation to environment means.

In 1964 Wolpert (1964, 558) wrote of central Swedish farmers:

> The concept of the spatial satisficer appears more descriptively accurate of the behavioural pattern of the sample population than the normative concept of Economic Man. The individual is adaptively or intendedly rational rather than omnisciently rational.

All three writers introduce the notion of satisfaction. Brunhes and Wolpert at the individual level and Sauer at the cultural level. Brunhes used the idea to counter the view of the environmentalist school which postulated a mechanistic response of man to his environment. Wolpert used the same idea to counter a similar mechanistic view of decision making—but in this instance it was the 'optimal' reaction of Economic Man. All three writers refer to the same psychological concept. Yet it could scarcely become explicit *without* reference to the work of psychologists. Thus Brunhes referred to Bergson, Wolpert to Simon. It is still not entirely clear how the concept of satisfaction can be defined and used with reference to human behaviour. The concept may well comprise a number of disparate psychological notions. Kates (1962), Gould (1966), and Saarinen (1966) have thus chosen to emphasise the *perceived* environment (as opposed to the *actual* environment) as the *action space* within which individuals make their decisions. Wolpert (1965) has used the idea of an *action space* to discuss migration-behaviour. Other writers have concentrated upon the stimulus-response mechanism as a descriptive device. Thus Golledge (1967) has adapted a stochastic learning model to describe the emergence of interpenetrating market areas.

It has been argued above (p. 95) that economic postulates may well be reduced to special cases of psychological postulates, and if this be so human geography will be correspondingly influenced. But a certain direct relationship between geography and psychology is emerging. From this we may conclude (i) traditional notions regarding the importance of individual and group behaviour in the creation of geographic patterns can be sharpened by reference to the psy-

chological literature, (ii) psychological postulates (particularly behavioural ones) can be employed directly by geographers with profit,
and (iii) much of the theory developed by experimental psychologists
applies only to very simple experimental situations, and the theory
does not, therefore, provide for all the needs of the geographer.

Similar conclusions may be reached with respect to the use of
sociological postulates in human geography. Thus the study of
diffusion—developed strongly by Ratzel as early as 1891 and subsequently elaborated by many geographers (see the reviews by
Brown, 1965 and Harvey, 1967A)—has drawn a great deal upon the
vast sociological literature (Rogers, 1962), to say nothing of the
stimuli received from anthropology, archaeology, epidemiology, and
so on. The general similarity of Hägerstrand's (1953) treatment to
that of Dodd (1950, 1953) with respect to the spatial diffusion of
information in populations is particularly interesting in that the
treatments appear to have been independently formulated. Again,
traditional geographical concepts can be sharpened and new ones
introduced by reference to the postulates of some other discipline.

(c) *Physical postulates* are of tremendous significance to research in
physical geography. Of all the sectors of geographic research, physical geography possesses the firmest grounding in the basic theory of
some other discipline and its concepts may thus be defined as derivations from physical postulates. To the degree that physics and
chemistry possess greater theoretical sophistication than other science
and a sounder empirical status than any of the social sciences, so we
may regard physical geography as possessing more soundly based
concepts than human geography.

Much of the work in desert erosion (Bagnold, 1941), coastal
erosion (King, 1961), glacial erosion (Nye, 1952), makes direct use
of the basic postulates and known relationships of physics. Scheidigger
(1961) has attempted to outline a purely theoretical geomorphology
starting from such postulates. Barry (1967) similarly relates work in
meteorology to the postulates of physics, while work on soil formation, weathering processes, and so on, refers to the concepts of
chemistry and biology. In fact *any* work on *process* in physical geography can be related, directly or indirectly, to the postulates of the
various physical sciences. In some cases it is difficult to achieve
integration simply because the deduction of, for example, geomorphological processes from physical postulates poses considerable
technical difficulty—indeed it often appears intractable. It is also
worth noting that many concepts in geomorphology, particularly
those formulated during the nineteenth century (Chorley *et al.*,

1964), were developed indigenously and have since been reduced successfully to special cases of physical law. Some concepts, such as that of grade, have not, apparently, survived this process intact.

But a great deal of research in physical geography has not been directly concerned with process. It has been concerned with spatial pattern, morphometry, and, most important by volume of activity, genetic morphology. The Davisian system in geomorphology has been particularly important within the last category since it suggested that an adequate explanation of landscape form lay in the interaction between 'structure, process, and stage'. The problem with the practical application of the Davisian system was 'the assumption that we know the processes involved in the development of land-forms. We don't and until we do we shall be ignorant of the general course of their development' (quoted in Wooldridge, 1951, 167). Wooldridge (1951, 166) considered that 'geomorphology includes or should include, both the comparative study of land-forms and the analytical study of the processes concerned in their formation.' C. A. M. King (1966, 328) similarly characterises the morphometric approach as 'an objective method of establishing fundamental empirical relationships between various attributes of the landscape', but points out that 'this type of analysis is one stage in the development of a theory of landscape development; by itself it cannot explain the significance of the relationships.' As King later points out, the explanation and the theory must rest on an understanding of process, and in this the postulates of physics play an all-important part. Thus although geomorphologists may have attempted to carry out research without specific reference to process, it now seems that this is the key to geomorphological explanation. Clearly that explanation must rest firmly upon physical theory.

These examples could be extended by examining the relationship between geographical concepts and concepts developed in many other disciplines. But they are sufficient to demonstrate that many of the concepts of use to geographers can either be taken from, or reduced to, the basic postulates of some other discipline. This demonstrable truth raises a number of questions.

Brookfield's (1964) major criticism of the Berkeley school is the failure, by and large, to seek explanation in depth and to pursue those explanations across interdisciplinary boundaries. Such pursuit is far from easy and may tempt the geographer into a form of 'intellectual dandyism' in which superficial explanations using notions from a poorly understood neighbouring discipline are offered.

Devons and Gluckman (1964) in their book *Closed Systems and Open Minds* have pleaded with their fellow-anthropologists for a deliberately cultivated intellectual naïvety. This involves closing the *domain* of their discipline and seeking explanations which can be provided within that *domain*, while retaining a certain humility regarding the effectiveness of the explanations offered. This plea for intellectual naïvety stems not so much from a wish to resist reductionism, as it does from a recognition of the dangers of amateur interdisciplinary dabbling. They conclude (p. 261) that

the different social and human sciences may be different realms, in whose borderlands trespass is dangerous save for the genius . . . a social or human scientist may profit by studying disciplines other than his own. It is dangerous to practise them without training and appropriate skills.

This is a serious issue. What degree of knowledge of other disciplines *do* we require to construct geographic theory? Should we simply assume the status of applied physicists or regional econometricians? And if so, does this imply the final dissolution of geography as a viable academic discipline? A number of points can be made in answer to these questions.

Firstly, crossing interdisciplinary boundaries in the search for concepts on which to base geographic theory imposes certain responsibilities. It involves a precise understanding of the concepts and postulates developed in another discipline and an understanding of the derived conclusions. It need not involve a precise understanding of the derivation. A meteorologist may thus work with the second law of thermodynamics quite adequately without necessarily understanding how that law is derived. It follows from this, however, that we should be fully aware of the empirical status of that law. We should be acutely aware, for example, that the laws of theoretical economics have an entirely different empirical status to those of physics. Thus a geographic theory of central-place systems based on economic postulates will have an entirely different empirical status to theories of slope-development based on physical postulates.

Secondly, it will prove very difficult for any individual discipline to resist the pressure towards integration in science as a whole. In this context it seems irrelevant if geography in particular disappears since the same fate will undoubtedly overtake many other disciplines. As Huxley (1963, 8) suggests,

In place of subjects each with their own assumptions, methodology and technical jargon we must envisage networks of co-operative investigation with common methods and terminology, all eventually linked up in a comprehensive process of enquiry.

If the choice simply lies between the over-all advancement of knowledge and the stubborn preservation of a disciplinary boundary, there can be no question which choice we should make.

Thirdly, there is no necessary intellectual inferiority involved in assuming a derivative position with respect to other sciences. If we took the limited view that the sole activity of geography was the deduction of spatial pattern or morphometry from the basic postulates of the social or physical sciences, then it can quickly be demonstrated that this task poses technical and conceptual problems of enormous complexity, the solution of which requires considerable intellectual ability. Dacey's (1965A; 1966A) geometric representation of central-place theory, for example, required considerable technical expertise. In some respects it could well be argued that geographers are faced with even greater difficulty in deriving the spatial consequences from a set of postulates than were other disciplines in creating those postulates. There can thus be no question of attributing an intellectual inferiority to geographic research merely because the basic postulates on which that research is based are essentially derivative.

Finally, although a discipline may ultimately resort to derivative concepts there is no reason why it should not play an important role in promoting their formulation. Explanation in astronomy, for example, must rest firmly on physical theory and physical postulates, yet astronomy has posed physics with a series of problems throughout history the solution of which has required a radical revision of the basic postulates upon which physics is itself based. Perhaps the greatest single measure of our 'isolationism' in geography over the past half-century has been our failure to ask any challenging questions of the other social and physical sciences. There are exceptions, of course, but by and large we have failed to develop hypotheses and concepts within geography which challenge accepted theory in other disciplines.

All these comments assume that geographers have not, on the whole, developed satisfactory *indigenous* concepts for explanation. Such indigenous concepts will now be considered.

(2) *Indigenous concepts*

There are plenty of 'concepts' and 'principles' developed by geographers which could function as postulates for theory. But few have been developed in such a manner. It is difficult to say, therefore, if the concepts and principles developed in the exposition of *Type 4* theories can be transformed into basic postulates that are not derivable from some other discipline. We have not sufficient experience of theory-construction in geography to discuss indigenous postulates with any certainty. But the limited experience we do possess, together with some *a priori* notions regarding the nature of geographical enquiry, provide some clues as to the nature of such indigenous postulates. Geographic concepts are of three types.

(a) In many cases concepts are developed to help explain and describe geographical phenomena. When introduced into a rigorous theory these concepts often require modification, and it frequently occurs that the postulates are either actually or potentially derivable from some other discipline. Concepts which turn out to be so derivative usually relate to temporal processes. This observation fits the generally accepted view of geography as being concerned with spatial systems rather than with temporal processes *per se*. It is tempting to conclude that all aspects of theory in geography which relate to temporal process are actually or potentially derivative of some other discipline.

(b) Some concepts in geography play a more ambiguous role. They sometimes play an explanatory role, but on other occasions they may be interpreted as procedural rules for conducting geographical research. This point can be demonstrated by a brief discussion of the role played by one of the central concepts of geography—the region. The region has sometimes been accorded the status of a 'theoretical entity' rather like an atom or a neutron which could not be precisely observed but whose existence could be inferred from its effects. The areal differentiation of the earth's surface could thus be 'explained' with reference to this theoretical object which governed human spatial organisation. Later writers denied such a mystical interpretation of the term 'region' and came to regard it as an essential mental construct for the organisation of geographic data (Hartshorne, 1959, 31). Bunge (*1966*) and Grigg (1965) have since indicated that the concept performs the same function as the concept of a class in any science and that therefore regionalisation is nothing more than a special form of classification. The double role which the concept of the region has played in the history of geography can be confusing. In particular the danger of tautology in this double interpretation of a single concept has not always been avoided. Once spatial phenomena are classified into regions it is not useful to explain the existence of such classes by reference to the regional concept itself.

The region is not the only concept of this type, nor is this problem simply confined to the more traditional formulations. Thus Berry and Garrison (1958) have pointed out that the concept of a hierarchy

of settlements is sometimes used as a 'natural' classification and sometimes treated as a derived theorem of central-place theory. Many
concepts in geography are classificatory in function (and therefore
act as procedural rules for examining data) but it has often appeared
as if classification acts as 'explanation' in geography (R. H. T. Smith,
1965A). It is a difficulty of classification that it can either be derived
from a theory or it may be regarded as a natural grouping which
needs explanation and, therefore, precedes hypothesis formation and
theory construction. Again, the danger of tautological argument is
evident.

(c) Some geographical concepts may well form a set of indigenous
postulates for the development of geographic theory. These concepts
are related to what are often called 'spatial processes'. Such a title
for them tends to be rather misleading since the processes are not
temporal and therefore not strictly processes at all—they are, rather,
sets of spatial relationships. These concepts are essentially to do with
location, distance, 'nearness', pattern, and morphology. They make
reference, therefore, to the rather special relationships which exist
between geography and geometry. These concepts are not simply
associated with those geographers who subscribe to the 'geometric
theme' nor have they simply emerged as a result of that particular
'directive' regarding geographic research. The common problem of
map-projection, cartography, map-transformation, exists, no matter
what the view regarding the 'nature' of geography, and these problems have been at certain times in the history of geography the main
object of research effort. The current status of the geography-
geometry relationship has been most fully reviewed in Bunge (*1966*,
chapters 7, 8, and 9).

The relationship between geography and geometry is special in
the sense that geometry is not usually regarded as an empirical
science but as a branch of mathematics. Geometry is essentially
abstract, analytic, *a priori*, knowledge, even though it does make
reference to terms such as 'point' and 'line' which have some kind of
intuitively reasonable empirical interpretation. Geometry therefore
provides an abstract language for discussing sets of relationships.
Geography maps many of its problems into this abstract language
without deriving its postulates from any empirical science *en route*.
It is no more derivative of geometry than marginal economics is of
calculus. Mathematics is a language in which we can theorise. The
various forms of geometry appear to be a peculiarly appropriate
language for theorising about spatial relationships, about morphometry, and about spatial pattern. From this language we may

derive the 'morphological laws' which help to explain geographical distributions (Bunge, *1966*, 249). We shall return to this problem of the geometry-geography relationship in chapter 14 since it is of considerable significance to explanation in geography.

(3) *General theory and synthesis*

A tentative conclusion can be extracted from the preceding discussion; geography is likely to possess an interesting dichotomy with respect to the postulate sets upon which theory may be based. Indigenous morphometric postulates on the one hand contrast with derivative process postulates on the other. It is possible to use our understanding of this dichotomy to throw fresh light, and perhaps even give a new interpretation to, the traditional notion of *synthesis* in geography. This notion has its roots in the emphasis which geographers have tended to place on what Brunhes termed *connexité* in landscape or in geographic area (Wrigley, 1965, 15). The interrelationships of a whole multitude of factors within an area create a 'unique' personality for a given area and have traditionally provided criteria for distinguishing regional units. From this has grown the idea that geography is concerned with the *synthesis* of everything within an areal context. This was, indeed, one of Hartshorne's (1939) key conclusions regarding the nature of geography. But given the dichotomy in the basic postulates of geographical theory, it might well be suggested that *synthesis* in geography can be interpreted as a matter of linking the theory governing processes (which is mainly derivative) to theories about spatial structure and form. This link demands a space-time transformation which is difficult to provide. It is evident, however, that it is largely governed by the degree to which the 'friction' of space influences the trajectory of some temporal process—hence the great emphasis in much recent geographical work on deriving various measures of the 'cost' of movement. But there are numerous examples in geographic research of this kind of approach. Hägerstrand's discussion of the way in which the spatial form of a diffusion process will depend upon the cost of moving information over distance, the degree of contact between individuals, and the degree of resistance within the population to new information, is a classic example of temporal processes leading to certain spatial patterns. Diffusion theory (Hägerstrand, 1953; Brown, 1965; Harvey, 1966B; 1967A) is a classic case of the interaction between process and spatial form. Lösch (1954) develops a similar kind of argument for he takes a particular temporal process (the demand-supply relationship), assumes it to be in equilibrium, and then derives the spatial form which must result. We need not confine our examples to human geography for some of the best examples of the interaction between morphometric laws and process laws can be found in geomorphology. Meander geometries have been examined by Leopold and Wolman (1957) who found that the development of meanders seemed 'to depend on a fairly simple multiple relationship between bankfull discharge and channel slope' (Chorley, 1965, 27). Leopold, Wolman, and Miller (1964) have also examined the interaction between fluvial processes in general and the morphometry of drainage basins. But an even more striking example is provided by Curry's (1967) recent treatment of central-place theory.

We have already indicated that a logical argument can be constructed which runs from economic postulates through to a geometric presentation of central-place theory (as given by Dacey, 1965A; 1966A). The problem with this approach, however, is that the results are geared to a set of postulates which make unwarrantable assumptions about human behaviour. Curry (1962A) attempted to break away from this framework by developing a more 'operational' approach in which economic postulates were replaced by random variables describing shopping behaviour, store inventory policy, and so on. The aim of using such random variables was to *describe* the general pattern of behaviour rather than to assume any underlying economic rationale. Curry (1967) then postulates certain statistical regularities in shopping behaviour by consumers over time. The description he furnishes is in terms of stationary stochastic time series where the behaviour of consumers is assumed to be periodic, and in which different goods are purchased in varying amounts at different time intervals. The behaviour of consumers may in very general terms be described by a characteristic spectral-density function. Several writers (Tobler, 1966B; Casetti, 1966) have suggested that point patterns of settlements can also be described in terms of a spectral-density function (Bartlett, 1964). Curry then invokes the ergodic hypothesis to link temporal behaviour with spatial form. The ergodic hypothesis amounts to assuming that the statistical properties of a time series are essentially the same as the statistical properties of a set of observations of the same phenomenon taken over a spatial ensemble (see below, p. 269). An ergodic process is thus a special type of stationary stochastic process. The assumption of ergodicity is often an act of faith, but it is nevertheless an extremely useful and necessary assumption (Kinsman, 1965, 328-9). Curry's application of the hypothesis to central-place systems is particularly interesting since the statistical properties of consumer behaviour over time become the same as the statistical properties of the spatial pattern of retail locations. Above all else, the assumption of ergodicity facilitates the necessary space-time transformation. When developed in this explicit manner the assumption of ergodicity may appear unacceptable, but it is worth noting how frequently geographers make this assumption in practice (e.g. when information culled from the historical record of one area is used to identify the 'stage of development' that other areas are in, or when information from many areas is used as evidence for some general historical process).

Curry essentially proposes one theory with a double text—one which relates the theory to consumer behaviour over time and another which relates the theory to spatial pattern. Such a device suggests the possibility of general theory in geography and, further, indicates new possibilities for synthesis. The interaction between what Bunge (*1966*) terms 'flows and movement' on the one hand and spatial pattern on the other may well emerge as a focal point for a new kind of geographic synthesis. Another example of this is Berry's (1966) proposal for the construction of a 'general field theory of spatial behaviour'. Berry tackles the traditional dichotomy between nodal and uniform regions in an interesting way. He constructs an *interaction matrix* for describing movement between places and an *attribute matrix* for describing the characteristics of regions. He then goes on to examine the relationship between these two matrices using data on the Indian economy as an example. Berry (1966, 192) states:

> The essential postulate of the general field theory is that the fundamental spatial patterns that summarise the characteristics of areas and the types of spatial behaviour that are the essence of the interactions taking place among areas are interdependent and isomorphic.

From this general discussion we can extract the sense of general theory in geography as follows: *it will explore the links between indigenous theories of spatial form and derivative theories of temporal process.* The links run in both directions. The usual direction in location theory, for example, is to attempt to deduce spatial form from process postulates. But it is possible to explore the links in the other direction by examining how a fixed spatial form affects the operation of a process (e.g. the relationship between the geometric form of a city and the process of communication within it). In practice we may conceive of feedback for process affects spatial form which may in turn affect process. We may also conceive of multiple sets of process postulates all of which interact in the context of some spatial form. Thus general theory can be extended to the total complexity of the 'man-environment system in space'.

C Formal statement of theory in geography

General theory, of the kind outlined above, demands some powerful logical development if it is to be of use. In practice geographical theory varies a great deal in its degree of formalisation. It was suggested above (pp. 96–9) that theories range along a continuum from completely formal *Type 1* theories, through the *Type 2* and *Type 3* theories which involve presupposition and quasi-deduction respectively, to the more nebulous *Type 4* theories which scarcely conform in any respect to the standards of scientific theory. Given the nature of geographic concepts the development of formal theory in geography appears to be a very restricted possibility. For the most part we must at best rest content with varying degrees of partial formalisation. In most cases the systematic presupposition or the quasi-deductions involved are not of the 'harmless' variety in which a full proof or full theory is available but not stated. In practice, most of our traditional theorising has been within *Type 4* theoretical structures.

There are two ways in which we may develop formal theory in geography. The first is by a clarification of the basic postulates and the development of an appropriate calculus. The second is by the direct application of model-building techniques to geographic problems. This second approach by way of models will be examined later and the discussion here will centre upon the first method.

In the previous section we examined the potential postulates for geographic theory in the hope that some clarification of the general form of that theory would emerge. It is difficult to think in terms of articulating that structure by way of formal logical argument. Such an argument can be verbal however, and there is no reason why it should not be logically sound. There is no *absolute* necessity to resort to some kind of formal logic to derive adequate and accurate conclusions. It is necessary to bear in mind logical considerations however. Let us consider a simple example of verbal theorising by examining the environmental determinist theory in relation to Hempel's covering-law statement.

We may begin by taking a very restricted view of environmental control and, for the sake of brevity in the exposition, restrict consideration to agricultural systems. Let us begin by stating two laws:

(i) If a farming system in any environment possesses natural advantages lacked by other farming systems, then that system will (is likely to) survive and develop.
(ii) If a farming system in any environment lacks natural advantages

possessed by other systems then that system will not (is unlikely to) survive and develop.

It should be noted that the derivation of these two law-like statements is unclear (and therefore we are dealing with the rump-end of a presupposed theory) and that we have allowed both a fully deterministic or a probabilistic interpretation. For convenience we will abandon the probabilistic statement from now on.

The first point that needs to be clarified is the concept of 'an advantageous characteristic'. Natural advantage cannot be defined in terms of the characteristics that have led different agricultural systems to survive and develop in different environments. To define it so is to indulge in tautologous argument and immediately invalidates the application of the theory. $L(1)$ and $L(2)$ would thus become trivially true statements. The notion of 'an advantageous characteristic' needs to be defined independently of the actual distribution of the cropping system. This amounts to:

(i) establishing the physical conditions necessary for the development of every farming system, and then
(ii) showing that an environment possesses those characteristics and
(iii) showing that there is one farming system the physical conditions for which can best be met from a particular environment.

It becomes apparent, therefore, that a theory of environmental control requires a prior statement of the physical conditions necessary for the development of a particular farming system. This specification is very difficult to provide, as various studies have demonstrated. In the present state of knowledge it must seem that any theory of environmental control must involve systematic presuppositions which are very far from being 'harmless'. Certainly the empirical verification of such a theory cannot be accomplished without such specifications.

But there is another respect in which the theory could be turned into a trivially true set of statements and, hence, protected against any empirical refutation. This is by defining the relevant environmental characteristics in terms of the cropping systems that have in fact survived. Again, we require an independent definition of environment such that:

(i) the 'relevant' physical conditions in any environment are defined with respect to the total set of all possible farming systems. This may involve listing exact physical states suitable for every crop, or it may involve establishing a set of variables (e.g. temperature, rainfall, etc.) each of which has a defined relationship with crop response.

(ii) these states or measures can be identified for any particular environment.

The theory of environmental determinism may then be stated in accordance with the standard form of scientific explanation. We have a set of initial conditions (measures of the relevant conditions in a particular environment), a set of law-like statements ($L(1)$ and $L(2)$), from which an *explanans* necessarily follows (in this case a particular cropping system must necessarily be found in that environment). We are then in a position to test the theory. If we assume that *all* the relevant variables and all the necessary physical conditions can be independently identified, then it is simply a matter of making a series of predictions for individual environmental situations and showing that these predictions are in fact correct. If the predictions are incorrect then we may reject the theory (unless we give it a probabilistic formulation in which case we require a certain proportion of correct predictions). The great danger, of course, lies in altering the relevant variables and the necessary physical conditions until the theory does work. In which case the independence of the measures is lost and we are dealing with a trivially true theory.

This example has not been given in order to discredit a theory of environmental determinism, although it does demonstrate how a long and tedious argument might well have been cut short by the application of logical principles of argument. The main aim, however, is to show how a theory can be stated and its assertions clarified simply by developing the theory in a way that conforms to a standard form of logical analysis. Many of the *Type 4* theories which we possess in geography require this kind of treatment. But the example given above requires further analysis. It should quickly be evident, for example, that there are numerous assumptions and concepts implicit in the theory which need further clarification. In particular there are behavioural notions implied. The idea of 'natural advantage', for example, involves advantage for something. In fact it implies some way of defining an optimal land-use system. Such an optimal system cannot be identified without reference to the psychological or economic concepts so important to our analysis of human decision-making. We quickly find ourselves, therefore, in the kind of theoretical structure which *explicitly* includes behavioural notions. Perhaps the first axiomatic treatment of this problem in geography was given by Garrison and Marble (1957); and it is worth examining their treatment in some detail.

Garrison and Marble begin by discussing the nature of the postulates and assumptions usually made with respect to the location of agricultural activities. They then construct an axiomatic system to discuss the problem. The system consists of two *definitions* of (1) R*— the set of all positive real numbers—and (2) R^{**}—the set of all non-negative numbers. They then state a set of primitive notions (together with a *text* to identify the empirical meaning of these notions). These were as follows:

C a finite set of crops

M a finite set of markets

d a real valued function defined on M; $d(m)$ is the distance of market m from the farm unit

y a function defined on $C \times R^*$; for $c \in C$ and $x \in R^*$, $y(c,x)$ describes the yield of crop c when x units of 'at site' inputs are employed in the production of the crop

a a function defined on $C \times R^*$; for $c \in C$ and $x \in R^*$, $a(c,x)$ describes the production cost of one unit of crop c when x units are directly employed by the farmer

p a function defined on $C \times M$; for $c \in C$ and $m \in M$, $p(c, m)$ is the price of one unit of crop c delivered at market m

t a function defined on $C \times R^{**}$; for $c \in C$ and $d(m) \in R^{**}$, $t(c,d(m))$ is the cost of transporting one unit of crop c one mile when the producing farm unit is $d(m)$ miles from the market.

Garrison and Marble then use these primitive notions to state eleven axioms. The first five of these are designed to exclude the possibility of negative prices, yields, and costs appearing in the system. The remaining six axioms describe the form of the functions explicitly and state, for example, that $y(c,x)$ is strictly increasing (i.e. constant marginal yields are derived from increments of input). These primitive notions and axiomatic statements are then used to define the rent received by the farmer as gross receipts, less production costs and transport costs to market. Stated symbolically:

$$R(c,m,x) = y(c,x)[p(c,m) - a(c,x) - t(c,d(m)).d(m)]$$

Garrison and Marble then show mathematically that for every location there exists some combination of crops, intensities of cultivation, and markets which will permit the agricultural entrepreneur to maximise this rent function. This is shown to follow logically and rigorously from the axioms.

We may use this analytic statement as an explanation in conjunction with the covering-law model. The initial conditions in this case are the full set of environmental characteristics, a set of markets (with associated demand for products), and a specified set of transport-cost functions, production-cost functions, and so on, and from this it follows that the land-use system *must* be of a certain kind, provided the farmers all maximise their profits.

It is not, however, the intention to prove substantive conclusions here. It is sufficient to demonstrate the utility of a rigorous statement of theory. This ensures rigorousness of treatment and it also ensures the precise specification of the assumptions necessary for the elaboration of the theory. By such treatment the foundations for geographic explanation will become much clearer. As Haggett (1965A, 310) has pointed out

> In the long run the quality of geography in this century will be judged less by its sophisticated techniques or its exhaustive detail, than by the strength of its logical reasoning.

There are several examples of the use of rigorous logic in geography in recent times. Kansky's (1963) axiomatic treatment of transport networks, Isard and Dacey's (1962) discussion of individual behaviour within regional systems, are examples of this kind of treatment. But formalisation is useful only provided there is some degree of certainty as to the nature of the basic postulates. It is quite possible to give formal expression and sophisticated logical treatment to a completely trivial theory. It may soon become evident, of course, that the theory is trivial, but the question still remains whether 'the disproportionate allocation of scientific energies available to this one facet of the scientific enterprise might result in the neglect of other equally important aspects' (Rudner, 1966, 52). In the present situation it may well be more advisable to develop clear heuristic treatments of geographic problems than to seek for full formal expression of theory. For this reason model-building techniques are much more significant in geographic research, and are likely to remain so in the near future. The problem of model-building will be left until later, however.

Most of our theorising is inexplicit and fuzzy. We can do a great deal about this state of affairs by simple verbal clarification of our notions. Thus the Sprouts (1965, chapter 8) have attempted a clarification of the man-milieu hypothesis 'in the light of more general theories of explanation'. The Hempel covering-law model is used to clarify both the traditional explanations given by geographers and the attempt to break away from those explanations by a number of geographers writing on problems of causality in the 1950s (Clark, 1950; Martin, 1951; Montefiore and Williams, 1955; Jones, 1956). This specific argument will be examined in detail in a later section, but the treatment of it by the Sprouts clarifies many of the issues involved. By such means our verbal theorising in the context of *Type 4* theories will at least be transformed to formulations which can be represented in a formal way without a great deal of difficulty. It is not easy to find examples of this in geographic thought. But it is intriguing to quote Simon's (1957, 99) formalisation of Homans' theory of interaction in social groups, which could be formalised because

first, although non-mathematical, it shows great sophistication in the handling of systems of interdependent variables; second, Professor Homans takes care with the operational definition of his concepts, and these concepts appear to be largely of a kind that can be measured in terms of cardinal and ordinal numbers; third, Professor Homans' model systematizes a substantial number of the important empirical relationships that have been observed in the behaviour of human groups.

The main problem of theory-development in geography is not a failure to formalise theory. It is, rather, a weak understanding of the role of theory in explanation and a failure to make verbal statements in ways which are 'explanatory' in some logically consistent way. This is not to say that these verbal statements are inherently uninteresting. Indeed they are not. Consider the following statement by Sauer (*1963*, 359):

The whole task of human geography, therefore, is nothing less than comparative study of areally localized cultures . . . But culture is the learned and conventionalized activity of a group that occupies an area. A culture trait or complex originates at a certain time in a particular locality. It gains acceptance—that is, is learned by a group—and is communicated outward, or diffuses, until it encounters sufficient resistance, as from unsuitable physical conditions, from alternative traits, or from disparity of culture level.

This statement is part a 'directive' for geographic study, it is part an explanation schema for understanding areally localised cultures, it is part a description of what happens. But all three notions are intertwined and it is not clear whether Sauer is proposing a theory or whether he is merely describing an average process. Yet the statement is surely a stimulating one and one that manages to sum up intuitively in very short space much of the activity of the human geographer. We could well pay more attention to statements of this type and attempt to show how they may function in an explanatory way. I have treated the above statement by Sauer as an explanatory schema elsewhere (Harvey, 1967A, 593–7).

The problem of formalisation is thus probably less significant at the present time than the failure to clarify our conceptual apparatus. We need to be much more certain what we are being rigorous about before resorting to full formal treatment of geographic theory. After all, 'axiomatization or even formalization as such does not add anything to the scope of what is being axiomatized' (Bergmann, 1958, 36). The failure to formalise geographic theory to any great extent is probably the least of our methodological worries. Indeed, it may be appropriate to heed Rudner's (1966, 52) warning that 'to insist on great rigor may be stultifying.'

D The empirical status of geographic theory

It is difficult to assess the empirical status of a theory which incorporates extremely fuzzy basic postulates, which is at best quasi-deductive, and which refers to an unspecified domain. It is thus contradictory to speak of the empirical status of a *Type 4* theory, because strictly speaking *Type 4* theories have no empirical status at all. Yet one of the aims of scientific explanation is to develop *testable* hypotheses, and to verify theoretical structures by testing the lower-level hypotheses against sense-perception data. Precision in the statement of postulates and in the definition of terms is a prior requirement for empirical testing. As Nagel (1961, 9) has it,

> Prescientific beliefs are frequently incapable of being put to definite experiential tests, simply because those beliefs may be vaguely compatible with an indeterminate class of unanalysed facts. Scientific statements, because they are required to be in agreement with more closely specified materials of observation, face greater risks of being refuted by such data.

A great deal of effort has been expended by geographers on verifying statements and testing the truth of propositions. Providing such confirmation is an exceedingly complex and laborious enterprise. It involves developing an experimental design, applying appropriate tests, and establishing principles of inference. These procedures will not be examined here. But from the accumulated experience regarding verification, two general points stand out:

(i) It is easier to test a hypothesis integrated into a theory than it is to test a completely detached hypothesis.
(ii) Verification is essentially an inductive procedure and it would be impossible for a scientist to provide evidence from every instance covered by a given proposition.

These two points are linked. As Zetterberg (*1965*, 154) has pointed out, to the theorist 'a huge accumulation of supporting evidence is hardly more impressive than a few strategically selected cases.' The advantage of a hypothesis embedded in a theoretical structure is that the evidence for that hypothesis comes partly from the confidence which we have in the theory itself, and partly from empirical evidence taken from 'strategically selected cases'. The evidence for a hypothesis embedded within a theory may be regarded as cumulative as more and more evidence accumulates for the theory itself. The evidence for a detached hypothesis is not cumulative in the same way. Scientific knowledge tends to progress by way of a circular and cumulative process. The deductive system of statements which characterises the standard-model form of a scientific theory is tested by testing the lowest-level hypotheses in the system (Braithwaite, *1960*, 13). Yet the decision as to whether we regard that hypothesis as being a law depends upon the confidence placed in the theoretical structure as a whole.

The aim of this section is to attempt a brief assessment of the empirical status of theories as a whole rather than to discuss the empirical status of individual propositions in geography. It is impossible to examine *all* theories in geography and to draw conclusions regarding empirical status from such an examination. If we did attempt such an examination we would probably find a continuum of theories running from those in which we have a high degree of confidence to mere speculations backed by minimal evidence. Such variability in empirical status will depend partly upon the amount of empirical evidence available, the quality of such evidence, and the degree to which the theory has been been given precise statement. We will thus find that theories of atmospheric circulation appear to have a far firmer empirical status than theories of continental drift. *Type 4* theories, for the most part, will fail to achieve a satisfactory empirical status.

The empirical status of geographic theories can, however, partially be evaluated analytically. If it is accepted that all postulates concerning process are either actually or potentially derivative from some other science, then we can use the experience of theory in that science as a guide. Here we may refer back to the discussion regarding the *text* of theories, and in particular draw attention to the practical difference there established between the idealisations of natural and social science (above, pp. 91–6, 117–22). The idealisations of, for example, physics can be derived theoretically from the requisite general theory. The idealisations of, for example, economics depend upon *introspection* for their justification since 'the non-rational and non-economic factors affecting human conduct' cannot be fully specified by way of some general theory. This difference leads to major differences in the empirical status of the theories developed. We must automatically expect derived geographic theories to have the same empirical status. This can best be demonstrated by an example.

Central-place theory, as we have already seen, is firmly based upon the postulates of economics. In particular, it is based on a theory of demand which Clarkson (1963) has shown is inherently untestable with reference to empirical evidence. Geographers have derived from this theory a number of hypotheses regarding the spacing and distribution of settlements of various sizes and have, in Dacey's presentation, provided a full geometric interpretation of the theory. Many geographers (Olsson and Persson, 1964; King, 1961; Thomas, 1962; Dacey, 1962, 1966A) have attempted to test the validity of central-place theory by examining empirical evidence of actual town patterns and spacings. Such testing has shown that actual spatial patterns do not conform to theoretical expectation. Numerous operational reasons have been adduced for this—non-isotropic surfaces, irregularities in population distribution, and so on. To get over this, various probabilistic interpretations have been given to the theory. Thus Thomas (1962) established the stability of population-size and distance relationships among towns by employing a normal curve of error to determine whether towns were of similar populations and at similar distances apart. More recently Dacey (1966A) has envisaged central-place theory as yielding an equilibrium solution for a spatial pattern of settlements, which is disturbed by other forces. The degree of displacement in Iowa was clearly quite considerable and therefore Dacey had to face two contradictory conclusions. Either (1) 'a displaced central place model describes the Iowa urban pattern though the Iowa pattern is in a strong state of disequilibrium, or (2) the utility of this stochastic interpretation of central place theory is limited to description of the urban pattern but has no explanatory implications to the locational process underlying the urban pattern' (Dacey, 1966A, 568).

The difficulty of finding empirical evidence for central-place theory has proved a major dilemma in human geography, given the generally acknowledged importance of the theory in geographic thinking. Yet it is surely evident that central-place theory will not and cannot yield empirically testable hypotheses since it is founded on a theory of demand which is itself inherently untestable except by introspection. This point has never been sufficiently understood in the argument regarding the empirical utility of central-place theory. Dacey's (1966A, 568) desire 'to incorporate these probabilistic notions within a more all embracive statement of urban systems than that allowed by the classical formulation of central place theory' is but a special case of the desire of many social scientists to provide the necessary general theory which will allow precise tests to be made. The only alternative (and this seems to have been tacitly accepted by geographers without much debate) is to regard settlement location as an economic process to which non-economic processes contribute a 'noise' or 'error-term' element. But as Dacey has shown, the error-term involved is so large that it is doubtful if it can reasonably be treated by the use of the classical theory of errors. At present theory derived from economics, psychology, and all the other social sciences, will be characterised by similar problems of empirical status. The possible exceptions to this statement may be the more descriptive formulations of behavioural psychology, econometrics, and operation research. Curry (1967) has used such formulations to break away from the traditional formulation of central-place theory and to provide a more general descriptive framework.

We may conclude, therefore, that derived geographic theory will vary in empirical status according to the empirical status of the theory from which that theory is derived. A clear practical difference is likely to emerge between theories in physical geography which can be referred to general theory, and theories in human geography which cannot in general be provided with an adequate *text*.

The empirical status of indigenously developed theory is usually less clear. When there is no clear derivation of the geographic theory from some other science we cannot make any *a priori* assessment of the empirical status, although we may make assumptions of some kind. Logically developed theories provided with the necessary *text*, which are not derivative, are, so far as I know, non-existent in geography. Such theories as do exist involve quite damaging quasi-deductive procedures and far from 'harmless' systematic presupposition. It is thus pointless to discuss the empirical status of such theories since, as we have already seen, *Type 4* theories cannot be said to possess any empirical status as theories although individual propositions contained in such theories may be verified by the use of careful experimental procedures. In the case of individual propositions, however, there is the constant danger of confusing the statement of the proposition with the statement of the evidence for it. The previous discussion of the regional concept and classification procedures referred precisely to this danger.

The empirical status of morphometric theories and theories of spatial pattern may be regarded as something special. In this case the problem we have been considering is reversed. Here we need to define the rules by which we map empirical situations into an abstract analytic calculus (such as geometry) and use the permitted manipulations of that calculus to obtain empirically meaningful results. Geometry and other mathematical systems are generally regarded as possessing no empirical status as such. But the calculus can be used for the analysis of empirical situations. The methodo-logical problem is here the problem of *model use*, and we shall there-fore consider the methodology of model use in the next section.

Basic Reading

Harvey, D. W. (1967B).

Chapter 10
Models

The term *model* has been avoided as much as possible in the discussion so far. Yet it has become very fashionable in geographic research. In this respect geography is but a short distance behind the social sciences in general. As Brodbeck (1959, 373) has pointed out, model-building has a kind of 'halo effect' upon the research worker since 'Models are Good Things'. Yet there is no consensus of opinion among philosophers of science as to what is meant by a 'model', or as to its function in scientific research. The general confusion surrounding the term is reflected in geographic research where very different opinions can be found. The meaning and function of the term *model* thus requires some careful methodological investigation.

I THE FUNCTION OF MODELS

In introducing a recent volume, *Models in Geography*, Chorley and Haggett (1967, 24) emphasise the many different functions which a model may perform in scientific investigation. They suggest, for example, that a model may act as a *psychological* device which enables complex interactions to be more easily visualised (a kind of 'picturing' device); as a *normative* device which allows broad comparisons to be made; as an *organizational* device for the collection and manipulation of data; as a direct *explanatory* device; as a *constructional* device in the search for geographic theory or for the extension of existing theory; etc. Given that models in fact perform all of these functions (and more), it is extremely difficult, as Apostel (1961) has emphasised, to provide any formal definition of the role of the model in scientific research. Such a definition requires, first of all, that the 'undeniable diversity' of scientific models should not be ignored, and secondly, a recognition that a single model is not always appropriate for all different functions. Model-use therefore poses two vital methodological problems:

(i) How to establish clearly which of the many possible functions a model is performing;

(ii) How to establish the appropriateness of a given model for the particular function we have in mind.

It should be made clear at the outset that these problems have not been solved. Apostel (1961) has considered them explicitly, but for the most part these two major problems have been ignored in the philosophical literature. Yet confusion regarding these issues lies at the root of most methodological controversy. Beach's (1957, 9) remark that 'the history of economic thought could be thought of to a very large extent as a history of misapplied models' could be generalised for all academic disciplines to some degree or other—and geography is not in any way an exception.

But the failure to solve these problems should not be regarded as a pretext for ignoring them. Models may be used to connect theory and experience, experience with imagination, theories with other theories, imaginative creations with formal theory, and so on. We may represent these functions schematically. Let a theory be represented by T, a hypothesis by H, a law by L, an actual data-set by D_o, a predicted data-set by D_p, and a model by M. Let T', H', and L', represent initial theories, hyotheses, and laws, and T'', H'', and L'', represent new theories, hypotheses, and laws. Then we may schematise a number of model functions as follows:

	Relationship	*Function*
(i)	T', H', or $L' \to M \to T''$, H'', or L''	Extending or restructuring T', H', or L'
(ii)	T' or $L' \to M \to D_o$	Validation procedure for T' or L' Establishing the domain of T' or L'
(iii)	$H' \to M \to D_o$ ($\therefore H' \equiv L'$?)	Validation of a hypothesis and creation of a law
(iv)	T' or $L' \to M \to D_p$	Prediction
(v)	$D_o \to M \to T'$, H', or L'	Discovering T', H', or L'
(vi)	T', H', or $L' \to M$	Representation of T', H', or L' (for teaching purposes, etc.)

Such schematic representations are probably much too simple however. In practice the relationships are much more complex, with one model being used to extend a theory and then another model being used to test the new theory against some set of data, and so on. The tautological case is interesting, however, and a couple of examples of it are:

(vii) $D_o \rightarrow M \rightarrow T' \rightarrow M \rightarrow D_o$

(viii) $L' \rightarrow M \rightarrow L'' \rightarrow M \rightarrow D_o$

The protection against mere tautology is provided by making the system as follows:

(ix) $D_o \rightarrow M_1 \rightarrow T' \rightarrow M_2 \rightarrow D_o$⎫where M_1 and M_2 are

(x) $L' \rightarrow M_1 \rightarrow L'' \rightarrow M_2 \rightarrow D_o$⎭independently formulated.

Most important, however, is the fact that different models are appropriate for different functions. Consider, for example, the class of relationship given in (i) above in which the model functions as a device for extending or restructuring a given theory. Apostel (1961, 5–7) has suggested how the relationship between a model and a theory should differ between these two situations. In extending or completing a theory, the model should satisfy all of the requirements of the theory but should possess, in addition, properties not contained in the theory. Suppose for an example, we theorise that migration into a town from all other towns is a function of the population of those towns and their distance away. This we may represent by a model:

$$_iM_j = \frac{P_j}{d_{ij}^b}$$

where: $_iM_j$ is the volume of migration into town i from town j

P_j is the population of town j, and

d_{ij} is the distance between town i and j, and

b is an exponent.

Suppose we now attempted to complete this theory and devised a new model of the following form:

$$_iM_j = \frac{w_j.P_j}{d_{ij}^b}$$

where: w_j is the average income of town j, per head of population.

In this case all the requirements of the initial theory are met, but the model we are now proposing contains an addition which, if it proves useful, may function to extend the initial theory.

In restructuring a theory, on the other hand, the model should not satisfy all of the requirements of the theory, and if none of the requirements is satisfied then a complete restructuring of the theory may take place. A partial restructuring of the migration theory given above might be accomplished by means of the following model:

$$_iM_j = \frac{P_j}{k_{ij}}$$ where: k_{ij} is some measure of the intervening opportunity between i and j.

If such a model can be demonstrated to be a successful predictor, then some restructuring of the theory is called for.

This example demonstrates in rather an obvious way how the relationship between a model and a theory varies according to the function which the model is performing. We might similarly demonstrate how a model designed as a test procedure for some theory should be different from a model designed for restructuring a theory. This is not to claim, however, that a particular model cannot perform different functions. Precisely the same model construct may be used for many different functions, but the way in which that particular construct is used will vary a great deal. This creates a very confusing situation and one which requires great care. In particular, models which may perform a multiplicity of functions should have that function fully identified, since the model can perform that function adequately only in a particular way. As we shall see later, failure to identify the function of a model construct and failure to proceed accordingly has been at the root of many methodological controversies in geography.

II THE DEFINITION OF MODELS

The multiplicity of functions of models makes any definition of a model extremely difficult. Definitions have, of course, been formulated. In some cases these definitions have been perfectly adequate since the function of a model in a particular academic discipline may be a very specific one. In mathematics and logic the meaning is highly specialised and complete definition is possible. Thus Tarski (see Suppes, 1961, 163) has stated:

A possible realization in which all valid sentences of a theory T are satisfied is called a model of T.

This example of a rigorous definition of the term *model* has been examined by Suppes (1961) and contrasted with the far less rigorous interpretations given to the term by practising scientists. Such a rigorous interpretation does not permit the model to function in any of the ways outlined above, and it therefore has little to recommend it from the point of view of creative empirical research. On the other hand, empiricists frequently adopt such a loose definition of the term that it almost loses all meaning. Thus Skilling (1964, 388A) writes that

A model can be a theory, or a law, or a relation, or a hypothesis, or an equation or a rule.

A more moderate statement, although it appears almost as all-embracing, is provided by Ackoff (1962, 108-9):

Scientific models are utilised to accumulate and relate the knowledge we have about different aspects of reality. They are used to reveal reality and—more than this—to serve as instruments for explaining the past and present, and for predicting and controlling the future. What control science gives us over reality we normally obtain by the application of models. They are our descriptions and explanations of reality. A scientific model is, in effect, one or a set of statements about reality. These statements may be factual, law-like, or theoretical.

But writers such as Braithwaite (*1960*), Nagel (1961) and Brodbeck (1959) have emphasised that a model should be regarded as distinct from a theory. If such a difference of opinion were simply a matter of semantics it would be pointless to pursue the matter. But the latter group of writers have convincingly argued that there are serious dangers in confusing what they call 'a model for a theory' with the 'theory' itself. The confusion which they portray is characteristic of methodological controversy in geography. This point will be taken up later, but for the moment we shall simply attempt to review the case as put forward by those who have emphasised the difference between models and theories.

A model may be regarded as a formalised expression of a theory. This formalised expression may be developed in terms of some other theory. Thus Brodbeck (1959, 379) states:

Two theories whose laws have the same form are isomorphic or *structurally similar* to each other. If the laws of one theory have the same form as the laws of another theory, then one may be said to be a *model* for the other.

Braithwaite (*1960*, 225) similarly states:

A model for a theory T is another theory M which corresponds to the theory T in respect of deductive structure.

According to these writers, all that is required of a model is that it should possess the same formal structure as the theory which it represents. There is no need for the physical properties of the phenomena described in the model and the theory to be the same. Physical similarities between model and theory may be of interest, but they are not essential. The crucial requirement is, then, similarity in formal structure (Braithwaite, *1960*, 93; Nagel, 1961, 110–11).

An important function of the model, according to this view, is to provide an *interpretation* of the theory 'in the sense that every sentence occurring in the theory is then a meaningful statement' (Nagel, 1961, 96). The model does not function as a *text*, however, since the *text* refers the abstract statement of the theory to the complete domain of that theory. A model can refer to only a small part of that domain, or it may refer outside the domain proper. In the first case, the model construct may be regarded as a kind of experimental design procedure by which the abstraction of the theory is brought to bear on a relatively small section of the reality which comes under the domain of that theory. In the second case, the model serves to transfer the theory into realms which are more familiar, more understandable, more controllable, or more easily manipulated. This is typically the case with analogue models which transfer the manipulations of some complex theory into some medium more convenient for analysis. A model is thus just one out of many possible interpretations of a theory.

From a logical point of view a model for a theory 'is an ordered set in which the postulates of the theory are satisfied' (Achinstein, 1964, 329). It is possible to construct many ordered sets in different media. Garrison and Marble (1957), for example, tackled their agricultural location problem by way of a symbolic model in which the ordered set $<C, M, d, y, a, p, t>$ satisfied the eleven axioms. The definitions of the variables were given above (p. 133). They thus provide a partial theory of agricultural location with a model which is the same in formal structure, and the model is then manipulated to give the necessary proof.

But the advantages of thinking about a theory by way of a model for it must be weighed against a number of disadvantages. According to Braithwaite (*1960*, 93–4) there are two significant dangers:

The first danger is that the theory will be identified with a model for it, so that the objects with which the model is concerned will be supposed actually to be the same as the theoretical concepts of the theory. . . .

But there is a second danger . . . that of transferring the logical necessity of some of the features of the chosen model on to the theory, and thus of supposing, wrongly, that the theory, or parts of the theory, have a logical necessity which is in fact fictitious.

In the first case we might assume wrongly that a population mass which we are examining by way of the gravity model has the same properties as any physical mass, and in the second case we may assume that the inverse-distance law is a logical necessity in the description of the population's behaviour.

But these views of a model, characteristic of writers such as Braithwaite, Nagel, and Brodbeck, do not agree with the presentations of other logicians and philosophers of science—such as Hesse (1963) and Ramsey (1964)—and have been openly attacked by writers such as Achinstein (1964; 1965) and Spector (1965). From this controversy no agreed view as to the definition of a model emerges. This is scarcely surprising since, as Apostel (1961) and Suppes (1961) have pointed out, the *definition* of a model partly depends on its *function* and there are a multiplicity of uses of the term in scientific explanation which are difficult to subsume under one all-embracing definition. As a result, a number of important epistemological problems remain unresolved. In the absence of an agreed definition, the most we can hope to do is to draw attention to the nature of some of these problems. They are part *logical* and part *procedural* and we shall attempt to separate them out in the following two sections.

III LOGICAL PROBLEMS OF MODEL USE

Most of the controversy regarding the use of models has revolved around the precise nature of the logical relationship between theory and model. Thus the view of Braithwaite is simply that model and theory have the same logical structure whereas the properties of the objects described may be different. An important criticism of this view rests upon the failure to differentiate between 'models' and 'analogies'—this difference is worth pursuing.

(1) *Models of an* x

The term 'model' is frequently used by the scientist in the expression 'model of an *x*' to refer to a set of assumptions or postulates describing certain physical objects, or phenomena, of type *x* . . .

Accordingly, the propositions comprising the model of an *x* are the same ones as those constituting what may also be called a *theory* of an *x*. . . . The terms 'model' and 'theory' are often used to refer to the very same set of propositions, and . . . hence the model object . . . is thought of as being identical with the theoretical object. (Achinstein, 1964, 330–1.)

In this case the model and the theory possess identical formal structures and the physical properties of the phenomena described in both model and theory are identical. But not all models are theories nor are all theories models. In general it is held that the model exhibits some underlying structure of the theory. A model may thus be thought of as a kind of skeletal representation of the theory, but still, it should be noted, the structural characteristics and the physical properties described are contained in the theory as well as the model. A model is not held to be unique, and many different models can be built to represent the same theory. On the other hand it is not generally admissible for radically different theories describing the same phenomena to coexist side by side. Achinstein (1965, 105) summarises the situation:

> To propose something as a *model* of (an) *x* is to suggest it as a way of representing *x* which provides at least some approximation of the actual situation; moreover, it is to admit the possibility of alternative representations useful for different purposes. To propose something as a *theory* of (an) *x*, on the other hand, is to suggest that *x*'s *are* governed by such and such principles, not just that it is useful for certain purposes to represent *x*'s as governed by these principles or that such principles approximate those which actually obtain. Accordingly, the scientist who proposes something as a theory of (an) *x* must hold that alternative theories are to be rejected, or modified. . . .

Given this view of a model, Achinstein (1964; 1965) is able to discount the dangers of misinterpretation discussed by Braithwaite, and is also able to dismiss the particular notions of Nagel as regards the interpretative function of a model.

In Achinstein's view, therefore, a model may be regarded as a simplified structural representation of the theory and several different models may be developed to represent the same theory. Thus a theory of spatial equilibrium might be represented by a regional input-output model, by a linear programming model, by a statistical equilibrium model (say a markov chain formulation), and so on. Some writers maintain that such representations are bound to be inferior to using the theory, and that therefore a model is a redundant device (see the discussion in chapter 1 of Hesse, 1963). The theory itself will provide a much 'deeper' framework for analysis, but a model may prove useful for a number of reasons. For example, a model may simplify calculation. Deliberately simplifying the assumptions of the theory and giving a structural representation to the theory may allow complex deductions to be reduced to fairly simple manipulations. Modelling a complex theory in this way may allow insight to be gained into the theory itself. From the point of view of teaching, of course, simplified structural representations of complex theories are extremely valuable. It is interesting to note that theories may become models with the advance of scientific understanding (Achinstein, 1965, 106). Thus Newton proposed his system as a theory (i.e. as if motions were truly governed by precisely the forces he proposed) and the system operated as a theory for many years. Now we may more correctly speak of the Newtonian model since it is clear that the structure which Newton proposed (i.e. the equation system) is a simplified structural representation of a far more complex system. Thus in the same way that the basic postulates of today may become the idealisations of tomorrow (above, p. 93), so the theories of today may become the models of tomorrow. But the Newtonian case illustrates how a structural representation of a theory can still be useful—after all, studies in meteorology and oceanography can safely rest their analysis on the Newtonian model and, by and large, do not need to refer to the far more complex mathematical system representing quantum mechanics.

In such cases there is apparently no need to heed Braithwaite's warnings. The only criterion for guiding the use of models in this context is that the model should be appropriate for the particular operation required of it. This criterion is provided through the *text* of the theory which the model represents. In the same way that an idealisation can be defined theoretically, so the domain of a model can be precisely defined if the appropriate general theory exists. Thus the domain of the Newtonian model can be precisely defined and it can be shown that the model is appropriate for a wide range of problems in, for example, oceanography and meteorology. But there are many instances in which the appropriate general theory does not exist; and, again, it is important to note an important practical difference between the natural and social sciences.

In the absence of general theory we may use a model as a temporary device to represent what we think the structure may, or ought, to be. There is, for example, no completely specified theory of interregional equilibrium of economic activity. Therefore we use input-output models, markov chain models, and so on, as structural representations of a theory which is not yet fully specified and which has no full *text*. Thus the domain of the model cannot be properly defined, nor its appropriateness for the particular purpose be fully established. Under these conditions a model may be used to indicate the theory. If so, it would seem that the model does not qualify for consideration as Achinstein's 'model of (an) *x*'. It appears to function more as an analogue.

(2) *Analogue models*

Achinstein (1964, 332) states:

It is important not to confuse the scientist's model (or theory) of an *x* . . . with analogies which he may invoke in explaining features of his model or theory, and which also may have aided him in its construction.

The term 'analogy' or 'analogue model' refers to the transfer of one model or theory to another model or theory. We may represent

atoms as billiard balls, population masses as physical masses, transport systems as electric circuits, and so on. Now the view developed
by Braithwaite, Nagel, and Brodbeck, is that all that is required of
the model-theory relationship is a similarity in formal structure. In
the case of analogue models Achinstein maintains that there is no
reason why similarities in the physical properties referred to should
not also be of logical importance. All writers admit, of course, that
similarity in properties may be provocative. Hesse (1963, 9–16)
divides the phenomena of model and theory into those which
exhibit *positive* analogy (i.e. the physical properties of the model
object and theory object are the same), *negative* analogy (the physical
properties are known to be different), and *neutral* analogy (the relationship between the model object and the theory object is not yet
established). Of these the last presents the most interesting challenge
to research. We have yet to establish precisely, for example, the
respects in which population masses and physical masses are similar.

It is clear that in analogue-model situations the properties of
the model object and theory object cannot be precisely the same.
Achinstein (1964) argues from this that the model and theory
cannot be precisely the same in all respects as regards formal structure. Thus the correspondence between model and theory can be
only partial and will depend upon the respects in which the physical
properties are the same and the respects in which the deductive
structure is the same. Thus Achinstein again concludes that the
analysis provided by Braithwaite, Nagel, and Brodbeck is deficient.
But Achinstein's analysis is not free from criticism either (Swanson,
1967). We have not space, however, to discuss all the ramifications
of an argument which has scarcely yet begun to take shape.

We may use an analogue model to deduce consequences of relevance to a theory. But these consequences will be relevant with respect only to those conclusions which depend upon properties and
formal structure which are identical between model and theory.
Clearly, it is important to be aware of the dangers of model use discussed by Braithwaite. The dangers envisaged apply with great force
to the use of analogue models.

The analysis so far has suggested that we can differentiate between
two types of situation. The first is a 'model' which can contain only
physical elements and structural characteristics which are already
contained in the theory. The second is an 'analogue model' which
contains physical elements and structural characteristics which are
not contained in the theory. Now Achinstein's analysis rests on the
assumption that we can clearly distinguish between these two types
of model situation. But, as Achinstein (1965, 116–20) freely admits,

there are substantial practical difficulties involved. It is often not clear whether scientists are proposing a new theoretical concept or invoking an analogy. Entities originally conceived of as analogies may, over time, quite legitimately become theoretical concepts of considerable power. It should be apparent that the logical distinction between *models* and *analogue models* can be drawn in practice only when a general theory, provided with an adequate *text*, can be used to show that a model is functioning in one of the two ways. We have already drawn attention to the lack of such general theory in the social sciences. It is, therefore, impossible to distinguish precisely between these two model situations in large areas of research. This raises some procedural problems which need separate examination.

IV PROCEDURAL PROBLEMS OF MODEL USE

The philosopher and logician may regard the problem of model use as being essentially soluble by logical analysis. The scientist, on the other hand, simply wishes to know how and under what circumstances a particular research procedure is or is not justified. In fact the dispute among philosophers of science and logicians reflects a number of acute procedural difficulties in the use of 'models' (of all types) in scientific investigation. Further, there is some suspicion that many of the controversies in empirical science over seemingly substantive issues are in fact generated by failure to agree as to the correct procedure to be used in employing models in scientific investigation. This procedural problem can be demonstrated in relation to the two ways in which a formal theory may arise in science and the role which models play in such a process. In brief we may distinguish between *a posteriori* and *a priori* models.

(1) A posteriori *models*

We have already described the two ways in which formal theory may arise (above, pp. 32–6). The first route begins with empirical observation from which a number of regularities of behaviour may be extracted. To explain these regularities a theory is proposed which may contain theoretical abstract concepts and eventually the theory may be given axiomatic treatment and may be verified. This theory may then be represented by some structural model, which can be used to facilitate deductions and simplify calculations. In this case the model is developed in order to represent the theory. By choice, we can make the model either contain terms and structure which exist in the theory (or can be defined with reference to the theory) or use an analogue of some kind. In this case the function of the model is simply to represent something which is already known, and the only question which arises is that of the appropriateness of a model for a given purpose and this, as we have already noted, can be fully defined only if the appropriate general theory can be referred to. When no general theory exists there must be some doubt as to the appropriateness of the model representation of the theory.

If a general theory exists, and if the model contains terms and structures referred to in the theory, then the conditions for Achinstein's 'model of an x' are fully met. In such a situation the dangers inherent in model-use are minimised. But the question arises as to the effect upon model-use of varying sophistication in the theory which the model is designed to represent. Model representations of *Type 2* or *Type 3* theories which simply involve 'harmless' elliptical argument need not be seriously affected. On the other hand, failure to specify the theory fully must automatically mean a certain loss of control of the model-theory relationship. Imprecise or partially formalised theories may be represented by precisely specified models, but the nature of the model-theory relationship must perforce remain unspecified also.

One of the roles of *a posteriori* model-building is to allow the easier manipulation of relationships and to facilitate testing procedures. The less sophisticated the theory, the less control we possess over the model-theory relationship, and the less we shall be able to tell if (i) Model conclusions can be transferred to the theory or (ii) A successful test of the model indicates a successful test of the theory which the model represents.

With *a posteriori* models we are thus working along a continuum from theoretical statements of great certainty represented by eminently controllable models, to extremely fuzzy theoretical statements which may be given precise, but completely uncontrollable, model-representation. At the latter end of this continuum it will be extremely difficult to tell whether we are invoking 'model' or 'analogue' representations.

(2) A priori *models*

A particularly important form of the second route to theory construction lies through giving an interpretation to a completely abstract calculus (Brown, 1963, 174). This calculus may have arisen simply from abstract analytic argument, or it may have arisen as a response to a particular set of empirical problems in a different domain. But for the moment we shall simply treat that calculus as having a completely abstract formulation as, for example, a set theoretic development of probability-theory, or an abstract development of some geometric system. This abstract calculus is then found some real-world interpretation by giving empirical meanings to the terms contained in the calculus. By providing a set of interpretative sentences we 'map' some aspect of the real world into this ready-made calculus. In other words we begin with the calculus and then seek to identify a domain of objects and events to which it can be applied. If it can be shown that this mapping is successful, then the calculus may be accepted as a model-representation of a theory, and we may, from the structure of the calculus, infer the structure of the theory. The model is thus set up first and the theory is developed from the model.

A priori models are probably more common than *a posteriori* models. It is interesting to note that in Braithwaite's (*1960*, 89–90) view a major difference between a model and a theory is simply that the former is epistemologically prior to the latter. A model is thus an *a priori* analytic construct which is applied to reality, whereas a theory is a construct which grows out of experience of the real world. This view of a model is particularly common in the social sciences. Thus economists usually view a model as an *a priori* construct (usually mathematical, but sometimes graphical) into which economic theory is mapped for clarity of exposition or for set purposes of manipulation (Arrow, 1951; Beach, 1957; Koopmans, 1957).

A priori models have a number of different functions. In terms of the context of discovery, 'picture' models of reality can be of great importance and often play a key part in the psychological process of theory formation. For the moment we will restrict consideration to some of the logical properties involved in the use of *a priori* models. Thus two of the main functions of such models are to suggest theory, and to allow manipulations and conclusions to be drawn about some set of phenomena even in the absence of full theory. In both cases there are some difficult logical problems of inference and control involved. In the first case it will not be clear whether the model concepts are the same as those of the theory-to-be, or whether they are in fact analogues. It is extremely dangerous to infer from a successfully applied model that the theory which governs the behaviour of the phenomena described in the model *must* necessarily possess the same characteristics as the model. Similarly, predictions made with the aid of an *a priori* model construct must be open to doubt, since it is not possible to know in what respects the model represents a theory and therefore the respects in which concepts and structure are similar cannot be assessed.

In general, therefore, we need to be clear that all of Braithwaite's warnings apply to *a priori* models and, worse, we must heed them knowing that we cannot yet specify the respects in which theory and model are similar.

With the development of general theory, however, it may be shown that an *a priori* model has the characteristics of an *a posteriori* model and hence problems of inference and control may be minimised. Again, there are special situations in which an *a priori* model may achieve the logical status of a 'model of an *x*', but in general such situations are rare in empirical science, particularly in the social sciences where the requisite general theory does not yet exist.

We assume in the above analysis that the two procedural situations can be distinguished. Characteristically, research tends to slip from an *a priori* model of some kind, to the postulating of some theory, to the development of an *a posteriori* model which then has elements added to it so that it then functions as an *a priori* model for the creation of further theory. Similarly, a model may subtly change its function. Originally functioning as some 'picturing' device for stimulating ideas, the model may then function as a set of rules for ordering data (e.g., the regional concept), and, later, become a structural schema suitable for explanation. From a procedural point of view it is essential that we differentiate clearly between these different functions and note when a particular model changes its function. But, again, it is not always easy to do this.

The procedural problem and the logical problem overlap. Philosophers and logicians frequently ignore procedural difficulties, claiming that the context of discovery and the psychological process of theory-formation does not concern them (Popper's dismissal of induction and other procedural problems is a good example). They also claim that most of the procedural difficulties will be solved if the logical problems are solved. This is undoubtedly true, but it is of small comfort to the scientist who needs to make a decision on some procedural issue without the appropriate logical guidance. Some, such as Workman (1964), argue that it is not necessary to relate model and theory since it is sufficient to design models which work. Even if we took this extreme view there would still be the problem of setting up criteria to judge in what sense and in what domains a model 'works'. The advantage of the model-theory debate is that it asks very explicit questions about the degree of realism in a model, even if it does so in a rather confusing way.

V TYPES OF MODEL

In the same way that a model may have various functions and definitions, so it may perform its function through a multiplicity of media.
Again, certain academic disciplines *tend* to resort to certain *kinds* of
model. It is useful, therefore, to review some of the typologies of
models which have been developed.

Ackoff (1962), for example, differentiates between *iconic* models,
which use the same materials but involve changes in scale, *analogue*
models which also involve a change in the materials used in building
the model, and *symbolic* models which represent reality by some
symbolic system such as a system of mathematical equations. Each
type of model varies in its appropriateness for different functions,
but, again, there are no hard and fast rules. Ackoff's classification
seems most closely geared, however, to model-functions which relate
directly to reality. Given the catholic definition of the term as used
by Ackoff, this is hardly surprising. As we have seen, however, a
theory may itself be defined 'as a language for discussing the facts
the theory is said to explain' (Ramsey, *1960*, 212; see also above, p. 88).
A theory may itself, therefore, be a symbolic system. Given Ackoff's
typology it becomes difficult to discuss the crucial model-theory
relationship in a succinct way. But Ackoff's point is nevertheless an
important one. By choosing to translate from one medium to another
(say, representing a symbolic system by an electric circuit) we may
facilitate manipulation of theory or the development of theory or
prediction etc. Ackoff's views have been taken up in the geographic
literature by Berry (1963, 105–6).

Chorley (1964) regards all models as being analogues of some
kind, but suggests a classification of models into those which translate into analogous natural circumstances (this is similar to Ackoff's
analogue model), models which involve experimental procedures
(which may either involve a change of scale or translation to analogous natural circumstances), and mathematical models (which
appear to correspond to Ackoff's symbolic models). In a later
presentation, Chorley (1967) revises and extends this classification
system (Figure 10.1). Three main types of model are now envisaged
with a number of sub-types.

1. Natural Analogue System:
    a. Historical Analogue
    b. Spatial Analogue

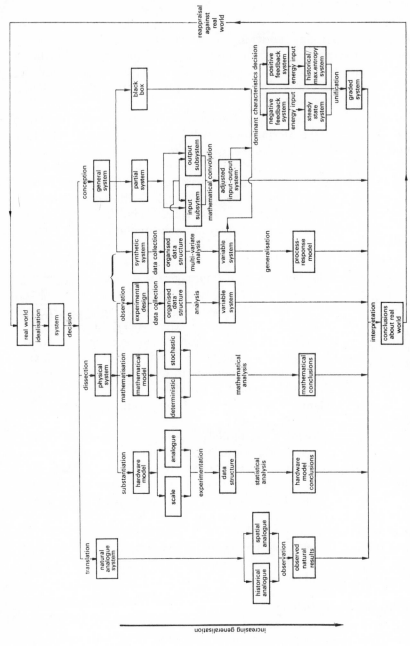

Fig. 101. Chorley's typology of models embedded within a 'map' of geomorphic activity (*from Chorley*, 1967).

2. *Physical System:*
   a. Hardware Model
      (i) Scale
      (ii) Analogue
   b. Mathematical Model
      (i) Deterministic
      (ii) Stochastic
   c. Experimental Design

3. *General System*
   a. Synthetic
   b. Partial
   c. Black box

Generally speaking the first group of models involves searching for analogous situations or events at different times or in different places, and drawing some conclusions. An example of such a procedure is Rostow's (1960) schematic representation of the economic growth process which is derived from historical analysis and searching for analogies between different countries at different times. The second group of models corresponds to the more conventional notion of a model in the sciences. The third is a newer concept which treats the structure of a landscape as an assemblage of interacting parts, and attempts to represent the processes as such. Synthetic systems are artificially built to simulate reality in a structural way and, as Chorley points out, such models may be similar to experimental design models. Partial systems are concerned with workable relationships and attempt to derive results without complete knowledge of the internal workings of the system. The 'black-box' approach attempts to derive results from a situation in which we have no knowledge of the internal workings of the system.

Chorley's discussion of these varied model approaches does not strictly amount to a classification of models, since the categories are not mutually exclusive. But it does suggest that different types of model have different properties and different aptitudes for representing different types of real world situation. Chorley's 'map of geomorphic activity' illustrates how these different types of model might be used in a comprehensive process of enquiry in geomorphic research (see Figure 10.1).

Typologies of this kind succeed in demonstrating the extraordinary breadth, flexibility, and potential, of the model-concept. They also may reveal a tendency for certain disciplines to be concerned with certain types of model (sometimes almost exclusively)—the dominance of hardware models and mathematical models in engineering, of natural analogues in history, of mathematical and graphical models in economics, are examples of this tendency. But such typologies do not solve the methodological problems of model use even though they may serve to clarify those problems to some degree. In practice the failure to develop a complete mutually exclusive classification of models depends on the prior failure to provide an adequate definition. Such a definition must, it is clear, embrace a multiplicity of types as well as a multiplicity of functions. Typologies succeed, therefore, only in exposing just one more facet of the complex problem of model use.

VI THE PROBLEM OF MODEL USE

The model-concept poses considerable methodological difficulty. There are a multiplicity of model-types performing a multiplicity of functions associated with a multiplicity of definitions. Each particular model exhibits a different logical capacity for performing the function required of it. As we have seen in the sections concerned with the model-theory relationship, the evaluation of this logical capacity and even of the way in which a model may perform a particular function, is extremely difficult. Yet such an evaluation is essential for the control of the model and for testing the validity of the inferences made from it. We can perhaps demonstrate the seriousness of this issue with reference to the all-important model-theory relationship.

A main aim of theory construction is to expose the 'order in seeming chaos' and consequently to allow information derived from certain individual cases to be brought to bear on other individual cases. In the search for theory we may begin by using some *a priori* model construct. In doing so we need to be able to say that order does exist in some form and has not been imposed merely by means of the *a priori* model. Further, we need to be able to show the legitimacy of transforming the physical properties of the model object and the logical form of the model, into properties and form of some theory. The legitimacy of this whole operation depends on an ability to differentiate between 'models' and 'analogues' and, as we have seen, such a distinction is sometimes very difficult to make. But without such a firm distinction we cannot tell how far the theory represents 'actual order in seeming chaos' or 'seeming order in actual chaos'.

Extending, restructuring, or creating, a theory invariably involves the application of *a priori* models. This means that the model pos-

sesses some characteristics or aspects of form which are not represented in the initial theory (if there is one). This may involve addition
(extending a theory), substitution (restructuring a theory), or invention (creating a theory) of model-concepts. Simply *representing*
a theory, on the other hand, allows the use of *a posteriori* models
which, if the appropriate general theory exists, can be fully controlled. The concept of a 'model of an *x*' used by Achinstein appears
to refer mainly to this kind of procedural situation.

In most of the social sciences and in many areas of natural science,
the requisite general theory does not exist. In geography, as we have
already seen, theory is weakly developed. It is certain, therefore,
that the use of models in geography (and in many other disciplines)
will involve *a priori* analysis. In such a situation the logical properties
of the model are obscure. But to insist on a rigid interpretation of the
term 'model', such as that proposed by logicians like Tarski, would
probably prove stultifying for research progress. In the search for
adequate theory, *a priori* models are indispensable. But their use
involves dangers—particularly when the model-theory relationship
can be only weakly controlled. The most serious problem is one of
identification. There are three kinds of situation:

(a) *Over-identified models*. We may postulate a model and then find
that the model has more than one possible theoretical interpretation.
Suppose, to use a simple example, we use a regression model for
analysis and get favourable correlations. Is this evidence of causal
inter-relationship, functional interrelationship, or is it simply spurious
correlation? The model, by itself, cannot tell us and the model is,
therefore, characteristically over-identified.

(b) *Unidentified models*. We may postulate a model and find that the
model gives excellent results, but that it is impossible to find any
firm theoretical interpretation. This kind of situation is characteristic
of some of Chorley's 'black-box' models, and may also be characteristic of some statistical models. The rank-size rule, for example,
appears to have no firm theoretical interpretation, and the gravity
model is, to a lesser degree, in this category. In each case several
theories have been developed to explain these regularities, but the
model itself does not indicate in any way what the nature of that
theory should be.

(c) *Identified models*. A postulated model may give rise to one and only
one theory. This circumstance is clearly the most desirable—but it
is not very common in geographic research.

It is characteristic of the status of geographic research that we are consuming models very rapidly in the search for geographic theory. Given the weak development of theory, however, we cannot establish the domain of the models we are using with any precision, nor can we identify the model-theory relationship with any precision. As Curry (1964, 146) has observed:

> That a given set of premises contains logical consequences which are in agreement with reality is no guarantee that a model is itself realistic. Several quite different models may give the same result. This type of work can only be judged over a period of years when separate models are being articulated and elaborated to the point where a whole theoretical structure exists to explain many seemingly isolated phenomena. In the meantime, 'explorations' are all that are possible with the considerable chance that they lead only up blind alleys.

In search for theory by way of the application of models the exploration of such blind alleys appears inevitable. Without taking such a risk it is probably unlikely that the development of theory will be rapid (if it develops at all). But there are certain risks which are justifiable and some that can be avoided by adequate research design and an adequate understanding of the role and function of models. Even though there are many unsolved epistemological questions surrounding model-building in the sciences, a number of broad procedural rules may be suggested as a guide so that the more obvious dangers associated with model use can be avoided:

(i) The proposed function of the model should be clearly specified; e.g., is it being used to represent a theory, suggest a theory, predict from a data-set in the absence of adequate theory, etc?

(ii) The function of a particular model should not change within a particular research design without adequate safeguards (otherwise the tautological case may be difficult to avoid).

(iii) A model used to infer or represent a theory should preferably be identified with one and only one theory.

(iv) Over-identified or unidentified models should be restructured so as to allow identification—otherwise alternative theoretical interpretations should be fully considered.

(v) Conclusions drawn regarding a theory from the manipulation of a model for that theory should not be automatically accepted unless

    a. The model is identified with the theory, or
    b. The domain of the model and the nature of the model-theory relationship can be fully specified.

(vi) Conclusions drawn regarding a particular subject-matter by way of some model-manipulation will be fully acceptable only in so far as the model will eventually be shown to represent a viable theory. Predictions simply based on model-manipulations are essentially uncontrollable. If it is necessary to accept predictions on the basis of a model only (as it often is in social science research directed to planning-ends) the predictions should be treated with care. Predictions accomplished with the aid of full theory are more controllable (although not necessarily correct if events outside the domain of the theory are admitted).

(vii) The multiplicity of functions, types, and definitions, given to the concept of a model should be fully appreciated in any research design.

These procedural rules are very general. We now need to consider more specifically some of the problems which have arisen in geography in relation to the use and abuse of the model-concept.

Basic Reading

Achinstein, P. (1964).
Apostel, L. (1961).
Brodbeck, M. (1959).
Hesse, M. B. (1963).

Chapter 11
Models in Geography

Given the very fluid state of philosophical opinion regarding the nature of models, it is hardly surprising to find a good deal of variation among geographers regarding the meaning of the term and the practical application of the model concept in empirical research. It is encouraging that geographers have recently become aware of some of the methodological problems surrounding the model-concept. Chorley (1964, 128) clearly recognises the model-theory dichotomy, stating that:

> A model becomes a theory about the real world only when a segment of the real world has been successfully mapped into it, both by avoiding the discarding of too much information in the stage of abstraction . . . and carrying out a rigorous *interpretation* of the model results into real world terms.

In later presentations this distinction becomes somewhat blurred. Thus Chorley and Haggett (1967) choose to emphasise a catholic definition of a model (notably that stated by Skilling cited above, p. 145) partly because, one suspects, the contributors to the volume of essays they were editing exhibited rather different attitudes to the model-concept. Thus George (1967, 43) and Harvey (1967A, 552–4) appear to regard a model as an interpretation or representation of a theory. Grigg (1967, 494–500) emphasises the confused meaning of the term and discusses the regional concept in relation to a number of different meanings. He considers the notion of isomorphism and analogy with reference to the regional concept, while in a later section the notion of a region is related to models which are essentially 'a step towards formulating a theory'.

The model-concept has, thus, been given some methodological attention in geography. Yet the explicit examination of the concept has exposed considerable confusion—a confusion of function and definition which, as we have seen, is also characteristic of the philosophical literature. Pahl (1967, 220) dismisses the debate surrounding this issue as a 'rather arid discussion about terminology'. If this assertion is true, then the whole problem could cheerfully be left to the philosophers and logicians to sort out at their leisure. But the debate is more than mere semantics. It is a debate over inference, over the organisation and validation of theory, over the very nature of the understanding which we can hope to attain regarding geographic phenomena.

In practice geographers have used (and sometimes abused) the model-concept in the course of their research. In many cases the use has been implicit rather than explicit, for in the same way that it is probably impossible to state a fact independently of some theoretical interpretation (above, p. 87), so it is also impossible to state a fact independently of a model for that theory. In sciences where theory is weakly developed—as in geography—the use of *a priori* models is inevitable whether or not such models are consciously used in the search for theory. The inexplicit use of such *a priori* models is particularly dangerous since the necessity for strict control over inference is not fully apparent. The application of *a priori* model constructs to suggest or extend theory, or to allow prediction in the absence of theory, requires that the model 'be used freely as long as it serves its purpose . . . be discarded without regrets when it fails to do so.' But we frequently become rather attached to a particular model and fail to heed Rapoport's (1953, 306) advice that 'the scientist, if he is completely a scientist, is unique among the users of metaphors in that he does not become an addict of a particular way of perceiving.' The addiction of geomorphologists to the Davisian model (Chorley, 1965), the more modern addiction of settlement-geographers to various central-place models, are examples of an attachment to models which are only partial representations of theories which are, in turn, far from being complete or explicitly developed. But addiction sometimes brings with it intellectual blindness. A curious transformation may take place from an *a priori* model which has no empirical justification, through the sudden acceptance of that model as *being* the theory (without any empirical evidence), to the ultimate canonisation of the model as the quintessence of reality itself. The history of the Davisian model exhibits some aspects of such an unwarranted transformation. But as Chorley (1965, 22) has remarked,

> The cycle is no more complete and exclusive definition of geomorphic reality than the pronouncement by the proverbial Indian blind man on feeling an elephant's leg that the animal is like a tree.

The cycle of erosion proposed by Davis is now 'recognised as merely one framework within which geomorphology may be viewed.' It would appear, then, that in geomorphology at least a past addiction is now being successfully overcome.

The history of geographic thought is full of the misapplication of *a priori* models. It has been a characteristic of the evolution of that thought that the master proposes the model (often fully aware of its temporary and *a priori* character), while the disciples rather blindly canonise it. The excesses of the Davisian school are attributable less to Davis himself than to the blind faith of his followers such as D. W. Johnson (Chorley, 1965). The excesses of the environmental determinists are, similarly, attributable less to Ratzel himself than to disciples such as Semple. In all such cases the disciples make one crucial mistake. The *a priori* model is suddenly accepted—without evidence —as a complete theory. It is for precisely this reason that a methodological understanding of the model-concept may be regarded as more than mere semantics. The history of geographic thought might thus be regarded as a 'history of misapplied models'.

Of course there are blatant examples of false argument by analogy in the history of geography. The analogy between political states and biological organisms failed to specify which properties were positively or negatively analogous. What could have been a stimulating view of geography—a view of the geographical system as *organism* and *ecosystem* (Stoddart, 1967A; Berry, 1964A)—fell into disrepute not so much because the model was itself invalid but because the model was misapplied to give spurious justification for a geopolitical theory which was both misleading and mischievous. The history of this particular model-concept is particularly interesting and has been dealt with at length by Stoddart (1967A). The impact of biological thinking on geographic thought has been strong—particularly through the two concepts of organism and ecosystem. But, as Stoddart shows, it is all too easy for a stimulating and perfectly valid analogy to fall into disgrace simply because the analogy is misapplied. The misapplication largely stemmed from a failure to specify in what respects regions possessed organic properties, and in what respects regional development could be explained by reference to the biological model.

But there are even more complex examples of the difficulty of applying model-concepts in geography. It is perhaps worthwhile examining some of these examples briefly.

Example 1

Consider the following logical schema:

(i) $A \rightarrow B$; $B \nrightarrow A$.

(ii)   If $A \rightarrow B$, and $B \rightarrow C$, then $A \rightarrow C$.

These two statements identify two important characteristics of cause-and-effect analysis. The first states the principle of irreversibility and the second of transitivity. This schema may be regarded as a logical model. Now a good deal of research in the natural sciences has been concerned with showing that this schema represented a universally valid theory. A good deal of evidence was collected to show that such a model was indeed appropriate. Now consider the position of the geographer who chooses to apply the model (i.e. give the schema an interpretation in geographical terms). The model is then being applied as an *a priori* model in the search for geographic theory. Events and interactions in the real world are mapped into the model and certain inferences regarding theory are made. It is not admissible, however, to infer that the theory must, of necessity, have the same logical structure as the *a priori* model which we have applied in the search for that theory. In particular, we cannot assume that the real world is *in fact* governed by cause-and-effect laws because we have successfully applied a cause-and-effect model to some aspect of reality. Such an inference would be true only if we could show that the model is uniquely identified with a theory, and if we could also show, independently of the evidence provided by the model itself, that real-world events were *in fact* governed by cause-and-effect laws rather than conveniently being analysed by the application of a cause-and-effect model. Certainly we cannot conclude, with A. F. Martin, that we must all, *of necessity*, be determinists. To this case we will return later (see chapter 20), but for the moment the moral can be made clear. The argument over determinism and possibilism in geography appears to be a classic case of two *a priori* models 'being transformed into two competing theories without the necessary safeguards' being observed.

Example 2

Probability models have become more fashionable in geographic research in recent years. Again, there is a danger that the successful application of a stochastic model will be used as direct evidence that the world is in fact governed by the laws of chance. Such an inference is not possible without evidence which is independent of the model itself. But probability theory provides us with even neater examples of some of the problems implicit in the use of models. Probability theory forms a vast and elaborate abstract calculus into which we can map geographic problems. The methodological difficulty which faces the geographer is to establish the rules which govern this mapping process. Without such rules it is difficult to judge whether a particular probability model is appropriate or if inferences made with the aid of the probability calculus are valid with respect to the real-world events we are examining. Consider, for example, the following probability model in which the probability of an event is occurring exactly i times is:

$$P\{X = i\} = p(i; k, p) = \binom{i + k - 1}{k - 1} p^k (i - p) \quad \begin{array}{l} 0 < p > 1 \\ k > 0 \\ i = 0, 1, \ldots \end{array}$$

This probability model has two parameters, p and k, and is usually called the *negative binomial* probability distribution. The model has been given a number of interpretations in geographic research (Dacey, 1967; Rogers, 1965; Harvey, 1966B; 1968A; McConnell, 1966). In particular it has been used as a device to summarise point-pattern measures derived by quadrat sampling. There are a number of assumptions in this procedure, but for convenience we shall assume that all of these assumptions have been met and that a good fit has been obtained between the model and a data-set. What inferences may we now draw?

The goodness of the fit indicates that the model is a reasonable description of the morphometry of that point pattern. Our primary interest, however, is to identify the particular process which leads to the creation of that pattern. To gain insight we may refer to the physical models which the mathematicians have used to derive the negative binomial distribution itself. Inspection of these physical models shows that mathematicians have derived the distribution from contrasting physical situations. Consider the following two models:

(i) 'If colonies are distributed over an area so that the number of colonies observed in samples of fixed area has a Poisson distribution, we obtain a negative binomial distribution for the total count if the number of individuals in the colonies are distributed independently in a logarithmic distribution' (Anscombe, 1950).

(ii) If the individual points are distributed according to random expectation (i.e. are described by a Poisson distribution) but the mean density varies from place to place, then we obtain a negative binomial distribution if the distribution of the mean values of the whole collection of Poisson distributions varies according to a Type III curve.

These models are not the only ones which give rise to the negative binomial distribution, but they are perhaps the most interesting, since they indicate that the same distribution can be derived from diametrically opposite processes. The negative binomial distribution is thus a classic example of an over-identified model. Without further evidence it is not possible to infer which of the theories is appropriate. Harvey

(1966B) assumed the contagious process to be the relevant theory, since a number of studies had shown this to be the case. On the other hand, Dacey (1967) has shown that the behaviour of the two parameters of the distribution under different sampling conditions (particularly with respect to quadrat size) may be used to infer which of the physical processes is occurring in the population distribution being studied.

In the case of the negative binomial model there are at least six different generating processes which can lead to the derivation of the distribution (Dacey, 1967). Knowing that a model is over-identified in this way is useful, if only because it indicates that inference is dangerous. But in many cases we do not know whether a model is over-identified or not. The general complexity of the situation is this: a number of different models may be built to represent a theory while each model may, under certain circumstances, represent several different theories. In the search for theory, the greatest problem is posed by the multiplicity of possible interpretations which can be given to any one structural model. Inductive statistical procedures —such as descriptive measures of pattern, principal components and factor analysis, regression analysis, and so on, all pose this same problem'. The negative binomial 'problem' is thus just one example of the overriding difficulty of constructing geographic theory with the aid of the model-concept.

These two examples indicate the kind of problem which we face when we use any *a priori* model in geographic research. Any logical schema or mathematical system (and in the latter category we may include geometry) may be regarded as an abstract calculus into which we can map geographic problems. Such a mapping process provides us with a powerful, if dangerous, device for generating theory or for making predictions in the absence of theory. The question arises, of course, why we should ever bother to use models when their use poses such difficulty and danger. The short answer to this is that we have no choice. With very weakly developed geographic theory and a highly complex multivariate subject-matter, it is inevitable that the model-concept should play a part in geographic explanation. In the absence of firm geographic theory, a model can provide a 'temporary' explanation or an objective (if often inaccurate) prediction. Such 'temporary' uses of the model concept are important, particularly in a world which demands some kind of answer to a whole range of complex socio-economic problems. But in terms of basic research the primary function of model-building in geography must be directed towards the creation of geographic theory. Models, it is true, can perform a multiplicity of functions. But the adequacy of performance of the model for a particular function can ultimately be judged with reference to the appropriate theory. Without such theory the appropriateness of a model-construct cannot be fully appreciated. It is heartening to know that *some* degree of control can be established over complex situations—indeed in human geography and many of the other social sciences 'the control science gives us over reality we normally obtain by the application of models' (Ackoff, 1962, 108). But as Lowry (1965, 164) has most cogently pointed out in the context of urban planning models, the model is 'a tool of unknown efficacy'. The employment of models is a *sine qua non* for the progress and application of geographic research, but 'the price of their employment is eternal vigilance' (Braithwaite, *1960*, 93).

Basic Reading

Chorley, R. J. (1964).
Chorley, R. J., and Haggett, P. (eds.) (1967), particularly the two introductory chapters and the chapter by Chorley on models in geomorphology.

Chapter 12
Theories, Laws and Models in Geographic Explanation—A Concluding Statement

Various threads in the argument of the preceding chapters need to be drawn together to form some kind of concluding statement regarding the function and implications of theories, laws, and models, for the explanation and understanding of geographic phenomena.

We began by suggesting that without theory of some kind the explanation and cognitive description of geographic events is inconceivable. The scientific method attempts to ensure that a theory is stated in a particular form; a form which is controllable, internally consistent, and verifiable. To this end philosophers of science and logicians have described a paradigm for a scientific theory. Academic disciplines vary in the degree to which they have managed to state theory in accordance with this paradigm, and it is well recognised that many theories in both natural and social science do not meet all of the requirements laid down by philosophers and logicians. Two things should be noted about this paradigm. First, it has itself been subject to change over time. Thus the views of Mill differ in many respects from those of Poincaré, and those of Poincaré differ markedly from those of Braithwaite, Nagel, and Hempel. Second, the paradigm does not tell us how to achieve theory. It merely tells us how we should present and write out a theory once it is achieved.

There is a world of difference between the way in which we set out a theory once it is created and the way in which we proceed in creating that theory. We can perhaps demonstrate this by way of an analogy. Philosophers of science have, on occasion, likened a theory to a map. Thus Toulmin (*1960*A, 105) states that

the problems of method facing the physicist and the cartographer are logically similar in important respects, and so are the techniques of representation they employ to deal with them.

Since many geographers probably know more about map-construction than they do about theorising, such an analogy may prove extremely instructive in the present context.

A map is drawn up according to certain cartographic principles.

These principles include consistency—the symbols should not change their meaning from place to place on the map—and logical coherence —we expect the map to make an even statement about whatever it is designed to represent. But the map in itself is an abstract system— a series of lines, marks, colours, signs, and symbols. In this respect the map is like an uninterpreted calculus—a theory without a *text*. The map can be interpreted only when the key is added, and the key should give us the relevant information for such interpretation. It should tell us what the signs, symbols, colours, and so on, mean, and it should also state in what respects the map is representative and in what respects it is not. Scale, location, map-projection and orientation are important statements which tell us the *domain* of the map in the same way that a proper *text* will tell us the *domain* of a theory. A theory without a complete text is like a map with an incomplete key. This incompleteness limits the utility of the map—it is dangerous to calculate real distances from a map surface if we do not know what map-projection was used in its construction. Many of the theories in the social sciences suffer from incompleteness and prediction from them is likewise affected.

Maps and theories have a similarity of purpose. We can use maps to supply information, we can use them to predict, and we can use them to analyse relationships. In the same way, says Toulmin, we can use theories 'to find our way around' phenomena, to predict, to explain, and to supply information. We can quickly answer enquiries such as 'Is Cairo north of Madrid?' or 'At what speed will a freely falling object hit the ground after a fall of 25 feet?' without actually making any direct measurements. Both maps and theories are, therefore, constructed for something. They are usually built to answer some significant need. Of course it is difficult to judge what is meant by significance here. Such a discussion would lead us only into the whole problem of the ultimate 'purpose' of human knowledge, and it is perhaps sufficient to state that maps and theories are constructed simply because someone regards the operation as being worth while. In terms of direct utility there are useless but beautiful maps and theories, inelegant but functional maps and theories, and so on.

Once the map has been constructed we may derive from it a number of models. We may simply sketch off the rivers, or the contours, or the transport network. The resultant sketch is rather like Achinstein's 'model of an x' in that it expresses only relationships and information contained in the master map. We may abstract from the map mathematical equations—fit trend surfaces to the contours, examine point patterns by way of nearest-neighbour measures, and so on. In this case we are translating the information on the map into some other medium for purposes of manipulation—we are constructing an 'analogue' model. In all of these cases we are, however, making *a posteriori* use of the completed map or theory.

But the way in which we draw up the map, and the rules which we follow in manipulating and using the map once it is completed, contrast with the procedures used to create the map in the first place. The method followed, the tools and implements used, the assumptions made, and the activities followed, are very different indeed. A scientific map or theory relates in some way to empirical phenomena. In drawing up a map which will allow us to 'find our way around' precisely we need to go out and survey the land—this involves measuring, sketching, and recording. The cartographer obeys one set of rules at his drawing-board and a different set of rules when he is out surveying the land. In the same way that we cannot judge the activities of the field surveyor by the criteria appropriate to the drawing-board, so we should not judge the scientist in the process of seeking for a theory by way of the criteria we use in drawing up that theory in its final form. This important point has been thoroughly emphasised by Gilbert Ryle (1949, 269–75). This is not to say that the field surveyor's activities are unrelated to the requirements of the drawing-board. A field surveyor ignorant of the rules of cartography will probably do a poor job; likewise, a scientist ignorant of the logical requirements for the statement of scientific theory will probably make a poor job of seeking out some theory. The two activities are different but their requirements are closely interrelated.

In the absence of detailed empirical knowledge the map-maker may be forced to make a number of *a priori* assumptions in drawing up his map. The medieval cartographers, for example, were forced to assume that the world had some shape before they could record anything. They may have assumed that the world was round, flat, or, as did Cosmas Indicopleustes, assume it to be shaped like a tabernacle. The point about such assumptions was that without them and without the knowledge of the correct shape of the earth no map could ever have been drawn. The assumed shapes thus functioned as *a priori* model-constructs. Predicting and explaining by means of *a priori* models is, thus, rather like trying to navigate to Jerusalem with the aid of the Hereford map.

The point of this extended analogy is to try and indicate simply and clearly the rather different ways in which we should approach the purpose of theory-construction, the rules which govern the statement of theory, and the procedures we follow in discovering a theory. By means of such an analogy it is also possible to make clear the rather difficult notion of the model-theory relationship. But it does not do to pursue the analogy too far. We shall therefore return to the substantive problems of methodology at issue and try to fashion some direct concluding statement on the *purpose, form,* and *strategy* of investigation in geographic research.

(1) *Purpose*

This book is mainly concerned with the way in which we make statements and the strategy of developing general explanatory statements of considerable reliability. Yet these two issues are not independent of purpose. It has thus been freely acknowledged that speculation regarding the 'nature of geography' forms the mainspring for the creation of geographic theory. Without the motivation and direction provided by such speculation, theory could not develop (see above, pp. 114–16). Nothing in the preceding chapters can be used to discredit the basic views which geographers have of the purpose of geographic investigation. In some instances it is hoped that these traditional views have been extended and sharpened. Consider, for example, the following statement made by Hettner in 1923:

> The thought of a general earth science is impossible of realisation; geography can be an independent science only as chorology; that is, as knowledge of the varying expression of the different parts of the earth's surface. It is, in the first place, the study of lands; general geography is not general earth science; rather, it presupposes the general properties and processes of the earth, or accepts them from other sciences; for its own part it is oriented about their varying areal expression. (Quotation from Sauer, *1963*, 317.)

This fairly traditional view of the 'nature of geography' fits, in a general sort of way, with conclusions regarding the nature of the basic postulates for geographic theory—conclusions which suggested an important dichotomy between indigenous morphometric postulates and derivative process postulates (above, p. 127). Although the analysis so far does not contradict such basic views regarding the nature of geography, it does suggest that there is an important case to be made for separating purpose from the form of statement. An important source of confusion in much of the methodological literature relating to geography (particularly Hartshorne's presentations) is the failure to differentiate between these two elements. The ultimate purpose of geographic analysis may be to understand individual cases; in the same way, to revert to our analogy for a moment, the ultimate purpose of a road map may be to allow individuals to choose their own routes in their own ways. This does not mean, however, that a separate map or theory must be created for every instance, nor does it mean that the principles which govern cartographic analysis or theory-statement can be altered or abandoned. Even unusual events require the application of such principles if they are to be explained. Above all, we cannot conclude, as many appear to do, that because geographers are very much concerned with particular cases, there is no possibility for formulating laws which can be applied to explain those particular instances. But this leads us to consider the form of explanatory statements in geography.

(2) *Form*

The basic concern of the analysis so far has been with the form of explanation in geography. Explanation in geography has, until recently, remained a process of applying intuitive understanding to a large number of individual cases. Scientific theories have not, by and large, been explicitly developed, laws have consequently not been formulated, and the usual requirements of scientific explanation have not been met. Such a situation can partly be attributed to the methodological position of many geographers (particularly Hartshorne) which rested on the false inference that because we are essentially concerned with particular cases we must necessarily seek for only particular explanations. Each game of chess may be unique, but the rules of the game are extremely simple and the field of play very constricted. Such a methodological position has led to the failure to investigate phenomena with the requirements of the scientific method in mind—we have been trying to create a map in ignorance of the rules of cartography. But the failure to achieve an adequate corpus of reliable theory must in part be attributed to the extreme complexity of our subject-matter.

There is no reason in principle why laws should not serve to explain geographical phenomena, or theories of considerable explanatory power be constructed. Explanations which conform to the rules of the scientific explanation as generally conceived of can, in principle, be offered. This is our central conclusion.

The main difficulty comes with the implementation of this conclusion. Given our lack of understanding and the extreme complexity of much of our subject-matter, it will be a long time before we possess relatively complete theories of any great explanatory power. Explanations which rest on partial and incomplete and inadequately specified theories are bound to be relatively weak and inefficient. In such circumstances it may well be that explanations offered on the basis of intuitive insight alone may be more effective. In the long term, however, such a procedure is inefficient. An intuitive explanation implies a model or a theory of some sort;

otherwise it would not explain. Why not, therefore, attempt to develop and expand such intuitive explanations with the requirements of scientific explanation in mind? We may, after all, successfully navigate to Jerusalem using merely a sense of direction and travellers' tales—but in the long run it is far more efficient to navigate by map and it is a pity, therefore, that our successful intuitive voyage was not recorded in detail for the purposes of map-construction. We are not, as yet, in the position to create powerful geographic theory, but it would be extremely useful if we proceeded in research with the requirements of such a theory broadly in mind. We have not yet achieved an adequate corpus of theory, nor does it seem that we will do so in the very near future. To attempt to apply any too rigid form of scientific explanation in the absence of adequate theory is to court disaster. Our explanations must, therefore, continue to resort very much to intuitive perceptions. But intuitive perceptions backed by a slowly developing corpus of theory are a more comforting sight than intuition grappling alone with complex problems.

It is pointless to judge current explanations or even the form of those explanations by the standards of the scientific paradigm. Such judgement will surely only discourage. If we cannot simply apply the paradigm itself, then this requires that we develop our own paradigm —one which is presumably weaker than, but not entirely unrelated to, the scientific paradigm itself. The rules of such a paradigm need to be flexible but firm. Perhaps the best way to approach such a geographic paradigm is to note those elements in the scientific paradigm which can most easily be discarded without great danger. Axiomatisation and formalisation of theory seem, for example, unnecessary and may even prove stultifying. Similarly applying rigid standards to the notion of a law appears pointless. Perhaps the central rule of some geographic paradigm should be the willingness to regard events *as if* they are subject to explanation by laws. This whole problem of setting up our own standards for research will ultimately be resolved by the geographer's behaviour with respect to the substantive issues he tackles. We may anticipate, therefore, that different groups of geographers will lay down rather different rules of behaviour. There will undoubtedly be conflicts between groups. But it seems to me that all rules *must* concede that the overriding aim of the geographer is to achieve a greater sophistication in the art of explaining the events he is concerned with. The strategy we adopt in seeking for this greater sophistication is a matter for debate.

(3) *Strategy*

We wish to navigate the world but we have no adequate map. How, under such circumstances, should we proceed? We cannot proceed without taking a risk, making a number of assumptions, in short, starting with a model which we regard as real. Such an *a priori* model allows us to navigate, and, further, it may allow us eventually to construct a more accurate map of the world than we currently possess.

Chorley and Haggett (1967, 33–9) have recently argued that geography should adopt a model-based paradigm of a new and significantly different type. One that is geared less to the activity of recording and classifying, and more to analysing and articulating theory. This amounts to a more directed use of *a priori* model concepts—directed, that is, to theory-construction. Given the present situation in geography there can be no doubt that the key to strategy is provided by the notion of *a priori* models. Such models have a dual advantage. In the first place they allow us to venture some predictions (even if rather suspect ones) in the absence of complete theory. Secondly *a priori* models can indicate the appropriate theory or an extension or modification of some existing but incomplete theory.

The use of *a priori* model-constructs poses dangers—dangers which have been emphasised in the preceding chapters. Awareness of the problems of identification, inference, and control, is an essential prerequisite for the responsible use of *a priori* model-constructs. Sailing the high seas without an adequate map is bound to be dangerous, but holding to a set course with breakers straight ahead is courting shipwreck.

Geography is probably in the stage of development when almost all explanation statements make some use of *a priori* models. Some of these models will be developed as a specific response to the geographical imagination. The quality and requirements of these cannot be established in advance. But more important, probably, will be the consumption of models from other areas of science which are more sophisticated in their theoretical development. We may adapt theories and models from some other area of empirical science, or from some area of logic and mathematics, to function as *a priori* models in geographic research. These theories and models can be examined in advance since the assumptions contained in them can be stated, and their appropriateness and applicability for certain types of problem be assessed in advance. There are vast ready-made calculi which can be given some geographic interpretation with great benefit. But in order to comprehend the results we must be prepared to learn the language of the calculus we are proposing to use, understand its properties, modify it where necessary, and consciously manipulate it to suit our own needs. There is a world of difference between mere unthinking model-consumption and model-use directed and controlled in a creative way.

The current situation in geography appears to call for a strategy of theory-creation through the interpretation of existing abstract calculi in geographic terms. This, as we have seen (above, pp. 152–3), corresponds to one of the major routes to theory-construction in science. Each of the calculi open to us has a certain set of characteristics associated with it—it possesses certain properties and certain limitations. The application of such calculi requires some prior assessment of the possibilities and limitations inherent in them. Since the general conclusion of the preceding chapters amounts to a plea for the application of controlled *a priori* model-thinking to geographic problems, it is appropriate that we should go on to consider in some detail the nature, form, and limitations, of the more important calculi which, on the face of it, appear to have great potential for interpretation in a geographic context. This will be the subject of the subsequent chapters.

PART FOUR

Model Languages for Geographic Explanation

Chapter 13
Mathematics—The Language of Science

Theory provides the key to explanation. A theory has been characterised as a 'language for discussing the facts the theory is said to explain.' It is relevant, therefore, to examine the nature of such languages and to indicate how theory in geography might be developed with the aid of specialised languages. We considered some of the general characteristics of languages in relation to perception and thought in chapter 2, and it will be useful to bear in mind certain aspects of that discussion in this and subsequent chapters. We shall, however, ignore the class of natural languages here, for our main concern will be to show how special artificial languages can (and should) be introduced in the search for adequate geographic theory. These artificial language systems provide an objective and universal language for discussing geographical problems, and in the multitude of mathematical languages we may well be able to identify some which are especially appropriate for the formulation of geographic theory. We shall be especially concerned with this latter aspect of the use of artificial languages in the next two chapters, on geometry and probability theory. In this chapter we shall consider the general nature of artificial language systems. We shall examine the internal logic of such artificial languages and their interpretation in a substantive empirical (and geographical) context.

A The structure of constructed language systems

We have already had occasion to refer to certain technical aspects of the structure of language systems (above, pp. 18–22), and to the way in which the terms 'theory', 'calculus', and 'language', are closely associated (above, pp. 87–91). We shall now endeavour to make some of these characteristics clearer. It is useful to begin by accepting Carnap's (1942; 1958) differentiation between *pragmatics* (the factors affecting the speaker or hearer of the language), *semantics* (the relating of symbols to concepts and terms which relate in some manner to experience), and *syntax* (the internal structure or 'grammar' of the language). A purely syntactical system, in which the elements and formation rules are not defined with reference to object-classes or process-postulates, is called a *purely formal language*. A system is called an *interpreted language* when the terms used are given some meaning. It is instructive to examine the nature of syntactical systems and their semantic interpretations, partly because such an analysis makes clearer the nature of scientific theory, but, even more important, because it clarifies the problem of finding and interpreting suitable languages for geographic research. The presentation that follows is based on Carnap (1942, 154–61) and on the account given in Rudner (1966, 12–18).

A syntactical system or, as we have hitherto referred to it, a *calculus* really consists of a system of formal rules. We may construct such a calculus, K, by first giving a classification of the signs which will be used in K. These signs form the *elements* or the 'vocabulary' for K. A set of *formation rules* are then stated as operating in K. These simply tell us which combinations of the elements are permissible and which combinations are not; i.e., whether or not a particular combination of elements forms a 'sentence' or not. In addition, a set of *definitions* may be provided which state how new elements may be formed by combining elements; i.e., new 'words' may be added to the vocabulary by permitting certain existing words to be combined. Proofs and derivations for the calculus may now be provided by furnishing a set of *primitive sentences* and a set of *transformation rules* (such as deductive rules of inference). The primitive sentences then function as the axioms from which other sentences, theorems, may be derived. These syntactical rules provide, it may be noted, a complete account of the structure of scientific theories (see above, pp. 88–9).

A semantical system, or an interpreted calculus, possesses the same form as a syntactical system, but in addition it requires a set of *designation rules* and a set of rules which determine the *truth condition* of the sentences contained in K. We thus have a semantical system, S, which may be related to a syntactical system, K. The essence of the logical problem which Carnap sets out to solve, is to show under what conditions S may be regarded as a true interpretation of K. The logical problem is extremely complex and we cannot, therefore, go into it in detail here. Briefly, the semantical system may be used to define certain sentences in K in terms of their content and it may also be used to show that certain sentences in K are also true by implication. If S contains all the sentences in K, then S may be termed an interpretation for K. The semantical system, S, thus provides the *text* or the *correspondence rules* (above, pp. 91–6) for a calculus, K. It is also interesting to note that we may provide K with

more than one S (i.e. we provide a calculus with more than one interpretation) and each S may be provided with more than one K (i.e.
several different calculi may be given the same semantical interpretation). This simply amounts to the basic problem of the model-
theory relationship and demonstrates the problem of identification
which the construction of artificial languages may lead us into.

It is also instructive to note the more general features of the
relationship between semantical and syntactical systems. Firstly, as
Carnap (1958, 101) points out,

> one who constructs a syntactical system usually has in mind from the outset some
> interpretation of this system. . . . While this intended interpretation can receive
> no explicit indication in the syntactical rules—since these rules must be strictly
> formal—the author's intention respecting interpretation naturally affects his
> choice of the formation and transformation rules of the syntactical system.

We may thus design syntactical systems specifically to deal with
certain types of empirical problem, but we ensure objectivity in the
process by observing the strict rules laid down for the construction
of such syntactical systems. In this case we are attempting to construct
a theory—i.e. develop an appropriate language—to deal with certain
sets of empirical phenomena.

Secondly, Carnap (1942, 203) points out that

> Interpretations of calculi play an important role in the method of science. In
> mathematics, geometry, and physics, systems or theories are frequently con-
> structed in the form of postulate sets. And these are calculi of a special kind. . . .
> For the application of such systems in science it is necessary to leave the purely
> formal field and construct a bridge between the postulate set and the realm of
> objects. . . . It is easily seen that this procedure, described in our terminology,
> leads from syntax to semantics and is what is called here constructing an inter-
> pretation for a calculus.

We shall discuss some examples of these two types of relationship
in later chapters.

B Mathematical languages

Symbolic logic provides us with the necessary tools to construct and
understand artificial language systems. Whitehead and Russell in
their *Principia Mathematica* (1908–11) succeeded in finally showing
that mathematical knowledge could be derived from logical principles and several mathematical systems have since been restated by
means of formalised languages. The publication of *Principia Mathematica* marked the end of an important stage in philosophical thinking
regarding the nature of mathematical knowledge. It did not, however, mark the end of controversy. The nature of pure mathematics

is still the subject of considerable argument. Körner (1960, 156) states:

It is logic, says the pure logicist; the manipulation of figures in calculi, says the formalist; constructions in the medium of temporal intuition, says the intuitionist; statements which we abandon more readily than some statements of logic and much less readily than empirical statements, says the logical pragmatist. And there are intermediate positions. The progress of mathematical logic since Boole and Frege has made little difference to the continuation of philosophical disputes about the nature of mathematics.

Yet it is universally agreed that mathematics is the language of science. The nature of the relationship which can be established between mathematical statements and perceptions becomes important, therefore, to our understanding of the use of mathematics in geographic research. This is a complicated and difficult subject and cannot be considered in detail here, but full accounts may be obtained in Körner (1960) and Beth (1965). From the general discussion on the philosophy of mathematics we may at least derive some clear notions as to what mathematics is not, and thus avoid some of the common pitfalls which hinder the application of mathematical methods to empirical research.

According to Hume, mathematical truths are analytic and *a priori*—that is, mathematical truths are true entirely by definition and such truths cannot be established by reference to experience. According to J. S. Mill, however, mathematics could be regarded as an empirical science in which the truth of the statements is established by reference to experience—that is, mathematical knowledge is, to use the philosophical terminology, synthetic *a posteriori*. According to Kant, mathematical knowledge is synthetic *a priori*. This complicated view of mathematical knowledge states that mathematical statements possess empirical content even though the statements themselves are formulated without reference to any empirically defined subject-matter. Thus the concept of the Euclidean triangle cannot be derived empirically—it is an *a priori* construct; yet the concept of the triangle succeeds in describing how points in space are in fact ordered.

The mathematical and philosophical developments of the nineteenth century resolved these conflicting views in a number of important respects. The development of new geometries such as those of Lobachevsky and Riemann (see below, pp. 199–202) demonstrated that the Euclidean system was but one out of many possible geometries. Peano and subsequent workers, such as Russell, succeeded in showing that every concept of mathematics can be defined by means of three primitive terms—'*o*', 'number', and 'successor'—and every proposition of mathematics can be deduced from five axioms with a set of definitions for non-primitive terms (Hempel, 1949, 228). It was also shown that geometry, algebra, and arithmetic, could all be reduced to the same basic set of postulates. The conclusion was inevitable. Mathematics was a formal language system—an uninterpreted calculus. It was, therefore, entirely *a priori* knowledge and pure mathematical systems had no empirical content whatsoever.

Such a conclusion leads, however, to an apparent paradox. Mathematical systems are used to gain information regarding empirical subject-matter. Euclidean geometry is applied to surveying problems, Riemannian geometry is applied to Einsteinian space, number theory is applied to engineering problems and economic problems, and so on. Such an observation merely amounts to saying that an *a priori* syntactical system, K (a mathematical calculus), may be given an interpretation by way of a semantical system, S. In one sense, therefore, Mill was correct in regarding mathematics as an empirical science. It is possible to give empirical interpretations to pure mathematical statements and thereby transform abstract concepts into empirical concepts and vice versa. Pure mathematics may be regarded as a syntactical system. Applied mathematics may be regarded as a semantical system. Hempel (1949, 237) thus concludes that the system of mathematics is 'a vast and ingenious conceptual structure without empirical content and yet an indispensable and powerful theoretical instrument for the scientific understanding and mastery of the world of experience.'

This view of mathematics has a number of important implications. Given that mathematics is *a priori* analytic knowledge, we cannot learn from it anything which we cannot otherwise know. But mathematics is an extremely efficient language and allows us to extract information which would otherwise be extremely difficult or psychologically impossible to attain. In its empirical uses mathematics functions as a 'theoretical juice-extractor' which can extract only information already there—but there may be much more information in a particular set of postulates than we can easily perceive and mathematics has an extraordinary power to extract that information (Hempel, 1949, 235).

It also follows that mathematics in its pure form has no necessary relationship with quantification and measurement. Much of the extraordinary power of mathematical thinking in the empirical sciences comes from the ability to handle quantitative problems with ease. But there are many areas of applied mathematics which have nothing to do with measurement. This area of 'relational mathematics', as it is sometimes called, has important practical applications in disciplines where quantification is difficult. Thus Radcliffe-Brown

(1957, 69–89) and Levi-Strauss (1963, 283) have both pointed out that mathematical logic, set theory, group theory, and topology, can be used to discuss many of the qualitative problems which anthropologists are concerned with. Many of the social sciences (and also human geography) are thus resorting to non-quantitative mathematics to discuss problems of structure and interaction. Mathematical thinking certainly does not necessarily involve quantification.

C The application of mathematical languages

According to the view developed in the preceding section, pure mathematics can have no logical relationship to the world of sense-perception data. Pure mathematics makes use of abstract terms and relational signs only. This severance of mathematics from sense perceptions poses a number of difficult problems (Figure 13.1). The most important of these can be collected together under two question marks. First, what role can perceptions play in influencing the choice of axioms and the rules for a particular mathematical language?

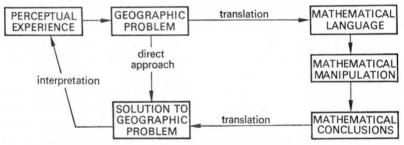

Fig. 13.1   The Use of Mathematics in Problem Solving.

Second, what are the rules which govern the use of such abstract mathematical languages in the context of sense-perception data?

As Carnap (above, p. 181) points out, a particular syntactical system may be chosen with an interpretation in mind. In two later chapters, for example, different axiomatic treatments of geometry and probability theory will be examined. In each case examples may be found where a particular interpretation of sense-perception experience has led to a particular axiomatic formulation. Klein's discussion of the foundations of geometry (below, pp. 203–6) and Nagel's discussion of the foundation of probability theory (below, p. 230) are perhaps the most eloquent examples. But the actual formulation of the axioms is an analytic exercise—it amounts to constructing a system with certain primitive terms, definitions, and rules. It is rather like inventing a new and elaborate game—a game played in symbolic terms.

The actual motivation of the mathematician in construction such games—be it the need to solve some empirical problem or be it a direct infatuation with constructing abstract symbolic systems—need not concern us. But the relationship between this symbolic system and real-world situations is of vital importance. Whatever Euclid's motivation, it is certain that the system he constructed possessed the power to systematise many of the empirical regularities which Babylonians, Egyptians, and Greeks, had discovered with respect to the mathematical form of plane figures. Euclid thus managed to develop a formal spatial language which could be so interpreted that known empirical regularities, and new regularities, could be predicted. This ability of Euclidean geometry undoubtedly did much to contribute to the belief of many mathematicians and philosophers, from the Greeks to Kant, that geometry was an empirical science. In so far as the interpretation of geometric systems involved (and presupposed) measurement, this view could be upheld. But later developments have shown that Euclidean geometry need have no direct empirical reference. The fact that the terms used in the calculus, K, can be given an interpretation by means of a semantical system, S (i.e. points, lines, angles etc., can be approximately represented), suggests that Euclidean, geometry functions as an *interpreted mathematical language*.

An infinite number of mathematical languages could be developed. The invention and elaboration of a particular language out of this infinite hypothetical set depends in part upon the inherent interest of a particular language form to the pure mathematician and partly upon the utility of that language for the scientist. Thus probability theory remained a curiosity applied merely to games of chance until problems in biology and physics necessitated the thorough exploitation of this particular mathematical language. As new applications for a particular language are found, so new problems in the language itself may emerge. The interaction between the extension of a mathematical language and the development of applications for that language is thus of some importance. But the application of the language depends on being able to interpret mathematical symbols in terms of sense-perception data and empirically based concepts.

This problem can best be approached by discussing the rules which govern the mapping of empirical problems into some existing mathematical language. This amounts to determining the rules by which we can tell when a semantic system, S, is a valid interpretation for an abstract calculus, K. It is possible to state a set of logical rules. Thus Carnap (1942) suggests that the semantical system, S, should be isomorphic with the calculus, K. This of course is the characteristic view of the logician regarding the relationship between theory (S) and model (K). But this view is not empirically very useful for it amounts to claiming an isomorphism between real-world structure and the mathematical language used to discuss that structure. It necessarily assumes, therefore, that we already know all about real-world structure—in other words K functions as an *a posteriori* model. In practice a number of steps are involved in mapping real-world problems into mathematical systems. These are described by Körner (1960, 182) as follows:

(i) the replacement of empirical concepts and propositions by mathematical, (ii) the deduction of consequences from the mathematical premises so provided and (iii) the replacement of some of the deduced mathematical propositions by empirical. One might add (iv) the experimental confirmation of the last-mentioned propositions—which, however, is the task of the experimental scientists rather than the theoretical.

The first step in this sequence is perhaps the most crucial. Empirical concepts have to be extracted from an inexact real world and these empirical concepts are then replaced by exact mathematical concepts and propositions. If the empirical concepts and propositions are formulated in a vague and ambiguous way it becomes very difficult to determine which mathematical language is appropriate, and difficult to translate the empirical propositions and concepts into mathematical ones. A prerequisite to the use of mathematical language, therefore, is precision in the formation of concepts and propositions. Failure in this respect is a common problem in many of the social sciences. Coleman (1964, 3) thus suggests that in sociology

the kinds of verbal theories and research results which have been set forth are so vaguely stated or so weak that it is difficult to translate them to mathematical language, and once translated they often fail to show an isomorphism with powerful parts of mathematics.

The mere use of mathematical notation does not amount to a proper mapping of some theoretical idea into a calculus. Thus Massarik (1965, 10) has criticised many social scientists for using 'shadowy systems of notation that lacked proper mathematical significance'. Simply stating that $Y = f(X)$ is not by itself very useful. Thus Arrow (1959, 149) has severely criticised Zipf's use of mathematical language in *Human behavior and the principle of least effort* (1949):

The fundamental postulates are nowhere stated explicitly; though mathematical symbols and formulas are sprinkled freely through a long work, the derivations involved are chiefly figures of speech and analogies, rather than true mathematical deductions; in some cases, they are simply wrong. Thus, as an attempt at a systematic social theory, Zipf's work can only be regarded as a failure.

On the other hand, as Coleman (1964, 3) has it,

Formal economic theories, with their sets of concepts precisely related to one another, have found mathematics extremely useful. They . . . have a partial isomorphism with algebra and calculus.

The attempt to mathematise large areas of social-science research is thus generally healthy, simply because it demands a prior clarification of concepts and propositions about empirical phenomena. In some cases this clarification may proceed by using some mathematical calculus as an *a priori* model and searching for concepts and propositions which can effectively be mapped into it. There are dangers in such a procedure. As yet only a few of the multitude of possible mathematical systems have been explored. The main motivation for their development has come from the physical sciences; thus the calculi useful to physics are probably better developed than any other calculi. The tendency has thus been for the social sciences to develop concepts which can be mapped into mathematical systems appropriate for the study of physical problems. It is hardly surprising that many of the concepts so developed in social science are analogous (perhaps spuriously so) to the concepts of physics. As with all *a priori* model-use, considerable care is required, but the rewards may be great. The impact of probability theory upon behavioural-science concepts should be sufficient example of that. Nevertheless, clarification of social-science concepts and postulates can proceed without the requirements of any specific mathematical system in mind. If the understanding of the empirical phenomena becomes sufficiently clear and precise, and no mathematical calculus yet exists appropriate for discussing the concepts and relationships set up, then conditions are right for the development of a new formal calculus, or the further exploration of some poorly-developed calculus. The growing interest in all the social sciences in relational mathematics is probably the most encouraging example of new developments in mathematics being provoked by the clear statement of empirical problems.

The reasonableness of a mathematical representation cannot be evaluated solely in terms of the validity of the transformation from empirical phenomena to conceptual idealisation to mathematical abstraction. If this were the only criterion then most mathematical treatments would be judged invalid. Although it is true that all mathematical systems can be manipulated and developed, given time and patience, it is also true to say that some systems are much easier to handle than others. Systems of linear equations are much easier to handle than systems of non-linear equations, Euclidean geometry is easier to work with than non-Euclidean, and so on. The methodological problem thus becomes more complex. It amounts to defining a mathematical calculus which does not do too much violence to the phenomena being discussed and which is itself simple and easy to handle. This general problem may be clarified by a simple example.

Suppose we use a simple regression model to discuss the relationship of two variables distributed over space. The model has the form $Y = a + bX + e$, where X is the independent variable, Y is the dependent variable, a and b are two parameters to be estimated from the data, and e is an error term. There are a number of theoretical interpretations which can be given to this mathematical model (i.e. it is over-identified) but let us suppose we are examining a causal relationship of the form $X \rightarrow Y$. The relationship we postulate is linear and it becomes very easy to handle if we can assume that the error term, e, is a normally distributed variate with zero mean and constant variance with every e term independent of every other. This is a very strong assumption, and it is undoubtedly very rare for this condition to be met in geographic data. But the question arises, before we abandon such a simple model and resort to more complicated techniques, how serious is the failure of the simple model to fit the data set? Our answer depends upon the purpose for which the model is being used. If, for example, tests of significance of the parameters are involved, it is absolutely essential that the conditions should be fulfilled. But it is often difficult to tell whether all the assumptions are fulfilled or not. Spatial auto-correlation in the error terms, for example, is very difficult to test for. If the assumptions are only weakly violated, the impact on the inferences may be small enough to be ignored. It is often difficult to judge when the advantage of simplicity in the model is offset by failure to fulfil all its assumptions. In this case we need to be able to distinguish the two variables, X and Y, to be confident that the relationship is a cause-and-effect one, and to be confident for purposes of inference that the assumptions of the regression model are *reasonably* fulfilled. This general problem of meeting the required assumptions and clarifying postulates and concepts is the most serious one that faces the geographer in his attempt to mathematise his study of phenomena. A specific example of this general methodological problem has been discussed elsewhere (Harvey, 1968A).

It is difficult to set up any absolute rules or criteria to govern the interpretation of mathematical calculi. Building the bridge between pure mathematical systems and empirical phenomena is bound to involve compromise. Thus Einstein (1923, 27) once wrote that

As far as the laws of mathematics refer to reality, they are not certain; and as far as they are certain, they do not refer to reality.

It is nevertheless possible to lay down some approximate rules to guide us down the rough road of mathematisation—

(i) A necessary prerequisite for the employment of mathematical calculi is (a) the development of empirically reasonable concepts which are precise and unambiguous, and (b) the precise statement of the relationships which bind these concepts together.

(ii) The mathematical calculus chosen to represent these concepts and relationships should (a) be as simple and easy to handle as possible, (b) represent the empirical concepts as accurately as possible, and (c) represent the structure and nature of relationships as accurately as possible.

(iii) The application of the mathematical calculus should take account of the assumptions made in the formulation of the mathematical model and ensure that, as far as possible, these assumptions are matched in the real-world circumstances being analysed, or that the method of describing those circumstances conforms to the requirements of the model.

(iv) If concepts and relationships are revised to meet the requirements of some mathematical model, the empirical validity of this procedure should be carefully evaluated. The use of mathematical models to formulate *a priori* models of real-world structure can be very rewarding, but the procedure also requires the 'eternal vigilance' which should automatically attend the use of such *a priori* models in the search for appropriate theory.

These rules are very general and in most circumstances it will be necessary to compromise. At this point the geographer, like every other scientist, has to be a careful decision-maker and try to optimise his choice, subject to the various constraints of simplicity, realism, and so on (Churchman, 1961). As theoretical analysis becomes more sophisticated, so the concepts and relationships tend to become clearer, and mathematical representation becomes easier and more fruitful. As geographic theory, together with the theory of disciplines from which geographers may derive concepts and relationships, becomes more sophisticated and more explicit, so we can expect a greater use of mathematical languages to formulate and discuss problems. It will also become easier to distinguish the kinds of mathematical language which are appropriate for discussing geographical problems and to establish the criteria for mapping geographical problems into those languages. The next few years will undoubtedly see a great deal of bridge-building between abstract mathematical languages and geographic reality. The next two chapters consider this process in detail by discussing the relationship which can be built between particular sets of mathematical language and the geo-graphical reality we are seeking to understand. The two examples chosen for this purpose are geometry and probability theory.

Basic Reading

Arrow, K. J. (1959).
Körner, S. (1960).

•

Chapter 14
Geometry—The Language of Spatial Form

The whole practice and philosophy of geography depends upon the development of a conceptual framework for handling the distribution of objects and events in space. At its simplest this amounts to defining some co-ordinate system (such as latitude and longitude) to give absolute location to objects and events. It may sound pretentious to call such a system a *language*, but this is in fact what it amounts to. In pursuing his objectives the geographer must necessarily resort to an appropriate language. The spatial language adopted should be appropriate for (i) stating spatial distributions and the morphometric laws governing such distributions, and (ii) examining the operation of processes and process laws in a spatial context (Nystuen, 1963).

For the most part geographers have assumed a particular spatial language to be appropriate without examining the rationale for such a choice. Like most other disciplines, geography was dominated by Euclidean geometry to such a degree that for many centuries it was never questioned as being the one and only spatial language suitable for discussing geographic problems. Some of the problems raised by location theory—the spatial expression of which is frequently discussed in Euclidean terms—have aroused interest in new ideas of *social space*. Such space often appears non-isotropic when judged by Euclidean standards and the processes operating in that space often seem to demand a different metric (or non-metric) system for discussing spatial relationships and spatial pattern; in short, geographers have begun to explore spatial languages other than Euclid in the belief that such languages could provide a more appropriate means for discussing geographical problems. In some cases an appropriate spatial language already exists, in others it is clear that new spatial languages need to be developed, while in some instances it appears that traditional Euclidean methods of handling problems are perfectly adequate. The geographer therefore faces the difficult problem of choosing between different languages for discussing spatial form and recognising that different languages may be most effectively used for different purposes. If this is so, then the geographer must needs be able to translate from one spatial language to another if, for example, morphometry can best be discussed in Euclidean geometry while the processes governing that form need to be discussed in terms of some non-Euclidean geometry.

The purpose of this chapter, therefore, is to examine explicitly the problem of developing a spatial language appropriate for discussing geographic problems, and to indicate the range of choice available to the geographer who is looking for a developed language to act as a model for geographic analysis. It should be clear from the preceding chapter, however, that the choice of language must depend on a dual understanding of the properties of a particular language and the empirical properties of the concepts which we seek to express in some abstract language. We therefore commence the discussion by examining the nature of spatial concepts.

I CONCEPTS OF SPACE

Concepts of space are founded in experience. In its most elementary form this experience is entirely visual and tactile. But there is a transition from such primary experience of space to the development of intuitive spatial concepts and, ultimately, to the full formalisation of such spatial concepts in terms of some geometric language (Cassirer, *1957*, 148). In the process of this transition, primary sensory experience, myth and image, cultural form, and scientific concepts, interact. As a result it is extraordinarily difficult to determine how concepts of space arise and how such concepts become sufficiently explicit for full formal representation to be possible. Certain phases in the emergence of spatial concepts can roughly be distinguished.

Psychologists have examined the physiological basis of space-perception. In part the evidence comes from an experimental approach to spatial perception which has attempted to distinguish what Roberts and Suppes (1967, 175) call 'primitive visual perceptions' from perceptions of space that are learned and thus in part culturally or physically determined. The basic result of this research appears to be that the visual perception of space is non-Euclidean and that it is in fact a Riemannian space of constant negative curvature (i.e. it conforms to the principles of Lobachesvkian geometry). This simply amounts to saying that we do not see in straight lines as defined by Euclid. Our ability to see in Euclidean terms is, according to Roberts and Suppes (1967), learned rather than innate. This learning process appears to be in part a response to direct tactile and motor experience and thus independent of any cultural conditioning. Piaget and Inhelder (1956) suggest that children automatically progress from perception of the topological characteristics of objects (characteristics such as proximity, separation, order, enclosure, and continuity), through perceptions which encompass perspective and projective relationships, to the ultimate ability to organise all objects in space in terms of some common spatial structure, such as a Euclidean system of co-ordinates. Most writers agree that the actual physical space which people experience and perceive is not measurably different from being Euclidean in structure. Euclidean geometry may thus be regarded as a natural outgrowth from tactile and learned visual experience, and certainly much of the initial justification for Euclidean geometry relied upon an appeal to the 'self-evident' nature of the Euclidean axioms.

Piaget and Inhelder are careful to distinguish, however, between the perception of space and the representation of space by means of imaginary concepts. At the representational level they suggest that children discover spatial concepts in the same order—that is, progressing from topological concepts to Euclidean concepts—but at a somewhat later age. The ability to represent space schematically is, however, undoubtedly influenced by the existence of signs and symbols designed to represent that space. It is influenced, therefore, by culture. In some cases the jump from perception to schematic representation is not accomplished. This is typical of many primitive societies. Thus Cassirer (1957, 153) writes:

> Reports on primitive peoples show that their spatial orientation, though very much keener and more precise than that of civilised man, moves wholly in the channels of concrete spatial feeling. Though every point in their surroundings, every bend in a river, for example, may be exactly known to them, they will be unable to draw a map of the river, to hold it fast in a spatial schema. The transition from mere action to the schema, to the symbol, to representation, signifies in every case a genuine 'crisis' of the spatial consciousness. . . .

Howard and Templeton (1966, 265–7) similarly suggest that it is dangerous to draw conclusions about the ability to move and act spatially from information regarding the ability to give schematised representations of that space and vice versa. The gap between the perceptual level and the representational level of spatial understanding is of the greatest significance. In particular it makes it extremely difficult to analyse individuals' actual spatial behaviour by means of the schemata they might use to represent that behaviour.

At the representational level the emergence of spatial concepts is inextricably bound up with the structure of the culture in which such spatial concepts are being developed. Anthropological studies have indicated considerable variation in the nature of spatial concepts from one society to another. This is scarcely surprising since the representation of space 'involves the evocation of objects in their absence' (Piaget, 1956, 17). It involves relating imagined concepts to other concepts and further it also involves concepts which have no empirical content—in particular it involves concepts such as 'empty space', 'infinity' and so on. The emergence of concepts of this kind is partly governed by language and partly by culture (Kluckhohn, 1954). In primitive societies it often seems that spatial concepts are rooted in the language developed to describe 'concrete and personal situations' (Lovell, 1961, 92). Similarly the cultural heritage:

limits or promotes the manner in which and the terms in which the individual deals with the spatial attributes of the world about him. If a culture does not provide the terms and concepts, spatial attributes cannot even be talked about with precision. . . . Without such instruments in the cultural heritage certain areas of action are excluded and the solution of many practical problems impossible. (Hallowell, 1942, 76-7.)

Eisenstadt (1949, 63) similarly suggests that 'every social structure lays different emphasis on different aspects of (or points in) time and space', and goes on to conclude:

The spatial and temporal orientation of social activities, their definite ordering and continuity are focused on the ultimate values of a given social structure.

Concepts of space thus vary from one cultural context to another, and within broad cultural configurations smaller sub-groups may develop a particular conceptual apparatus with respect to space geared to the particular role which they perform in society. Any individual within the society may likewise carry around with him a spatial schema (Lee, 1963) or a cognitive or mental map (Hallowell, 1955; Gould, 1966) which reflects that individual's cultural and physical experience and which, in turn, affects that individual's behaviour in space and, perhaps, his visual perception of spatial relationships (Segal et al., 1966).

The conceptual framework which a society develops to represent space is not, however, static. Spatial concepts have changed very substantially since antiquity. Cultural change often involves change in spatial concepts, but on occasion the sudden need to reappraise spatial concepts through scientific discovery has delivered a powerful jolt to an existing set of cultural values. The general history of scientifically based spatial concepts is worth considering very briefly since it highlights some of the general problems inherent in developing spatial languages as formal representations of spatial concepts. It also demonstrates how the solution of problems depends upon the development of spatial concepts appropriate for that purpose. Thus Childe (1948, 15–17) has noted how the Greeks were able to solve a number of problems which the Babylonians could not simply because they replaced the Babylonian concept of space (which was essentially additive in its metric) by the concept of a continuous space.

The historical evolution of scientific concepts of space is inextricably bound up with the progress of physical theory. Newton's laws of motion, for example, required the definition of a 'straight line' and such a definition could be provided only by assuming a given geometry—Newton thus assumed the Euclidean geometry (which was the only developed geometry at that time) as being self-evidently and a priori true (Nagel, 1961, 203–4). Any measurement system, according to Russell (1948, 282), 'presupposes geometry'. Since the history of physical theory is intricately related to the development of systems of measurement the connection between concepts of space and physical theory is a very intimate one indeed. At times, philosophical speculation regarding the nature of space has thus influenced the development of physical theory.

Jammer (1954) has reviewed in detail the history of concepts of space in physics. He contrasts two essentially different concepts of space. The first regards space as a positional quality of the world of material objects or events—i.e. space is a relative quality. The second regards space as a container of all material objects—i.e. it is an absolute quality. Both concepts, be it noted, are an outgrowth and abstraction from primitive concepts of place. Jammer suggests that the absolute concept of space was not developed until after the Renaissance (although in the foreword to the book Einstein disagrees with this view, pointing out that the Greek atomists appear to have assumed an absolute space). The real triumph of the absolute concept of space came with Newton, according to whom space

consisted of a collection of points, each devoid of structure, and each one of the ultimate constituents of the physical world. Each point was everlasting and unchanging; change consisted in its being 'occupied' sometimes by one piece of matter, sometimes by another, and sometimes by nothing. (Russell, 1948, 277.)

One simple pragmatic reason for adopting such a view of space was that the Newtonian laws could not work without such a concept. But as Jammer (1954, 108) points out, the concept of an absolute space was an important element in much of the theological writing of the time and in his later writings Newton came to identify absolute space with God, or with one of His attributes. The metaphysical overtones of much of Newton's writing about space led to a reaction. Leibniz thus contended that space 'was only a system of relations' (Russell, 1948, 277), and philosophical opinion at least became steadily relativistic in its approach to space. But physical theory still required the concept of an absolute space, simply because there was no alternative to Newtonian mechanics. The progress of mathematics, and in particular the development of non-Euclidean geometries in the nineteenth century, posed a further problem for 'there was no *a priori* means of deciding from the logical and mathematical side which type of geometry does in fact represent the spatial relations among physical bodies' (Jammer, 1954, 144). The physicist therefore became acutely aware that only appeal to experiment could solve the difficult problem of selecting that geometry which expressed the true spatial relationships between objects. Jammer gives an account of the experiments on terrestrial and astronomical scale which were designed to solve this problem, but as Poincaré (*1952*, 72–88) has pointed out, experiment could only indicate which was the most *convenient* geometry; it could not determine the *true* geometry. Measurement, he suggested, presupposed that the property of, for example, length of a rigid body remained invariant in all parts of the universe. Any deviation from Euclidean geometry could thus be interpreted in two ways. First, the geometry of space was genuinely non-Euclidean. Second, the measuring-rod was itself changing its properties over space. Einstein's theory of relativity, which demonstrated that space must under the conditions of the theory exhibit constant positive curvature (i.e. it was elliptic rather than Euclidean), did not put an end to controversy. Jammer (1954, 2) suggests that the theory has led to 'the final elimination of the concept of absolute space from the conceptual scheme of modern physics'—a conclusion which Grünbaum (1963, 421) regards as being inconsistent with Einstein's theories. Modern relativity theory replaces the concept of matter by the concept of the *field* which is specified by the 'properties and relations of ponderable matter and energy'. The metric (or geometry) of the field is entirely determined by matter. Unfortunately some prior specification of the boundary conditions at infinity still cannot be avoided. Thus Grünbaum continues:

the boundary conditions at infinity then assume the role of Newton's absolute space . . . and . . . instead of being the *source* of the *total* structure of space-time, matter then merely *modifies* the latter's otherwise autonomously flat structure.

The other great conceptual change which Einstein wrought was to replace the individual concepts of space and time by the single concept of space-time. This replacement was a necessary technical change in seeking to measure phenomena moving at the speed of light. In a sense, of course, it is a mere convenience and certainly, as Russell (1948, 291) has pointed out, the concept of continuous space-time necessary to relativity theory does not in any way affect the time and space of perception. It is nevertheless of philosophical interest in that it demonstrates how different theoretical frameworks may imply or require as a precondition for their development, new concepts of space. It is clear, also, that different concepts of space may be appropriate for different theoretical purposes. It may be realistic to regard the concept of space, therefore, as a 'multidimensional' concept in the sense that the concept has a different meaning according to cultural background, perceptual ability, and scientific purpose.

Given this 'multidimensional' view of the concept of space itself it is fortunate that mathematicians have been able to develop many different types of geometric system. This simply means that different spatial concepts may be represented by different, but appropriate, formally developed geometries. We shall therefore consider some of the properties of formal geometry.

II THE FORMAL REPRESENTATION OF SPATIAL CONCEPTS

The first formal development of geometry came with Euclid's attempt to axiomatise and synthesise the empirical observations and partial theories which the Babylonians, Egyptians, and early Greeks had accumulated to explain and describe relationships on a plane surface. The *Elements*, written about 300 B.C., was an amazingly sophisticated contribution for it is still a model of axiomatic thinking as well as a geometric system which has extraordinarily wide application. Euclid began by furnishing a set of definitions for the concepts he used. Thus *point* ('that which has no part'), *line* ('breadthless length'), *surface* ('that which has length and breadth only'), and so on, were all defined. The definitions are not, however, very useful; and so modern presentations regard these terms as *primitive*, and the need to define them is thus dispensed with. The geometry which Euclid devised flowed from five axiomatic statements.

(i) A straight line may be drawn from any point to any other point.
(ii) A finite straight line may be extended continuously in a straight line.
(iii) A circle may be described with any centre and any radius.

(iv) All right angles are equal to one another.

(v) If a straight line meets two other straight lines so as to make the two interior angles on one side of it together less than two right angles, the other straight lines, if extended indefinitely, will meet on that side on which the angles are less than two right angles.

It has subsequently been shown that these axioms are not complete and that there are some hidden assumptions in Euclid's development (Tuller, 1967, 3). It has also been shown that Euclid's axiomatic statements are not the only ones from which Euclidean geometry can be derived (Barker, 1964, 23–4). Nevertheless, the *Elements* effectively dominated geometric thought for over 2000 years. Any new development in geometry during this period was invariably regarded as a mere extension of the Euclidean system. Thus the study of projective geometry, which became particularly important from the fifteenth century onwards (Tuller, 1967, 34), was regarded for four centuries as a mere logical extension of Euclidean geometry to problems of perspective. Not until the late nineteenth century was it shown that projective geometry could be developed in an axiomatic manner independently of Euclid. Curiously enough, this independent development also demonstrated that Euclidean geometry could be regarded as a special case of projective geometry and that the latter was much more general than the former. This point will be taken up later.

The long period between the statement of the *Elements* and the development of non-Euclidean geometries was not entirely barren. Perhaps the most important development was the statement of geometric problems in algebraic form. This was not entirely new, for both Egyptians and Babylonians appear to have made use of a co-ordinate system for discussing spatial problems. But the Greeks appear to have largely ignored this method (Jammer, 1954, 23), and it was not until the seventeenth century that Descartes convincingly showed that 'every geometrical result could be turned into an algebraic result' (Sawyer, 1955, 103). Thus any point on a plane surface could be represented by two co-ordinates (x, y) which measured the distance of that point from two orthogonal axes (Coxeter, 1961, 108). This algebraic development of geometry has subsequently been of great importance. It allows geometric concepts and theorems to be extended to three, four . . . n dimensions, and it allows complex geometrical problems to be solved by way of analytic algebraic techniques. The later important development of analytic and differential geometry rested upon Cartesian concepts.

These developments did not challenge the supremacy of Euclidean geometry. Indeed, Euclid had been almost too successful. Designed originally to synthesise a set of concepts about space, it later became the rigid framework into which all concepts of space had to be fitted. Philosophers ceased to speculate on the geometry of space but sought instead to explain why spatial relationships which were observed could be so successfully described by the abstract axiomatic and deductive system postulated in the *Elements*.

The strength and appeal of Euclidean geometry was based on two major features. First, when the abstract calculus was given an interpretation—as Euclid envisaged—it proved extremely effective in predicting relationships. Second, the axiomatic statements seemed self-evident and in no way contracted perceptual experience—all, that is, except the fifth postulate. The attempt to resolve the difficulty of this fifth postulate led, ultimately, to the emergence of non-Euclidean geometry.

The fifth postulate—usually called the axiom of the parallel—implied that there is one and only one parallel to a given straight line which may be drawn through a point not on that straight line. This axiom is not as self-evident as it appears. There is

something unsatisfactory about it, because it contains a statement about infinity; the assertion that the two lines do not intersect within a finite distance transcends all possible experience. (Reichenbach, *1958*, 3.)

There was, thus, a long history of attempts to derive the fifth postulate from the other four. These attempts failed. A number of mathematicians therefore attempted to prove the necessity of the postulate by postulating either (i) more than one parallel could be drawn through a point or (ii) no parallel could be drawn through a point off the line and demonstrating that this led to a contradiction. In both cases it turned out that perfectly consistent geometries could be devised. The first system led to the geometry of Lobachevsky and Bolyai—later called *hyperbolic* geometry—while the second led to Riemann's geometry—later called *elliptic* geometry. Both of these geometries proved to be internally consistent (although Riemann was forced to replace Euclid's first axiom). It was clear, therefore, that different sets of axioms could yield different but just as consistent geometries as Euclid's. This simple conclusion dealt a shattering blow to *a priori* concepts of physical space, to the whole philosophy of mathematics as set up by Kant, and indeed to the whole traditional framework of rational scientific thought, which had been much influenced by the power and supposed uniqueness of the Euclidean system.

It was also clear that some rational grounds had to be found for choosing one axiomatic system in preference to another. In modern terminology this amounted to finding a semantic system to give a valid interpretation of an abstract calculus. The grounds for this choice, it was quickly realised, had to be empirical rather than mathematical. Thus 'the problem of mathematical space [was] recognised as different from the problem of physical space' (Reichenbach, *1958*, 6). The axiomatic method thus provided a convenient mechanism for generating non-Euclidean geometries and also separated out the rather different problems of mathematical and physical space. But the non-Euclidean geometries still required an interpretation.

One of the great advantages of Euclidean geometry was that it could be easily interpreted and that the interpretations were useful, and, as far as could be shown, were also valid. The Euclidean system could be applied with good results to a wide range of empirical phenomena. The initial difficulty with the non-Euclidean geometries was to provide them with any kind of interpretation (let alone any application) for they possessed properties far removed from direct perceptual experience. Poincaré (*1952*) provided some kind of intuitive interpretation of the geometries of Lobachevsky and Riemann

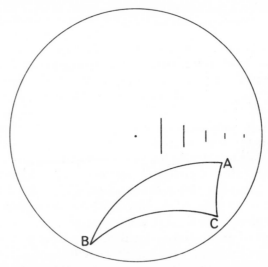

Fig. 14.1 A Euclidean diagram to demonstrate the properties of Lobachevsky's hyperbolic geometry. Imagine living within a circle within which the metric shrinks the closer one comes to the edge. Thus a man the height of the bar would shrink in size as he approached the edge of the circle, would take smaller and smaller places, and, thus, never reach the edge of the circle except at infinity. As a result distances (geodesics) are curved, the angles of triangles (**ABC**) add up to less than 180°, and most other geometric relationships are altered (*after Sawyer*, 1955).

by modelling some of the properties of these geometries in the Euclidean plane. Lobachevskian—or hyperbolic geometry—was, he suggested, rather like living in a self-contained universe with the properties of a circle in which the metric shrinks as the circumference is approached. An inhabitant of this world can therefore never reach its boundary since any object moving becomes smaller (and thus moves in smaller and smaller increments) as the boundary of the universe is approached (see Figure 14.1). If the inhabitants of this universe are unaware that the metric is shrinking as the boundary is approached, the kind of geometry which they would devise would be Lobachevskian. Straight-line distances to them would appear curved, the angles of a triangle would add up to less than 180°, an infinite number of parallel lines could be drawn through a point not on a given line, and so on. Riemann's geometry, however, would be the kind constructed by inhabitants living on a spherical universe. Straight lines to them would be equivalent to great circle distances, the angles of a triangle would add up to more than 180°, and since all great circles on a sphere intersect no parallel lines could be drawn. Intuitive interpretations of this kind are useful (see also Nagel, 1961, 238–41; Sawyer, 1955, 65–88), but they can also be misleading since we are trying to interpret non-Euclidean concepts and relationships in Euclidean terms. In fact most systems of geometry postulate plane surfaces. It is the relationships between objects placed on that surface which vary. This leads, therefore, to a second avenue of approach to non-Euclidean geometry—namely that *via* geodesics.

In the Euclidean system a straight line was defined as the path of shortest distance between two points. Such paths are termed *geodesics*. The mathematician Gauss had studied the properties of such shortest paths on curved surfaces during the early nineteenth century. He showed 'that given any surface, it is possible to express the equation of any figure on it in terms of a co-ordinate system wholly embedded in that surface' (Nagel, 1961, 241). Riemann was able to generalise these Gaussian ideas regarding geodesics and he conclusively showed that the hyperbolic, Euclidean, and elliptic, geometries were special cases (and very simple cases at that) of what came to be known as the geometry of 'Riemannian spaces'. The essential point of this approach is that 'the type of geometry required is a consequence of the rules adopted (or tacitly employed) for making spatial measurements' (Nagel, 1961, 246). It was not necessary, according to Riemann, to develop an axiomatic system in order to state the properties of different kinds of space. The properties of a space could be adequately defined in terms of the form of the geodesics embedded in that space. Thus on a flat plane surface the geodesics would be the familiar straight-edge lines of Euclidean geometry, on a pure spherical surface they would be great circle arcs. But the geodesics, and the co-ordinate system embedded in any surface, could assume an infinite variety of forms, depending on the shape of the surface, or, to use the rather misleading technical term for it, the *curvature* of the surface. The hyperbolic space of Lobachevsky and Bolyai was the space of constant negative curvature, the elliptic space of Riemann was the space of constant positive curvature, and the flat space of Euclid was the space of zero curvature. Every perceptual shape, from a flat board to a Henry Moore statue, could be given a co-ordinate system which would define the intrinsic geometry of that shape. But so general was the theory that Riemann devised that it could be extended to more than three dimensions. It was possible, therefore, to discuss *n*-dimensional space given Riemann's general theory.

Two- and three-dimensional Riemannian spaces could be given a direct intuitive interpretation but the physical application of Riemannian geometry was not immediately apparent. Riemann himself anticipated some of the central ideas of relativity theory by pointing out that the assumptions of a homogeneous metrical field in space were idealisations and that 'just as the physical structure of the magnetic or electrostatic field depends on the distribution of magnetic poles or electric charges, so the metrical structure of space is determined by the distribution of matter' (Jammer, 1954, 159). It was left to Einstein to provide the application of Riemannian general theory, since the

geometrical structure which relativity physics ascribes to physical space is a three-dimensional analogue to that of the surface of a sphere, or, to be more exact, to that of the closed and finite surface of a potato, whose curvature varies from point to point. In our physical universe, the curvature of space at a given point is determined by the distribution of masses in its neighborhood; near large masses such as the sun, space is strongly curved, while in regions of low mass-density, the structure of the universe is approximately Euclidean. (Hempel, 1949, 248.)

The study of different geodesic systems, with its implication of varying co-ordinate systems and varying shapes, formed a direct approach to new forms of geometry. It led, in turn, to a third approach associated in particular with the work of Felix Klein carried out towards the end of the nineteenth century. Klein's (*1939*, 159–60) stated intention was

to erect the entire structure of geometry upon the simplest foundation possible, by means of logical operations. Pure logic cannot, of course, supply the foundation. Logical deduction can be used only after . . . we have a system which consists of certain simple fundamental notions and certain simple statements (the so-called axioms), and which is in accord with the simplest facts of our perception. . . . The one condition which the system of axioms must satisfy is [that it] must be possible to deduce the entire contents of geometry logically from these fundamental notions and axioms, without making any further appeal to perception.

The parallel between Klein's system and Piaget's observations on the growth of spatial concepts in children is quite striking. Klein chose to found his geometry on the topological characteristics of objects. His geometry is therefore essentially non-metrical—which is in itself an important indicator of the *qualitative* nature of much mathematical thinking. Topology relies upon the statement of certain important properties. For example, the sphere is essentially different from the plane because it is closed and finite rather than open and infinite. A sphere can be distorted without tearing into a cube, into a shape like a potato, and so on. The property of such a distortion is that the transformation involved is *unique and continuous over all points*. The meaning of this statement needs further clarification. Suppose we map a series of locations on a spherical surface on to a flat plane surface (this is the basic problem of map projection). The relationships between points on the flat paper may represent spherical relationships *locally* (e.g. the positions of London, Birmingham and Bristol may not be distorted), but on a typical Mercator map Japan looks to be at the other end of the world from Seattle. At some point in the transformation of points from a spherical surface to a plane surface the neighbourhood relationship between points has to be disrupted. The transformation cannot be both *unique* and *continuous*. On the other hand, points on a spherical surface can be mapped on to a cube so that the transformation is unique and continuous. In general, the principle of such unique and continuous transformation is that the surface can be bent and stretched but it must not be torn (Figure 14.2). Klein's geometrical system is based entirely on the concept of transformation and the properties of figures which remain invariant under a group of transformations. His definition was

A geometry is the study of those properties of a set S which remain invariant when the elements of S are subjected to the transformations of some transformation group. (Tuller, 1967, 70.)

The precise meaning of this definition is difficult to convey (see Klein, *1939*; Tuller, 1967; Nagel, 1961, 246-8), but it is again possible, as Bunge (*1966*, 215-29) and Tobler (1963) have pointed out, to understand the general import of it by referring to the traditional geographic problem of map-projections. One way of classifying map-projections, for example, is by stating the properties of the spherical surface which are preserved on the flat projected surface. It is impossible to preserve all properties, and a characteristic transformation

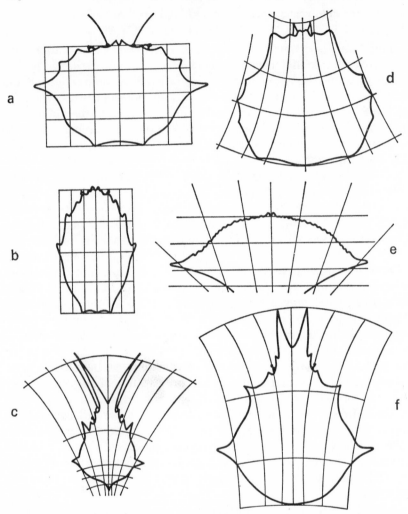

Fig. 14.2 Transformations employed by d'Arcy Thompson to show the relationship between the carapaces of various crabs. The same basic shape can be discerned but in each case it is embedded in a different co-ordinate system (*from d'Arcy Thompson*, 1954).

group would be the various projections which seek to preserve the areas of objects located on the spherical surface. Such equal-area projections regard all areas as the set S in Klein's definition and keep each of the elements of that set invariant under the transformation. Other types of projection, such as azimuthal, equidistant, etc., may be identified in the same way and each type forms a family of projections which satisfy the condition that some property or other should remain invariant under the transformation. Some of the most interesting work using transformations is that of d'Arcy Thompson (*1950*) who used transformation techniques to show relations among biological forms (Figure 14.2).

Map-projections form just one special case included in Klein's general system of geometry. Starting from a set of primitive terms (such as point, line, plane, etc.) and the principles of such transformations, Klein is able to progress from topology, through projective and affine geometries, to the axiomatic systems of Euclid, Lobachevsky, and Riemann. This analytic approach to geometry

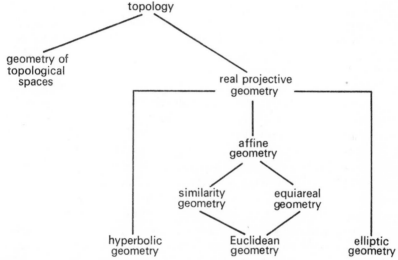

Fig. 14.3 Diagram to show the interrelationships among the different geometries with topology at the top of an all-inclusive hierarchy of geometries (*after Klein*, 1939; *Adler*, 1966, 351).

synthesises all geometries into one consistent and coherent system and provides another means for identifying the properties and form of non-Euclidean geometries.

These three approaches, through axioms, geodesics, and transformations, suggest certain conclusions regarding the structure and nature of formal geometric languages. The axiomatic approach gives a complete specification of the geometry since every theorem can be deduced from the specified axioms. The geodesic approach, on the other hand, is incomplete in the sense that the geometry can be completely specified only when all the properties of the space are known. The approach through transformations seeks to specify only certain properties and is therefore incomplete, although it may include axiomatically developed geometries as special cases. Klein's system also showed how one formal geometry could be related to another in a complete hierarchy of geometries (Figure 14.3).

Klein's unification of geometric systems was, however, only one element in a broader mathematical synthesis accomplished by Hilbert (*1962*) and by Whitehead and Russell (1908-11). This synthesis amounted to showing that all branches of mathematics could be developed out of mathematical logic and that spatial languages were subsets of a far broader class of syntactical systems. This point will be examined in a geographic context later (pp. 212–17). The general implication of this is, however, that different mathematical systems may be related to geometry. Thus the algebraic representation of geometric problems developed by Descartes was only one out of many possible relationships. It is thus possible to study geometry by means of the algebra of vectors. It is possible to apply calculus to the study of curvature (a branch of geometry known as differential geometry). It is possible to relate geometry and probability theory (Kendall and Moran, 1963), geometry and number theory, lattice theory, and so on. All these interrelationships allow formal geometric problems to be developed in some alternative and perhaps more readily manipulated syntactical system. It also implies that spatial concepts and ideas may be formally represented in a very wide number of possible calculi. The model languages available for geographic use are potentially infinite and are in fact very numerous at the present time. It is therefore pertinent to enquire how far geographers have resorted to such formal spatial languages both in the methodological development of their subject and in empirical research.

III SPATIAL CONCEPTS AND FORMAL SPATIAL LANGUAGES IN GEOGRAPHY

The concept of geography as a science of space has been exceedingly important in the history of geographic thought (Hartshorne, 1939; 1958). In part at least, the history of geography may be regarded as the history of the concept of space in geography, since space is a basic organising concept in geographic methodology. Although this general point has been accepted by most geographers, there has been little methodological debate on the nature of space as an organising concept. Yet the implications of the spatial concept for geographic method are frequently appealed to in discussion of the nature of geography, while geographers frequently make assumptions about space in empirical work and resort to formal spatial languages for discussing geographic problems. There seems, therefore, to be no very close relationship between the nature of space as understood in science, the methodological treatment of space in geography, and the actual use of spatial concepts in geographic research. In the following sections we shall consider some of the possible relationships which might be developed in terms of the philosophy of space in geography, the measurement of geographic distances, and the use of formal spatial languages in a geographic context.

A The philosophy of space in geography

One of the central conclusions in Hartshorne's (1939) *The Nature of Geography* was that the distinctive aim of geography as a science could be defined in terms of spatial concepts. The task of the geographer, it was claimed, was to describe and analyse the interaction and integration of phenomena in terms of space. This view of space as a basic organising concept in geography was traced from Kant and Humboldt through to Hettner's clear restatement of the idea at the beginning of the twentieth century. Hartshorne (1939, 1958) thus proceeds by classifying geography among the sciences according to Kantian principles. We shall therefore begin by stating Kant's philosophy of space.

Kant's initial concept of space was, according to Jammer (1954, 130), a relative view in which space consists of a system of relations among substances and 'spatial magnitude is therefore only a measure of the intensity of acting forces exerted by the substance.' In 1763 Kant appears to have been completely converted to the Newtonian notion of absolute space in which space has an existence of its own independent of all matter. By 1770 Kant had stated his 'transcendental idealist' view of space in which space is regarded as a conceptual fiction. Space is not a thing or event. It is 'a kind of framework for things and events: something like a system of pigeon-holes, or a filing system, for observations' (Popper, 1963, 179). Both space and time may thus be described 'as a frame of reference which is not based upon experience but intuitively used in experience, and properly applicable to experience' (Popper, 1963, 179). Geometry could be regarded as synthetic *a priori* knowledge (above, p. 182). Given this philosophy of space, Kant was able to classify geography in relation to the other sciences and this, according to Hartshorne (1939, 39; 1958, 98), Kant first did in 1775.

The Kantian view of geography has already been discussed (above, pp. 70–4), but briefly recapitulating, Kant suggested that geography and history were fundamentally different from other disciplines. Geography thus constituted the study of all phenomena organised according to the dimension of space, history constituted the study of all phenomena organised according to the time dimension, and both disciplines together filled the 'entire circumference of our perceptions' (Hartshorne, 1939, 135).

The first and most important point to note about this Kantian definition is that it postulates an absolute space. Kant's 'filing-system' or 'abstract-frame-of-reference' approach to space is a key concept in understanding the Kantian definition of geography. Much of the philosophy of geography thus stems from a 'container' view of space which is particularly associated with the concepts of Newton and Kant. Yet there has been little examination of the justification of this concept in geography, nor have all the implications of the concept been fully realised. Hettner and Hartshorne thus regard geographic space as being essentially absolute. This is one of the basic unwritten assumptions in their presentations of the philosophy of geography. The following excerpts from Hartshorne (1939) serve to demonstrate the dominance of this particular view of space and to indicate some of the implications:

> The whole of reality may be divided into sections in terms of either space or time. . . . The consideration of sections of reality in terms of space is the chorological point of view, represented by astronomy and geography. (p. 371.)

> The area itself is not a phenomenon, any more than a period of history is a phenomenon; it is only an intellectual framework of phenomena, an abstract concept which does not exist in reality. It cannot, therefore, be compared as a phenomenon with other phenomena and classified in a system of generic concepts, on the basis of which we could state principles of its relations with other phenomena . . . the area, in itself, is related to the phenomena within it, only in that it contains them in such and such locations. (p. 395.)

Many of the philosophic notions of Hettner and Hartshorne—particularly those to do with regionalism and uniqueness—stem from this 'container' view of space. The relationships between objects —i.e. relative locations—are similarly discussed through an imposed metric system inherent in the concept of absolute space. The objects being discussed at no point interfered with this absolute metric.

The absolute view of space has not been general in the philosophy of science for the last hundred years and the Kantian views on space and geometry were quickly discredited by the discovery of non-Euclidean geometries during the first half of the nineteenth century. Gauss, involved in a geodetic survey of part of North Germany, became interested in the more general problems of the geometric properties of curved surfaces, and in the process became aware of the possibility of non-Euclidean geometry—a possibility quickly confirmed by Lobachevsky and Riemann. Gauss regarded Kant's views on space and geometry as being either trivial or false (Bell, *1953*, 263). It is perhaps ironical that the main current of philosophical opinion in geography—particularly that associated with Hettner and Hartshorne—takes more guidance from Kant than it does from Gauss, who, partly by way of solving technical problems of map-projection, was led to initiate a set of important mathematical discoveries which culminated in the geometry of Riemannian spaces.

It may have appeared, of course, as if such geometries, closely associated as they were with advancing knowledge in physics, were inapplicable in geography. More recently, the investigation of location theory has led to the development of relativistic notions about space. Cities influence the properties of the space around them, varying patterns of human activity form fields of influence which distort the properties of space, and so on (Olsson, 1967). In such cases it is no longer feasible to take a container view of space. Activities and objects themselves define the spatial fields of influence. The empirical problem which geography thus faces is to select a geometry which can deal with the complexities of such fields and forces. The notion of an absolute space specified in Euclidean form no longer holds. The growth of this relativistic viewpoint regarding space, and some of the technical problems which it involves, will be examined in the next two sections. In terms of the philosophy of geography, however, the concept of relative space has scarcely been discussed. Whether or not the concept of absolute space will finally prevail scarcely matters at this juncture, since neither has the opposing viewpoint been directly stated nor have the implications been fully explored. It may well be that geographers will eventually decide upon the kind of compromise view suggested by Grünbaum (above, p. 196) in which matter modifies the autonomously flat structure of space. For the moment it is sufficient to point out that much of the philosophy of geography still relies upon the Kantian concept of absolute space—a concept that has been generally discredited for a century or more—while much of the practical work of geographers operates with relativistic views of space. These views are in open conflict. The opposition between Hartshorne and Bunge, for example, can be almost directly interpreted as an opposition between an absolute and a relativistic concept of space. Space may be the central concept on which geography as a discipline relies for its coherence. But the nature of space itself and the different interpretations which may be put on the concept have scarcely been appreciated.

B The measurement of distance

The importance of distance in geography is closely associated with the definition of geography as 'a science of space'.Thus Watson (1955) has termed geography 'a discipline in distance'. This view of geography may be associated with a particular operational problem which frequently occurs in practical geographic research. This operational problem simply amounts to defining a means for measuring distance.

In a later chapter the general problem of measurement will be considered. We have already established, however, that measurement either presupposes or implies geometry. The actual measurement of the distance variable in geographic research has, therefore, enormous implications. It not only helps to define the nature of geometric concepts in geography—it also has implications for the whole philosophy of geography simply because it is directly related to the concept of space itself.

Given the philosophy of absolute space, the metric in that space must remain isotropic and constant. To Kant and Humboldt the only metric available was that defined by Euclidean geometry. Relationships between objects on the earth's surface, the extent of areal units, and so on, could be measured by the direct extension of Euclidean concepts of space and distance to the surface of a sphere. Straight line distances were, thus, regarded as great circle routes, and so on. There were, it seemed, no problems in the measurement of distance which could not be solved by elementary trigonometry.

This view is no longer generally acceptable. Thus Watson (1955) has pointed out that distance can and must be measured in terms of cost, time, social interaction, and so on, if we are to gain any deep insight into the forces moulding geographic patterns. The general argument about the nature of distance in geographic research (Olsson, 1965A; Bunge, *1966*) has effectively been resolved. Distance, it seems, can be measured only in terms of process and activity. There is no independent metric to which all activity can be referred. In the discussion of the location of economic activity distance may be measured in terms of cost, in the discussion of diffusion of information distance is measured in terms of social interaction, in the study of migration distance may be measured in terms of intervening opportunity, and so on. The steady realisation in much empirical work that the metric appropriate for measuring distance was variable came to a head in the attempt to compare theoretical patterns, such as those derived in location theory for agriculture, industry and settlement, with observed patterns. The distance measures for observed patterns were derived using Euclidean concepts of distance. The theoretical patterns clearly referred to some other measure of distance, such as that determined by cost, convenience, time, social contact, or a mixture of such measures. The difficulty of comparing actual with theoretical patterns led, in the work of Tobler (1961; 1963) and Bunge (*1966*) to the notion of employing map-transformations in order to match patterns measured according to some different metric system (Figure 14.8). Getis (1963) provides a simple example of this procedure with reference to central-place theory. The theory postulates an even distribution of population. Given a non-isotropic distribution of population the hexagonal market areas derived by Lösch would clearly be deformed. Getis tackled the problem of matching reality with theory by stretching the areas of high-density population relatively to the areas of low-density population and hence derived an isotropic population surface on which the Löschian network could then be imposed. The method was crude, but it demonstrated how the general problem with which all location theorists are faced could possibly be solved. The solution, as we shall see in the next section, depends on the ability to use formal geometry to treat such complex problems.

The notion of distance in geography has thus changed its status very markedly during the twentieth century. The general conclusion which we may draw from such studies, is that distance cannot be defined independently of some activity. The metric is thus determined by activity and by the influence of objects. Such a concept of

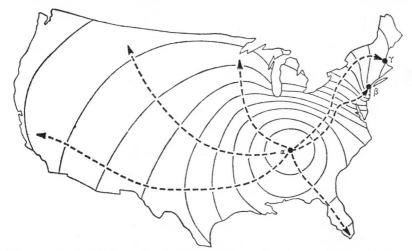

Fig. 14.4   Geodesic paths (shortest distance lines) on a non-Euclidean iso-cost surface (*from Haggett and Chorley* 1969; *after Warntz*, 1965).

distance is purely relative. Geographic distance can no longer be equated with great-circle distance. The significance of geodesic theory for geography has only recently been appreciated. Warntz (1965), for example, has studied least-cost paths and shown that such paths lie orthogonal to the iso-cost lines on a surface, while Haggett (1967, 620) has examined a number of other examples (see Figure 14.4).

Such geodesics, as we have seen, can be used to define a geometry. It is clear that the geodesics being introduced into geographic research are too complex to be regarded as mere extensions of Euclidean notions of distance to simple spherical surfaces. The geodesics are far more complex than that. The spherical surface of the earth is thus modified by activity to a complex shape. Each activity serves to define a rather different shape. To date, geographers have tended to content themselves with creating individual measures of distance for specific purposes and have not concerned themselves with the general theory of geometric surfaces and non-Euclidean co-ordinate systems. Such a general theory will undoubtedly call for the use and development of non-Euclidean geometries.

The implication of this for the philosophy of geography cannot be ignored. In particular it indicates that the naïve 'container' view of space postulated by Kant, Hettner, and Hartshorne, is no longer acceptable. We are bound to return to Kant's initial view of space in which 'spatial magnitude is . . . only a measure of the intensity of acting forces exerted by the substance' (above, p. 207). Such a view of space is contrary to the view on which Kant based his philosophy of geography. Thus space is no longer something which can encompass our perceptions of the world. It is, rather, a collection of measures determined by those perceptions. If space and matter can no longer be effectively separated and if the properties of space can no longer be regarded as given *a priori*, then the logical justification for the particular view of geography adopted by Kant, Hettner, and Hartshorne, can no longer be sustained.

C Formal spatial languages in geography

It has been emphasised in preceding sections that formal mathematical languages can be usefully applied in geography only if the spatial concepts existing in geography are precise and unambiguous and if an appropriate mathematical language can be identified for discussing these concepts. There are numerous cases where these conditions are reasonably satisfied and formal geometric theorems may be used to yield meaningful geographic results. Before examining some of these we shall describe a simple well-known case and attempt to derive from it some general conclusions about the relationship between geographic concepts on the one hand and formal geometry on the other.

Places on the earth's surface are designated by proper names such as Bristol, Birmingham, London, etc., and each place stands in some relation to every other place with respect to the physical distance which separates them. One of the major activities of geography over the centuries has been to replace such proper names by the use of some co-ordinate system and thereby enable the relationships to be generally stated by means of some convenient spatial schema. We shall call such a spatial schema a *spatial language*. The typical spatial language used by geographers is that of latitude and longitude (other local varieties are provided by national grid systems, etc.). This language allows proper names to be replaced by co-ordinate references, which allow point-to-point distances to be calculated by the application of certain computational rules. It is interesting to enquire, therefore, as to the nature of these rules and to ascertain whether any statements at all refer utterly and entirely to the earth and not to anything else. Bertrand Russell (1948, 243) has dealt with this case at length. He suggests that only two primitive terms are necessary in geography:

> The remainder of the words used in physical geography such as 'land' and 'water', 'mountain' and 'plain', can now be defined in terms of chemistry, physics, or geometry. Thus it would seem that it is the two words 'Greenwich' and 'North Pole' that are needed in order to make geography a science concerning the surface of the earth. . . .

Assuming the earth to be a spheroid, all points on the surface of the earth and all relationships between points can be stated by means of formal geometry. Again this requires abstraction to some degree. Thus the term 'point' in geometry needs to be represented by a 'dot' on the surface of the earth, and so on. The conceptualisation of places that have areas as geometric points (which have no dimension according to Euclid or which remain undefined in modern geometries) does not cause undue difficulty. The system of latitude and longitude thus constitutes a simple but extremely useful spatial language for discussing the distribution of spatial events and analysing the relationships between them. It also constitutes a general theory about spatial structure although the theory is of very limited scope. This example therefore demonstrates once more the sense in which a theory may be regarded as a language. It also demonstrates an interesting, although not very profound example, of the dichotomy already noted (above, p. 127) between derivative process postulates and indigenous spatial postulates in the construction of geographic theory since the language we adopt for discussing spatial form reflects certain assumptions about the nature of activity and process.

The main purpose of this example was to demonstrate how geographic information may be ordered and analysed by means of a spatial language. However, in the past 'owing to certain technologic constraints, such information was often ordered in a non-geographic or quasi-geographic format. Some of these constraints are fast disappearing since the introduction of the computer, but the methodology developed within their context lingers on' (Kao, 1963, 531).

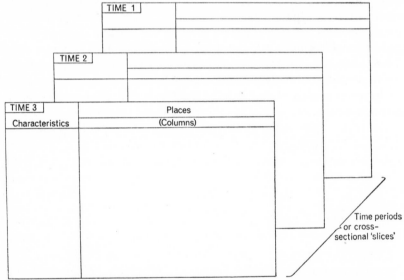

Fig. 14.5   Berry's representation of geographic information by way of a matrix of attributes by locations arranged in a series of time 'slices' (*after Berry*, 1964).

The current issue is not one of revolutionising geographic information systems—it is, rather, one of examining the underlying logic of any geographic information system, and formalising the method whereby geographers may spatially order the mass of information available to them. One example of such a procedure is Berry's (1964A) proposal for ordering all information by way of a three-dimensional array (Figure 14.5). The columns of this matrix represent places, the rows represent characteristics and the third dimension collects the same information in a series of time 'slices'. A number of geographers have taken up this general problem of ordering geographic information either in the context of general methodology (Chorley and Haggett, 1967, 28–32) or in the context of more specific data-gathering activity, such as that associated with census recording (Hägerstrand, 1967). In each case some spatial language is necessary for the rational ordering of data. It is therefore important to understand some of the basic properties of spatial languages.

The most instructive approach to spatial languages is to regard them as providing sets of rules governing the use of co-ordinate systems. In general the location of an event or object in space and time may be described in terms of a four-dimensional co-ordinate system (x, y, z, t). This system forms what Carnap (1958, 161–7) calls a *space-time language*. It contrasts with the non-spatial co-ordinate language (a *substance language*) which identifies an object or event by measures on a set of properties $(p_1, p_2, \ldots, p_n)$. Now Berry's data-matrix approach to geographical information (Figure 14.5) expresses that information in terms of both a space-time language and a substance language. It therefore collapses one complex language on to one dimension. The aim of space-time languages is, however, to state the location of objects and events (*things*—as Carnap terms them) in as much detail as possible. Carnap (1958, 158) writes:

a *thing* occupies a definite region of space at a definite instant of time, and a temporal series of spatial regions during the whole history of its existence. I.e. a thing occupies a region in the four-dimensional space-time continuum. A given thing at a given instant of time is, so to speak, a cross-section of the whole space-time region occupied by the thing. It is called a *slice* of the thing (or a thing-moment). We conceive of a thing as the temporal series of its slices. The entire space-time region occupied by the thing is the class of particular space-time points . . .

Carnap then goes on to identify three basic forms of language. The first conceives of the individual as constituting a space-time region (i.e. an individual is defined by the spatial extent of the thing over all time periods). The second conceives of the individual as forming a space region without any temporal extent (regionalisation at one point in time is a good example of this). The third regards the individual as a series of space-time points (this provides a way of handling phenomena varying continuously over space and time— a climatic region might be built up in this particular language). These three forms of language can be developed and expanded, but they have not been given a great deal of attention in the philosophical literature. To the geographer an understanding of their properties and relationships is vital, for, whether we like it or not, we are forced to make use of such languages. Frequently we try to use simultaneously two rather different languages possessing rather different characteristics and it may well be that many of our methodological problems arise from failure to understand the properties of such languages. This can be demonstrated with reference to geographic information systems.

Dacey (1964C; 1965B)[1] has reviewed much of this work and he points out the relationship between the logical languages developed by Carnap and the problem of ordering geographic information—particularly the problem of identifying what has come to be known as the 'geographical individual'. As Grigg (1967, 483–4) has pointed out, 'the geographical individual has presented problems to all who have attempted to classify features with areal expression.' The difficulty lies in identifying the *individuals* which geographers deal with. Is the *geographical individual* a farm, a farmer, a temperature-reading at a point? If all of these may legitimately be termed *individuals*, then grouping and treating them clearly poses problems since some individuals are discrete, some continuous. N. L. Wilson (1955), followed by Dacey (1964C), has distinguished between two different languages for identifying individuals. What is termed the *individuation* of an object may result from (i) the properties which the object manifests or (ii) the position occupied by an object. The first identification is provided by the *substance language* (e.g. the p_1, p_2, . . ., p_n co-ordinate language), and the second is provided by a space-time language (e.g. the x, y, z, t, co-ordinate language). These languages identify two different kinds of individual possessing different properties. The notion of similarity or 'sameness', for example, relies upon the properties of two individuals being the same in the substance language, but relies upon two individuals occupying the same position in the space-time language. Geographers, in attempting to order information, have never quite sorted out which language they are using, and in regionalisation frequently appear to mix the two. This is a classic case where prior failure to conceptualise the problem has led to a great deal of confusion in the whole process of analysing geographic data. On the one hand geographers usually insist that a region has to be spatially contiguous (this suggests that some space-time language is appropriate), yet on the other hand the elements (or 'individuals') contained within a region are frequently required to exhibit similar properties. Deriving 'individuals' in one language (say, the space-time language) from 'individuals' specified in another language requires an adequate translation procedure. Simply mixing up two very different languages will only yield garbled results—it is scarcely surprising, therefore, that so much controversy surrounds the regional concept in geographic thought. On a lesser scale of complexity is the specification of the 'individual' in the variations of the basic space-time language. Carnap's first

[1] I am indebted to Dr Dacey for the loan of certain unpublished materials.

space-time language conceives of an individual as forming a space-time region and is therefore most appropriate for discussing the location of discrete objects. His third space-time language which views individuals as a series of space-time points appears more suitable for discussing phenomena which vary continuously over the earth's surface. Again, the problem of ordering geographic data amounts to the difficult logical problem of working with two different language systems in the same context since characteristically geographers deal with both discrete and continuous data.

It is therefore vital that geographers should be clear about the logical properties of the formal languages which are available to them for ordering areally distributed data. An understanding of these logical properties of the formal languages yields an insight into the dimensions of the problem which must face geographers in ordering data. Thus the argument by Grigg (1967) and others that the logic of regionalisation must be followed in depth inevitably leads to examining the nature of spatial languages. Yet so far, with the exception of Dacey's largely unpublished work, there has been little work done by geographers in this vital methodological field. We will examine some of the practical problems involved in chapter 19.

Simple spatial languages are nevertheless used with considerable success in geographic research (e.g. the latitude-longitude system). It is thus quite possible to apply such spatial languages without necessarily understanding all their formal properties. Much the same observation can be made of the application of formal geometry to geographic problems. Formal geometry (in the usual sense of the term—e.g. topology, Euclidean geometry and so on) is regarded by Carnap (1958) as an extension of the simple spatial languages he developed. He thus shows by way of the axiomatic method how the various formal geometries which have already been described may be related to simple space-time languages. Again, geographers are intimately concerned with the application of such geometries to geographic problems. We shall consider, therefore, some of the applications of formal geometries to geography. Since the number of actual applications is large and the number of potential applications infinite, we shall proceed by way of examples.

(1) *Topology*

Topology is qualitative geometry. It is concerned only with 'the continuous connectedness between the points of a figure' (Hilbert and Cohn-Vossen, *1952*, 289). Topology remained a relatively undeveloped form of geometry until the late nineteenth century. Yet it is a very basic form of geometry since it represents some of the simplest and earliest perceptual concepts of space and, as Klein showed, it is the form of geometry from which all other geometries may be derived. Topology has now become a highly developed branch of geometry, and it has also been shown that 'the theorems of topology have been found to be connected, despite their apparent indefiniteness, with *the* most precise quantitative results in mathematics, that is, with the results of the algebra of complex numbers, the theory of functions of a complex variable, and the theory of groups' (Hilbert and Cohn-Vossen, *1952*, 289).

Since topology deals with the holistic properties of objects and in particular is concerned with connectedness, we may expect topological theorems to be applicable to geographic problems if the geographical problem itself can realistically and successfully be stated in terms of connectedness. Fortunately, there are numerous problems which have arisen in geography which can be stated in such terms. The simplest example of this comes from examining the connections between settlements by a set of transport links. But there are numerous other problems—such as the connections across boundaries in studying contiguity (Dacey, 1968)—which may be regarded as topological problems. The particular branch of topology which has been most invoked to date is undoubtedly that of *graph theory*, and a number of geographic problems can be identified with this particular branch of topology. The primitive terms of graph theory— terms such as 'edge', 'path', 'node', and 'vertex'—can easily be related to actual geographical objects (in much the same way that Euclidean terms such as 'point' and 'line' could easily be operationalised in surveying problems). It is thus relatively easy for actual geographic relationships to be mapped into graph theory. Little abstraction or generalisation is called for, yet graph theory provides a framework for analysing and manipulating geographic data-sets in a fairly sophisticated way. It is clear also that the solutions which can be found by using graph theory—such as measures of connectivity in the network structure, measures of network capacity, shortest-path solutions, and so on—are of considerable utility.

Haggett (1967) reviews in detail the application of graph theory to the study of network problems in human and physical geography. He identifies four major types of geographic phenomena that may be examined using topological notions—paths, trees, circuits, and cells. Geographic phenomena that can be conceptualised in these terms may be described and analysed by way of topology. The large literature dealing with network structures in geography—from the analysis of transport networks to the analysis of complex cell structures—undoubtedly reflects the reasonableness of conceptualising geographic problems in topological terms and the utility of topology in describing and analysing geographic problems (Haggett, 1967; Kansky, 1963; Garrison and Marble, 1965; Medvedkov, 1967).

(2) *Projective geometry and transformations*

It has been suggested (p. 215) that a spatial language may be regarded as a set of rules defining a co-ordinate system. It is possible to specify rules which transform one co-ordinate system into another. These rules provide a translation from one spatial language to another. This general statement has direct implications for the application of formal geometry to geographic problems.

Perhaps one of the oldest technical problems facing the geographer in ordering information is that of map-projection, for the parallels and meridians of the sphere need to be delineated on a plane surface. Various approaches may be adopted to this problem. Ptolemy and Mercator considered it specifically and attempted to derive geometrical solutions. In other cases—such as the *T-O* maps and the portolan charts of the medieval period—no conscious attempt is made to solve the problem, but, as Tobler (1966A) has shown, such maps imply a projection even if they do not specify one.

The modern treatment of map projections relies upon formal projective and analytic geometry. The history of this approach probably begins with Lambert in 1772 and after his analytic treatment several mathematicians—including LaGrange, Euler and Gauss—wrote treatises on the problem. Tobler (1966A)[1] has reviewed this development. Thus:

> Gauss' contributions were fundamental. Characteristically, he does not attack so simple a problem as representing a sphere on a plane, but rather considers the general question of representing one arbitrary surface on another arbitrary surface in such a manner that similitude relations are maintained. He thus gives a general solution, using complex variables, to the problem, posed by Lambert, of conformal representations. He continues by creating the subject of differential geometry, now basic to advanced work on map projection.

At this point in history the problem of map-projection links up with formal geometry to stimulate the latter to new formulations. These developments enabled map-projections to be discussed analytically—in other words the rules specifying the transformation from one co-ordinate system to another could be specifically stated. Melluish (1931, 2) puts it as follows:

> When a map is being drawn, each point on it is fixed according to some given law which expresses the co-ordinates of that point on the map in terms of those

---

[1] I am indebted to Dr Tobler for the loan of an unpublished manuscript and for permission to quote from it.

of the corresponding point on the earth. Such a law is called the Projection on which the map is drawn; and the equations of the projection are those which give the relation between the terrestrial co-ordinates and those of the point on the map.

Solving such equation systems requires formal geometric methods. Further, the properties of these equation systems can be analytically investigated. Tissot (1881) was thus able to make a general study of deformation involved in map projection. Tobler (1966A) writes:

> If the objective is to preserve some spherical property on a flat map, what happens to the other properties when this is done? How are the angles distorted on equal-area projections, and so on? Tissot's indicatrix allows such questions to be answered, in a local sense, with comparative ease.

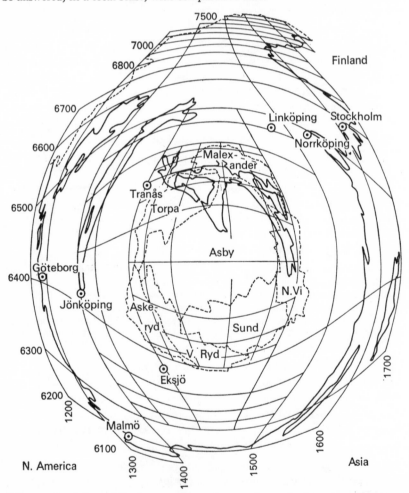

Fig. 14.6 A logarithmic transformation to show the location of important places in the migration field of Asby, Sweden (*from Hägerstrand*, 1953).

The indicatrix developed by Tissot (Melluish, 1931, 98) allows the distortion of a number of chosen properties to be measured and it therefore allows a rational choice of projection to be made according to chosen criteria. Suppose, for example, a conformal (angle-preserving) projection is chosen, then it is possible to choose that particular projection which for a given area minimises the distortion of areas and so on. Analytic formal geometry has important applications.

The detailed study of map-projections, however, has tended, until recently, to decline in importance in geography. Thus Hartshorne (1939, 398) does not regard it as being an integral part of geography. The lack of interest in projection problems may be partly attributed to the quite correct assertion that the problems of projecting the spherical surface on to a plane have been essentially solved. But if, as we have already suggested, the earth in activity terms is a far more complex shape than a sphere, then the whole problem of map projection is raised anew in a much more complex form. This leads us to consider the modern problem of map-transformation.

It has been shown in a variety of studies that distance may be measured in a variety of ways and that areas may be distorted according to the volume of activity occurring within them. To get round these problems rough transformations have been constructed, such as Hägerstrand's (1953) logarithmic map (Figure 14.6) and

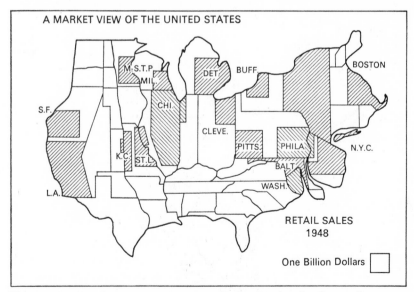

Fig. 14.7 An elementary map transformation in the form of a cartogram of United State retail trade with the area of each state equated with its volume of retail trade (*from Harris*, 1954).

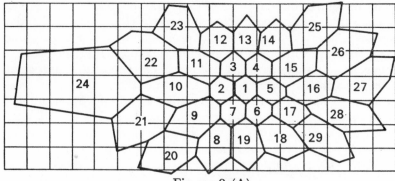

Fig. 14.8 (**A**)

Fig. 14.8 Comparison between theoretical and actual map patterns by way of map transformations. **A:** An approximation to Christaller's theoretical model in an area of disuniform rural-population density.

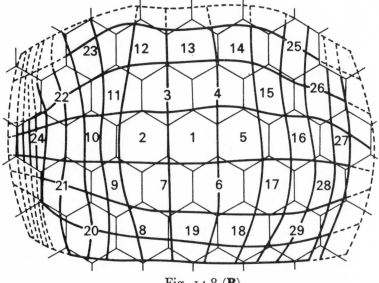

Fig. 14.8 (**B**)

B: A transformed map to produce even population densities and the superimposition of the theoretical Christaller solution (*from Bunge*, 1966).

Harris's (1954) cartogram of United States retail sales in which the size of each state is proportional to its volume of retail sales (Figure 14.7). It has also been shown that theoretical systems, such as Christaller's central-place model, cannot be applied to geographic circumstances without some transformation (Figure 14.8). Tobler (1963) has shown that all of these problems may be discussed analytically by way of projection and transformation, for each requires that one co-ordinate system be changed by some specified law into another co-ordinate system (Figure 14.2). As with all map-projection problems the difficulty is to choose the most appropriate transformation from an infinite set, and in this Tissot's analytic work is again of considerable value. Tobler (1963) suggests, for example, that if the basic aim of the transformation is to produce a map of uneven density in such a manner that density becomes uniform, then the appropriate transformation is one which is as conformal as possible (i.e. one which gives the least angular distortion). Tobler concludes:

valuable map projections can be obtained that do not conform to the traditional geographic emphasis on the preservation of spherical surface area but rather distort area deliberately to 'eliminate' the spatial variability of a terrestrial resource endowment. In many ways these maps are more realistic than the conventional maps used by geographers. . . . The important point, of course, is not that the transformations distort area, but that they distribute densities uniformly.

Such transformations involve translating a three-dimensional co-ordinate language (x, y, z) which may be used to describe a surface of hills, hummocks, and depressions (according to the density of activity) to a two-dimensional co-ordinate language (x, y). This can be done by using formal analytic geometry, and it is Tobler's unique achievement to have demonstrated the utility of a formal geometric approach to this exceedingly complex problem.

(3) Euclidean geometry

It may appear from the preceding discussion as if Euclidean geometry could have little application to geographic problems. Certainly classical Euclidean geometry appears 'almost childishly simple by comparison to . . . geographic geometry, which takes into account the realities of transportation' (Tobler, 1963). In many ways the simplicity of Euclidean geometry may be regarded as its great strength. Many of the non-Euclidean geometries are exceedingly difficult to work with and however much we may wish to work with 'periodic surfaces carried over into the nth dimension of (some) strange and unknown geometry' (Olsson, 1967), it has to be recognised that Euclid still provides us with a well-developed simple geometry which has tremendous utility in geographic research.

The simplicity of Euclidean geometry can be demonstrated by considering the measurement of distance between two points. In analytic geometry the position of two points is given by their respective co-ordinate references. Let us suppose we are considering a two-dimensional surface and the co-ordinates are $x_1 x_2$ and $x_1' x_2'$. On the Euclidean plane the pythagorean theorem gives the measure of distance between these points as $d = \sqrt{(x_1 - x_1')^2 + (x_2 - x_2')^2}$. If we assume the distance between the two points to be very small (ds) and if we simplify the terms $(x_1 - x_1')$ and $(x_2 - x_2')$ to dx_1 and dx_2 we have $(ds)^2 = (dx_1)^2 + (dx_2)^2$ which may be extended to the n-dimensional Euclidean space as $(ds)^2 = (dx_1)^2 + (dx_2)^2 + \ldots$ $(dx_n)^2$. Now it is clear from this statement that the special nature of the Euclidean case is that there is no interaction between subsequent terms. Thus there is no relationship between e.g. dx_1 and dx_2. We may regard the Euclidean space, therefore, as being characterised by the following distance measure system (Adler, 1966, 282–3)

$$(ds)^2 = 1(dx_1)^2 + 1(dx_2)^2 + \ldots 1(dx_n)^2 + 0(dx_1)(dx_2) \ldots$$

Riemann introduced the more general formula for distance in an n-dimensional manifold:

$$(ds)^2 = a_{11}(dx_1)^2 + \ldots a_{nn}(dx_n)^2 + a_{12}(dx_1)(dx_2) + \ldots$$

in which the coefficient of each term is a function of the co-ordinate system itself. In fact, Riemann regarded any system in which such a measure of distance was possible as defining the Riemannian space. It is clear that Euclidean geometry is a special case in which all $a_{ij} = 0(i \neq j) = 1(i = j)$. It is possible to measure the distance between two points in any Riemannian space, but it is far simpler to measure that distance when Euclidean conditions hold. Euclidean geometry thus forms one of those special situations where the ease of manipulation and calculation is so great that any problem tends to be mapped into the Euclidean calculus if that calculus provides a 'reasonable' representation of the structure of the empirical problem. We shall consider some important cases where this is true in the chapter on measurement. For the moment we will confine attention to the problem of measuring geographical distance.

In terms of spatial concepts in geography the Euclidean system appears to provide reasonable approximations in a number of situations. A simple demonstration comes from navigation, where navigating around the surface of the earth requires spherical geometry (a particular form of Riemannian space) but on which Euclidean geometry provides reasonable approximation for the calculation of distances and angles, provided the distances involved are not more than, say, 250 miles. Similarly, for ordinary purposes we may use Euclidean geometry to discuss the perception of space and physical distribution of objects in physical space. The curvature of space postulated in Einstein's theory of relativity only becomes measurable at a great distance.

The measurement of distance over complex geographical surfaces, such as those defined by transport cost, travel time, social interaction, or individual mental maps, is more difficult to discuss, largely because we do not yet know enough about the properties of such surfaces to be able to say with certainty whether a Euclidean approximation is reasonable or not. Locally, for example, transport-cost relationships may be so structured that Euclidean approximation is reasonable. In other situations we may find aggregate human behaviour may be approximated by Euclidean space but individual behaviour cannot reasonably be so approximated. In still other situations it may be perfectly obvious that the Euclidean metric is inappropriate (Figure 14.4). In such cases it is possible to use a non-Euclidean geometry directly, but it may prove technically simpler to transform the relationships between objects and events so that they can be analysed by the use of the simpler Euclidean geometry. The latter option is the one most frequently chosen because most non-Euclidean geometries are either unfamiliar or analytically difficult (and in some cases they are both).

Most specifications of spatial theory also resort to a Euclidean geometric framework. The assumptions of isotropy and homogeneity in the specification of location theory thus allow the analytic tools of Euclidean geometry to be used in the process of theory construction. The familiar assumptions of location theory (e.g. flat plane surfaces, equal transport facility in all directions, and uniform resource endowment) are assumptions which are specifically designed to allow a Euclidean treatment of the problem. The solution of the problem may later be disturbed to take account of non-Euclidean properties of actual spaces. Thus the familiar Thünen rings become distorted and disturbed with variation in transport cost. In demonstrating this graphically von Thünen is in fact developing a rough map-transformation (Tobler, 1963).

Perhaps the most sophisticated development in this area has been Dacey's (1965A) specification of central-place theory. In this case Dacey uses many of the tools associated with Euclidean geometry— vector analysis, lattice systems, and some of the important links which have been built up between Euclidean geometry and number theory —to specify the geometry of central-place systems. Dacey views the central-place system as 'a motif constructed on a plane lattice' with the motif repeated an infinite number of times over an unbounded plane. The motif is obtained by equating the geographic concept of the market area in central-place theory with the mathematical concepts of the *Dirichlet* region. The primitive Dirichlet region is defined as 'a polygon that contains a lattice point at its centre and every point of the plane closer to that lattice point than to any other lattice point'. The competitive process involved in central-place theory then becomes a packing problem.

Dacey's treatment of central-place theory demonstrates how geographic problems can, once they are unambiguously conceptualised (and Dacey found some ambiguities in the usual presentation of central-place theory), be represented in an appropriate mathematical language. In this particular case the intention is to describe mathematically the nature of central-place systems as they are normally described. There is no attempt to derive the geometry from process postulates (e.g. the theory of supply and demand). Dacey's stated intention is limited to discussing the mathematical properties of the diagrams usually used to demonstrate the spatial form of a system of central places. He goes on to state that 'a suitable mathematical system is available, so this task only requires placing the appropriate geometry in juxtaposition with central place concepts.' The appropriate geometry is provided by a special arithmetic and algebraic development out of Euclid.

(4) Space-time problems and Minkowskian geometry

The examples discussed so far refer to situations in which formal geometric concepts have been shown to be applicable to actual geographic problems. This section considers a potential rather than a realised development. For the most part formal geometry has been applied in connection with one- two- or three-dimensional spatial languages. Consider the following conceptual development of a four-dimensional problem in which time becomes an important co-ordinate; an individual person constantly moving about in a two-dimensional co-ordinate system (x, y) with a varying amount of money to spend (z) over time (t). Hägerstrand (1963) presents a simplified version of this space-time conceptualisation in discussing the movement of individuals over space and in time (co-ordinate z is omitted). This movement Hägerstrand terms a *life-line*:

every individual can be represented by a *life-line*, starting at the place of birth, now and then making a jump over to a new station, and finally ending at the place of death. The life-lines of a population in a block of time-space are twisted together in a very complicated way.

This conceptual arrangement has considerable importance for the ordering of geographic information and the construction of dynamic spatial theory in geography. It also bears an interesting resemblance to the analytic and graphical treatment of Einstein's space-time concepts by Minkowski. The co-ordinate language appropriate to relativity theory defines the simultaneity of two events x and y in time as a state in which it is impossible for a signal to go from x to y or from y to x (Carnap, 1958, 203). Reichenbach (1958, 145) thus points out that the notion of simultaneity of events is reduced to the concept 'indeterminate as to time order' and that it implies 'the exclusion of causal connection'. In terms of physical signals (determined by the speed of light) one second on one axis of the space-time graph may be equated to 186,000 miles on the distance graph and therefore on a terrestrial scale there is no measurable difference from ordinary Euclidean geometry (d'Abro, *1950*, 467–81). But Hägerstrand's concern is with the diffusion process, and here the speed of movement of signals is socially rather than physically determined. This necessarily makes the analytic treatment of time-space problems in human geography very complex since the social process is far from being invariant over time. Nevertheless it seems technically feasible to deal with notions about spatial causal connection, spatial diffusion and interaction, and so on, by way of the analytic geometry which Minkowski developed.

The four examples discussed of the interrelationships and 'bridges' which can be built between geographic problems and formal geometry are brief illustrations rather than exhaustive analyses. The myriad applications of topology, projective and Euclidean geometry, would require extensive treatment. In Klein's sense 'projective geometry is all of geometry', and a detailed examination of the use of different forms of transformation in geographic research would require a book in itself.

What I have sought to show in the preceding sections is the methodological possibility of treating geographic problems of spatial form and spatial pattern by way of formal geometry. At the moment the bridges between geometry and geography are rather weak and, to use the topological term, the network for the interplay of formal mathematics and empirical problems is poorly connected. Undoubtedly, in the general process of bridge-building between mathematics and geography, the special bridges between geography and the formal spatial languages that constitute the whole of geometry will be some of the first to be fully explored and strengthened.

D Space, culture, geometry, and geography

Geographic concepts of space are founded in experience. In part that experience is common to the whole society in which the geographer works. It is thus dependent upon actual physical experience and upon the accumulated cultural experience of a particular society. It is impossible to comprehend geographic concepts of space without reference to the concepts of space developed in the language, art, and science, of a particular culture. Particular geographic notions about space are thus embedded in some wider cultural experience. But in part the geographic concept of space is special to geography. It has developed and evolved out of the special experience which geographers have of working with actual spatial problems. In this respect the special interpretation of the notions of 'space and distance' forms one of the key identifiers of the disciplinary sub-culture that forms geography itself. Special concepts of 'activity spaces', 'process distances' and so on, may in turn be used to elucidate spatial form and the processes that create such forms. Concept formation may be quickly followed by formal representation in some space-time language, and formal abstract languages may allow the geographer to rise above his own cultural background and examine spatial form and processes in different cultural complexes. Formal geometric languages, already developed in response to the analytic needs of other disciplines (such as physics), may in turn be used to clarify geographic formulations. Thus formal geometries provide useful *a priori* model-calculi or model-languages for discussing geographic problems.

The interrelationships are complex. Geographers cannot seriously hope to understand their own notions about space, let alone comprehend spatial behaviour or give it formal representation, in academic isolation. This chapter has sought to sketch in some of the cultural, scientific, and mathematical, background necessary for an understanding of the nature of spatial concepts in geography. It has shown that the term 'space' may be treated in a variety of ways and that the concept of space is itself multidimensional. Yet many formal languages are available to discuss different facets of this multidimensional concept. The lesson that should be learned is that there is no need to take a rigid view of the spatial concept itself either for philosophical purposes or for purposes of empirical investigation. The concept itself may thus be regarded as flexible— to be defined in particular contexts, to be symbolised in particular ways, and to be formalised in a variety of spatial languages. Such flexible use requires care. But it also provides the challenging opportunity to develop geographic theory in a new and creative way.

Basic Reading

Adler, I. (1966).
Hallowell, A. I. (1955).
Hempel, C. G. (1949).
Jammer, M. (1954).

Reading in Geography

Bunge, W. (*1966*), especially chapters 9–11.
Dacey, M. F. (1965A).
Gould, P. (1966).
Hartshorne, R. (1958).
Nystuen, J. D. (1963), reprinted in Berry and Marble, 1968.
Tobler, W. (1963), reprinted in Berry and Marble, 1968.
Warntz, W. (1967).
Watson, J. W. (1955).

Chapter 15
Probability Theory—The Language of Chance

The mathematical theory of probability provides a language for discussing a variety of empirical problems in a variety of ways. In recent years the use of this language has increased very rapidly throughout the whole of science, and geography has been no exception. If we were to select any one mathematical language as dominating the current *Zeitgeist* in academic research, it would almost certainly be that of probability theory. Yet in spite of its enormous importance, the notion of probability turns out to be an extraordinarily confused one.

Savage (1954, 2) has written:

> There must be dozens of different interpretations of probability defended by living authorities, and some authorities hold that several different interpretations may be useful, that is, that the concept of probability may have different meaningful senses in different contexts . . . It is surprising . . . to find that almost everyone is agreed on what the purely mathematical properties of probability are. Virtually all controversy therefore centres on questions of interpreting the generally accepted axiomatic concept of probability, that is, of determining the extramathematical properties of probability.

Savage thus suggests that there is little or no disagreement regarding the *syntax* of probability theory but considerable confusion regarding its *semantics*. One version of the theory of probability—the axiomatic statement by Kolmogorov in 1933—has dominated the mathematical treatment of the subject. But other axiomatic statements are possible (e.g. Koopman, 1940; Lindley, 1965, I). Thus Nagel (1939, 39) writes:

> As in the case of geometry, the probability calculus can be formalised in different ways, depending on what terms are selected as primitive, on the mathematical apparatus which is to be employed in developing it, and also upon the use to which it is to be put subsequently.

A syntactical development of probability theory may proceed with a particular semantic interpretation in mind (see above, p. 181). But the various mathematical developments of probability theory do bear a close relationship to one another since it is required that the standard theorems of probability (such as the addition and multi-plication theorems) should be derivable from the axioms. The choice of one axiomatic development instead of another, however, is un-doubtedly influenced by the interpretation given by the scientist to the term 'probability' itself.

The considerable confusion surrounding the meaning of 'prob-ability' in science as a whole has important implications for geo-graphic research. The use of the term in geography may itself be confusing and ambiguous. Words such as 'probable', 'luck', 'chance', 'random', and the like, are not foreign to geography. In some cases such words appear to have an everyday meaning, to be culled merely from ordinary speech. On other occasions such words appear to have some kind of technical meaning, but the *precise* technical mean-ing is not always clear. In other situations a mathematical model based on the probability calculus may be used to analyse a particular geographical problem, without its being entirely clear which of the multiple interpretations of the calculus is being implied.

The probability calculus, when used as an *a priori* model lang-uage, thus appears to be characteristically overidentified (above, pp. 165–8) in that the successful application of the language may be interpreted in rather different ways. It seems important, therefore, to examine the general nature of these various interpretations put upon probability, to examine the structure and development of the language of probability (in both its deductive and non-deductive aspects), and, finally, to examine the importance, both actual and potential, of this particular language in geographic context.

I THE MEANING OF PROBABILITY

The tremendous variation in meaning attached to probability in the sciences is demonstrated by the following list of terms, taken from Fishburn (1964, 134), which seek to clarify the meaning of probability:

Degree of confirmation	*Degree of conviction*	*Degree of rational belief*
Empirical probability	*Geometric probability*	*Impersonal probability*
Inductive probability	*Intuitive probability*	*Judgement probability*
Logical probability	*Mathematical probability*	*Objective probability*
Personal probability	*Physical probability*	*Psychological probability*
Random chance	*Relative frequency*	*Statistical probability*
Subjective probability		

Such qualifying terms demonstrate the range of the concept of probability but do not entirely succeed in clarifying the meaning.

Most writers on probability theory identify three major groupings, within which there may be considerable variety of interpretation, but between which there are major philosophical differences (Nagel, 1939; Savage, 1954; Fishburn, 1964; Churchman, 1961).

(i)   Frequency views 'are characterized by defining the magnitude of a probability as the relative frequency with which a stated property occurs among the elements of a specified set of elements, called the reference set or reference class' (Fishburn, 1964, 139).
(ii) The logical view of probability is concerned with the logical relation between hypotheses and the evidence for such hypotheses. Thus 'probability measures the extent to which one set of propositions, out of logical necessity and apart from human opinion, confirms the truth of another' (Savage, 1954, 3). This view is thus closely bound up with the problem of induction and has application to the problem of confirmation of hypotheses (above, pp. 38–40).
(iii) The subjective view of probability 'measures the confidence that a particular individual has in the truth of a particular proposition' (Savage, 1954, 3). Such a view involves defining normative decision procedures, but it allows for different initial judgements regarding the probability of an event's occurring.

These three major views of probability will be examined in detail later. But before such an examination it is convenient to introduce a further distinction which stems from the elaboration of probability theory itself. It is frequently held that 'probabilistic' is the polar opposite term to 'deterministic'. The sense in which this is true needs to be stated. The axiomatic development of mathematical theories of probability relies entirely upon the logic of deduction. The inferences are certain, the theorems are absolutely determined, given the axioms, and therefore, this deductive elaboration of mathematical probability theory must be regarded as *deterministic*. The deductive elaboration of probability theory contrasts with the inductive inferences involved in the theory of statistics. Here the development, which amounts to an extended and elaborate theory of the application of probabilistic concepts to decision-problems, is radically different in form and does involve uncertainty. Statistical inference thus involves applying certain rules of decision in non-determinate situations. These rules are deterministic in the sense that they are derived from certain axiomatic statements. The application of such rules may or may not involve making statements which are probabilistic in the sense of being 'non-determinate'.

A The classical view of probability

The historical development of probability theory is a neat illustration of the way in which mathematicians frequently rely upon some model of a real-world process in order to develop mathematical theory, while physical and social scientists concerned with constructing theory in their respective disciplines frequently turn to the mathematical theory as a convenient model for formulating theory.

Cramer (1955, 12) suggests that the modern mathematical theory of probability has its origins in the famous correspondence between Pascal and Fermat in the second half of the seventeenth century. This correspondence concluded that problems in assessing a 'fair' wager in games of chance 'could be reduced to problems in the mathematical theory of permutations and combinations' (Nagel, 1939, 8). This combinatorial approach—which is still of great importance today (David and Barton, 1962; Feller, *1957*)—led to the development of numerous mathematical theorems which could be applied to any kind of problem where an element of chance was involved. These theorems were brought together and synthesised in the work of Laplace (*1951*). Probability theory was necessary, in his view, because of our own ignorance and lack of knowledge. Laplace did not regard the world as being governed by probabilistic laws since 'all events are regulated by "the great laws of nature" which a sufficiently powerful intelligence could use to foretell the future in the most minute way' (Nagel, 1939, 9). Given our vast ignorance, it was necessary to develop some theory which could be applied to situations in which we were uncertain. But the theoretical foundations of this theory were themselves rather ambiguous. In particular, 'the principle upon which Laplace assigned numerical values to probabilities was that of analysing the possible outcome of a situation into a set of alternatives which could be judged as 'equally possible' (Nagel, 1939, 8). This meant an arbitrary *a priori* assignment of a probability value to some outcome. This view in its most rigid form led to a number of contradictions and ambiguities. To resolve these the *principle of indifference* (or the *principle of insufficient reason* as it is sometimes called) was invoked. This amounted to saying that equal probabilities could be assigned to the various outcomes of an event provided there was no evidence to suggest that some other assignment of probabilities ought to be made. Thus the automatic assignment of probabilities on the throw of a six-faced die was $\frac{1}{6}$ to each face unless there was definite evidence that the die was biased in some way.

This rather unsatisfactory way of assigning probability did not inhibit the mathematical development of the theory. The mathematician simply assumed that a probability could be assigned and regarded the 'how' of such an assignment as an empirical rather than a mathematical problem. In any case it was possible for the mathematician to specify simple real-world situations where 'equally possible outcomes' seemed a perfectly reasonable assumption. Games of chance, urn models, and so on, provided (and still provide, e.g. Feller, *1957*) elementary physical models suitable for mathematical exposition. Yet the application of probability theory also expanded during the nineteenth century. The development of statistical mechanics brought together probability theory and theoretical physics to begin a long and fruitful interaction between physical and mathematical theory. Later in the century the development of genetics—in particular the Mendelian theory of inheritance—gave a considerable impetus to mathematical development since this problem appeared inherently probabilistic in form.

Perhaps the most important development of all was the fusion of statistics and probability theory. Statistics—originally conceived of as 'the factual study of society'—made use of certain descriptive measures—mean, variance, and correlation, for example—in classifying and tabulating information. The distinctive method of statistics was thus 'to accept variability and try to study it, in contrast with the traditional scientific method of investigation, which involved trying to eliminate variability' (Anscombe, 1964, 157). This difference is of tremendous significance, for it sheds light on many aspects of the use of probability theory in empirical work. It is interesting to contrast Gauss and Galton in this respect. Thus Gauss developed the theory of least squares and explored the properties of the normal curve (sometimes called the Gaussian curve) in order to develop a method for eliminating observation error in making observations which required a high degree of accuracy. In geodetic work, for example, observation error in measuring angles and distances was of great significance in his attempt to decide empirically whether physical space was Euclidean or not. Galton, whom T. W. Freeman (1966) has claimed as one of the important contributors to geographic knowledge in the nineteenth century, in developing regression analysis (which in modern statistics uses the Gaussian method of least squares) was less interested in the true value of some measure than in the degree of variability inherent in genetic processes.

But statistical investigation was not merely descriptive. Inference was also involved. It was necessary, for example, to know something about the properties of such measures as the mean and variance in order to gauge the relationship between such measures and the data-set being examined. A similar but more sophisticated problem concerned the testing of hypotheses—when, for example, was a correlation coefficient significantly different from zero? Hypothesis-testing required some appropriate statistical theory and this theory was derived from the Laplacian theory of probability. The application of Laplacian theory to problems of statistical inference raised a number of difficult problems which emphasised the inadequacy of the Laplacian method of assigning probabilities on an *a priori* basis. The Laplacian method appeared to many to be lacking in rigour, and the attempt was therefore made to be more objective about the assignment of probabilities to events. This attempt led, in the work of Pearson, Fisher, and Neyman, to the foundation of a school of 'statistical orthodoxy' based on a relative-frequency interpretation of probability.

B The relative-frequency view of probability

The relative-frequency view of probability is undoubtedly the most important of the various views of probability in terms of the volume of literature produced and the variety of applications which have been found for it. There are a number of variants of the view, but essentially it rests on the belief that there is some ratio between the actual number of times a particular outcome of an event is recorded and the total number of events. Given a total set of events, R, and a subset of R exhibiting a certain property a (we will term this subset A), then the frequency, r, of A in R is given by:

$$r(A, R) = \frac{n(A)}{n(R)}$$

which can be empirically determined, given any reasonable number of events R. The relative-frequency view goes on to postulate that r stabilises as n is increased and that the meaning of the term 'probability' can be defined by stating that the probability p, is the value of this ratio at the limit. Thus:

$$p(A, R) = n \overset{\lim}{\to} \infty \frac{n(A)}{n(R)}$$

Simple experiments have been used to demonstrate how r does stabilise (see Figure 15.1), and in certain respects the relative-frequency view is undoubtedly intuitively appealing.

It replaces the *a priori* assignment of probabilities by an empirical method for determining those probabilities. Such a method minimises individual judgement and therefore the relative-frequency view sometimes goes under the name of objective probability. But the actual operational method is not free from awkward assumptions. It assumes the existence of some hypothetical infinite population and it also assumes that the value of p in the limit can be estimated. This is possible only if it can be shown that the particular sample of outcomes of events is representative of the population as a whole. The relative-frequency view thus involves a definition of a random sample or a random experiment.

Consider the following urn model, in which one urn contains 750 red balls and 250 black balls, and a series of drawings without replacement is used to estimate the probability of obtaining black.

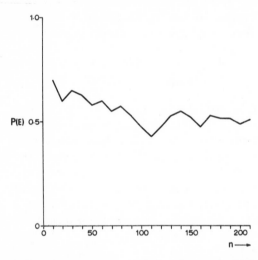

Fig. 15.1 An experiment in which the observed number of outcomes of a particular type (e.g. obtaining heads on the toss of a coin) as a ratio of the total number of outcomes stabilises with increasing sample size.

Since the sequence is finite the probability can be estimated as ·75. But suppose the red balls weigh twice as much as the black and therefore tend to settle at the bottom of the urn. The estimates of the probability would not oscillate about the limiting value as in a simple coin-tossing experiment but would change progressively. In an infinite sequence there is similarly no reason why the extension of the sequence should not converge on any value between 0 and 1 (Figure 15.2).

The relative-frequency view thus replaces the *a priori* assignment of probabilities by some operational method for determining probabilities based on the notion of 'randomness'. Yet Cramer (1954, 5) suggests that 'it does not seem possible to give a precise definition of what is meant by the word *random*'. This may seem rather like defining one undefinable term by way of some other undefinable term. This is only partly the case since it is far easier to give an operational definition of the term 'random' than it is to give definitive meaning to the far broader notion of probability itself. Most writers on mathematical probability (such as Cramer, 1955; Parzen, 1960; Feller, *1957*) who take the relative-frequency view define randomness by example. Such definitions rely heavily upon specifying a simple urn model or a simple game of chance. The specification usually contains the statement that each face of a die has an *equal* and *independent* chance of occurring on a given play. This heuristic definition of randomness is important because from it the mathematical properties of a given sequence of events can then be deduced. These properties then provide a mathematical norm against which

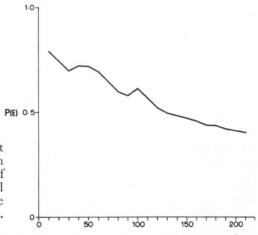

Fig. 15.2 An experiment in which the ratio between the number of outcomes of a particular type and total outcomes fails to stabilise with increasing sample size.

a given sequence can be measured in order to determine whether that sequence is random or not. In the latter situation, however, it is necessary to employ methods of statistical inference in order to determine whether a given sequence is significantly different from a random sequence. Thus 'the definition of probability in terms of relative frequency apparently presupposes the definition of probability in terms of judgement' (Churchman, 1961, 161).

Randomness may, however, be defined by way of sampling-theory. The aim of such a theory is (i) to ensure as far as possible that the estimate which is made of the probability of a given events occurring converges in the long run to the correct value, and (ii) to ensure that the estimate based on a particular sample comes as close to the truth as possible. Sampling-theory is, of course, extremely complex (see, e.g. Cochran, 1953), and it will not therefore be considered in detail here. But it is important to recognise the way in which an operational definition is given to randomness and, hence, to relative-frequency probability. A random experiment may be defined as an experiment whose outcome is independent of the outcome of any other experiment. Hacking (1965, 129) points out that 'in practice it is very hard to make a set-up, experiments on which are independent.' The solution to this difficulty 'is to have a standard chance set-up and anchor all random sampling theory to it.' The necessary set-up is provided by random-number theory. This provides a sequence of digits which can be shown to conform to the calculus of probability, assuming that each digit has an equal and independent chance of occurring.

A table of random numbers, such as RAND's million digits published in 1955, thus provides a basic chance set-up for sampling. In general a table of random numbers possesses the properties which may be predicted from probability theory, assuming each digit has an equal and independent chance of occurring. The difficulty with such a set-up is, however, that certain blocks within the set-up may not conform (Hacking, 1965, 131). This difficulty has been dealt with by Kendall and Stuart (Vol. 1, *1963*) who point out that in a table of $10^{10^{10}}$ digits blocks of a million zeros should occur fairly frequently in order to satisfy the requirements of the probability calculus.

Thus, it is to be expected that in a table of random sampling numbers there will occur patches which are not suitable for use by themselves. The unusual must be given a chance of occurring in its due proportion, however small.

There are extreme difficulties in defining infinite random sequences and in generating specific examples of portions of such sequences (Churchman, 1961, 158). Yet 'the notion of randomness as it appears in connection with sampling procedures is basic to a vast body of research being carried on today', and unfortunately it is altogether too often 'that one is tempted to assert that any sample that happens along is a random sample, simply for the sake of being able to use techniques of statistical analysis that may be applied to random samples or their closely related brethren.' Fishburn's (1964, 159-60) comments should provide a salutary warning to the would-be geographic quantifier.

The relative-frequency view of probability therefore amounts to conceptualising real-world processes as repeated experiments conducted under essentially the same conditions and thus independent of one another. Assuming this conceptual representation of real-world processes is reasonable, these processes may then be mapped into the powerful calculus of probability for analytic treatment. The reasonableness of such a mapping of real-world events into the syntactical system of probability theory depends, however, upon an effective sampling design which ensures that each event is randomly selected from an infinite number of possible events. This operational procedure for ensuring randomness in sampling is not without difficulty, for it relies upon a probabilistic test for randomness in a given sampling structure (such as a sequence of random numbers). The latter procedure is essentially inductive and, therefore, involves the usual difficult logical problems inherent in such a form of argument.

C The logical view of probability

Carnap (1950; 1952) is undoubtedly the most important philosopher to expound the logical view of probability. Carnap does not object to the frequency view, which he regards as an important theory in its own right (terming it 'probability$_2$') but suggests that an alternative view of probability (termed 'probability$_1$') may be employed which explores the relationship between certain aspects of mathematical probability and inductive logic. Carnap (1950, v) introduces his study as follows:

> The theory here developed is characterised by the following *basic conceptions*: (1) all inductive reasoning, in the wide sense of nondeductive or nondemonstrative reasoning, is reasoning in terms of probability; (2) hence inductive logic, the theory of the principles of inductive reasoning, is the same as probability logic; (3) the concept of probability on which inductive logic is to be based is a logical relation between two statements or propositions; it is the degree of confirmation of a hypothesis (or conclusion) on the basis of some given evidence (or premises).

This logical relation between a hypothesis and the confirming evidence for it is conceived of by Carnap as being wholly analytic and therefore entirely independent of personal belief or judgement. Probability statements according to the logical view are entirely formal and have no empirical content.

The attempt to furnish an inductive logic which measures 'the extent to which one set of propositions, out of logical necessity and apart from human opinion, confirms the truth of another' (Savage, 1954, 3) is undoubtedly of tremendous significance, for it holds out the prospect of solving the problem of induction once and for all. There are, however, serious difficulties in developing such an inductive logic—difficulties of which exponents of the logical view, such as Keynes (*1966*), Nagel (1939), and Carnap (1950; 1952) are acutely aware. Savage (1954, 61), in reviewing these attempts, comments:

> . . . there is no fundamental objection to the possibility of constructing a necessary view, but it is my impression that the possibility has not yet been realised, and . . . I conjecture that the possibility is not real . . . Keynes indicated . . . that he was not fully satisfied that he had solved his problem . . . and Carnap regards

(his work) as only a step toward the establishment of a satisfactory necessary view, in the existence of which he declares confidence. That these men express any doubt at all . . . after such careful labor directed toward proving this possibility, speaks loudly for their integrity; at the same time it indicates that the task they have set themselves, if possible at all, is not a light one.

The problems associated with developing a logical view of probability have not yet been solved. We shall not, therefore, consider this particular aspect of probability theory further.

D The subjective view of probability

The subjective view of probability is similar to the logical view in that it accepts that probability represents a relation between a statement and a body of evidence. The main difference is that the subjectivist denies that this is (or ever could be) a purely logical relationship and conceives of the relationship as representing a degree of belief. Clearly this degree of belief may vary from person to person, and there is, therefore, no unique relationship between a statement and the evidence for it. Holders of the subjective view—such as Ramsey (*1960*), Savage (1954) and Jeffrey (1965)—'put the individual back into the picture by giving him center stage' (Fishburn, 1964, 167).

It should not be thought, however, that the subjective view does away with logic. In fact the subjective view amounts to a normative theory of rational choice in the face of uncertainty. There is a close relationship between this theory and the rapidly growing literature on decision theory and utility. The reason for such a connection appears to be that the only way of making subjective probability an operational concept is to measure a person's degree of belief by way of his behaviour. Most subjectivists, such as Ramsey and Savage, identify degree of belief with a particular behaviour pattern—such as making a wager or staking money on the occurrence of one event rather than another. Hence the extremely close connection which exists between utility theory as developed by von Neumann and Morgenstern, decision theory as developed by Fishburn, and subjective probability theory as developed by Savage.

The normative theory of choice inherent in the subjective view of probability contains two important logical conditions (Kyburg and Smokler, 1964). The first is that of *consistency* which ensures that only certain combinations of degrees of belief in related propositions are admissible. The degree of belief in an outcome must be compatible with the degree of belief in the non-occurrence of that outcome. The second logical condition is that of *coherence*. Thus:

No set of bets on a proposition is allowable which insures that whatever the outcome of events, the bettor will lose money. It is possible to show that conformity of degrees of belief to the rules of the probability calculus is a necessary and sufficient condition of coherence in this sense. (Kyburg and Smokler, 1964, 11.)

If a person is coherent and consistent in his behaviour, his betting behaviour will conform to predictions derived from the axioms of probability theory. *Bayes' theorem* is one important prediction of behaviour. The theorem defines a way of transforming a set of prior probabilities (which may be determined subjectively as a degree of belief in something) into a set of posterior probabilities given that information or evidence becomes available about the events in question (below, p. 249). Bayes' theorem thus provides a consistent way of changing beliefs as evidence becomes available. There is nothing particularly controversial about the rule itself, since it is common to all versions of probability theory. The peculiarity of Bayesian thinking, however, is that the prior distribution does not necessarily have to be objectively determined. Subjective degrees of belief are admissible and therefore the prior probability may be assessed differently by different people. This view is unacceptable to most frequentists and Fisher (1956) totally rejects it—a decision which has had very important implications for the development of statistical theory (Anscombe, 1964, 161).

Acceptance of the Bayesian view has the virtue—regarded as being very doubtful by some—of allowing a far wider range of phenomena and situations to be examined by way of the probability calculus. The trouble with the frequency interpretation is that it restricts the application of the calculus to repetitive, independent, events observed under strictly controlled sampling conditions. It is thus not possible to refer to the probability of obtaining an ace on a drawing from a pack of cards unless a number of drawings have already been made. Further, it is not possible to take the sense of probability contained in a statement like 'I believe the probability of iron ore costing $z per ton next year is y', and treat it analytically. The Bayesian viewpoint allows such topics to be analysed by the use of probability theory. The extension of the probability calculus to cover 'degrees of belief' is thus intuitively appealing even though it does involve a number of awkward assumptions. But subjectivists also argue that frequency views are not free from awkward assumptions either, and that all they are really doing is to replace a set of rather awkward assumptions about the behaviour of variables when taken to the limit by one very explicit assumption which is intuitively reasonable (Savage, 1954, 62). In addition they can point to the far greater applicability of the Bayesian method to concrete situations. Statisticians, mathematicians, and philosophers appear, however, to be divided on the acceptability of the Bayesian viewpoint.

We must conclude, therefore, that all three interpretations of the term 'probability' suffer from logical difficulty. The attitude taken to the scientific meaning of probability, as opposed to the everyday meaning of probability, thus depends partly on which set of assumptions seems most reasonable and partly on which set of difficulties seems the least insuperable. It is also possible to take a more pragmatic approach and attempt to assess each version of probability by way of the successful applications of a particular version. In this respect there can be no doubt that the relative-frequency view has been highly successful, although this success has been restricted to the analysis of certain types of phenomena. The greater success of the frequency interpretation, however, must partly be attributed to the fact that many more studies have been undertaken with this interpretation in mind (Anscombe, 1964, 165). The Bayesian approach has been relatively neglected until recently, and the achievements are less easy to assess. Certainly, it may be applied to rather different kinds of phenomena from those analysed in frequency probability. The application of the probability calculus to provide a normative decision-theory under conditions of uncertainty has certainly been moderately successful, especially where the decision problem can be clearly and unambiguously formulated. Thus Anscombe (1964, 164) comments:

> Generally speaking, vagueness is reduced by previous experience in the same field of observation; in a well-developed field we can legitimately feel more confidence in proposing specifications and prior distributions than in a new field. Naturally, it is when the vagueness is least that we find the analysis most satisfying. So Bayesian decision theory is rather like a hothouse plant—it needs a favourable environment, it flourishes best in a rather well-developed field of inquiry.

Of all the interpretations of probability, the logical view has the least to recommend it in terms of actual achievement and application. This is not to suggest that this interpretation is to be abandoned entirely. The problem of induction—and hence probability in the sense proposed by Carnap—is perhaps the most significant of all problems when it comes to understanding the nature of scientific knowledge. It has significance in the context of verification and confirmation. The solution to Carnap's problems would have tremendous implications for scientific philosophy. It is hardly surprising that this—the most ambitious interpretation of probability—has yielded the least in terms of practical applications.

The scientific interpretation of the term 'probability' is open to doubt and is the subject of wide disagreement among scientists. The interpretations are not contradictory but rather complementary. This at least allows the scientist to choose one interpretation without necessarily denying the possibility of choosing another. It appears that different interpretations are concerned with rather different things. The only common ground therefore appears to be the probability calculus itself and to this aspect of probability theory we shall now turn.

II THE CALCULUS OF PROBABILITY

Whatever contention may surround the semantical interpretation of probability theory, there can be little doubt regarding its syntactical structure. It is true that different axiomatic formulations may be given to the theory and that the choice of axioms is usually influenced by a particular interpretation of probability, but the derived theorems of probability are the same. The development of a theory of statistical inference, which employs the concept of probability inductively, is much more contentious. From the point of view of the calculus of probability, therefore, it is important to differentiate between the deductive development of the theory and statistical inference which employs the theorems of probability to construct a rational decision theory.

A Deductive development of probability theory

The diverse theorems of probability were finally brought together into a unified axiomatic system by Kolmogorov in 1933. Since then a number of axiomatic treatments have been devised. Good accounts of contrasting axiomatic treatments are given in McCord and Moroney (1964, 20) and Lindley (1965, I, 6–11) while Feller (1957) gives a rigorous but non-axiomatic account. It is not intended to discuss these axiomatic presentations here. However, it is important to remember that probability theory, in common with all axiomatic mathematical systems, is concerned solely with relations among undefined things. The axioms thus provide a set of rules which define relationships between abstract entities. These rules can then be used to deduce theorems, and theorems can be brought together to deduce more complex theorems. These theorems have no empirical meaning although they can be given an interpretation in terms of empirical phenomena. The important point, however, is that the mathematical development of probability theory is in no way conditional upon the interpretation given to the theory.

Like any axiomatic system, the mathematical theory of probability relies upon the statement of a set of primitive notions and postulates. Geometry used terms such as 'point' and 'line'—probability theory uses terms such as *event* and *sample space*. The primitive statements of probability theory are drawn from the theory of sets. These statements can be given an interpretation in the same way that the primitive terms of geometry can be represented by dots and ruled lines. This requires a general interpretation of probability theory, however, and for the purposes of exposition a frequency representation will be given.

The basic primitive notion in the mathematical theory of probability is that of the *sample space*. Consider an experiment in which we toss a coin. All possible tosses may then be characterised as forming a sample space. The sample space contains what are called *members of the sample space* or *elementary events*. In this example each elementary event may be equated with the outcome of an experiment (the toss of the coin). In set theoretic notation we can then define the sample space as $A = (a_1, a_2, a_3, \ldots)$. We may then consider any *subset of the sample space* which we may call a *compound event* or sometimes just an *event*. In our example an event might be all those tosses that are recorded as heads. If there are only two events in the sample space (e.g. heads and tails) then we have the condition that $A = (A_1, A_2)$. It is important here to introduce the notion of mutually exclusive events since this is also basic to the mathematical development of probability theory. Two events may be defined as mutually exclusive if the set formed by their intersection contains no members (or $A_1 \cap A_2 = \phi$, the empty set). It is easy to see in our example that heads and tails are mutually exclusive events. All of the terms we have used so far are primitive to probability theory and they can be put in the language of set theory. All that remains is to define probability as a *set function defined on subsets of the sample space* which translated into simple English amounts to a rule for assigning values to the events contained in the sample space. This rule then gives a number to the probability of obtaining heads, say $P(A_1)$, and the probability of obtaining tails, say $P(A_2)$. We can then state three basic axioms which the numbers assigned must conform to.

(i) $P(E) \geqslant 0$ for every event E (i.e. no negative numbers).
(ii) $P(S) = 1$ for the certain event S.
(iii) $P(E \cup F) = P(E) + P(F)$ if $E \cap F = \phi$ (the empty set), which is the addition rule which states that the probability of the union of two mutually exclusive events is the sum of their probabilities.

The above statement provides foundation for the mathematical theory of probability. There are, of course, a number of different versions available but they are all very similar (Parzen, 1960, 18; McCord and Moroney, 1964, 20; Lindley, 1965, I, 6).

It is useful at this point to digress a little regarding the nature of the sample space and in particular to note that the application of probability theory to any empirical phenomena is dependent upon finding an adequate operational definition (or interpretation) for this abstract theoretic concept. In the example chosen above the sample space contains an infinite number of members since we can carry on tossing coins indefinitely. Suppose, however, that the space is thought of as containing the 52 cards from an ordinary pack. In this case the sample space contains a finite number of members although we can, if we sample with replacement, generate an infinite series from it. Suppose, in this case, we ask what suit a card is when we draw it? There are four possible answers to this question and we may think of the space as containing four mutually exclusive events. If on the other hand, we ask what number or picture the card is, there are 13 mutually exclusive events in the sample space. The sample space can therefore be changed at will according to how we wish to conceptualise it. Therefore the mathematical concept of the sample space only has an interpretation by way of the question we are asking. As Parzen (1960, 11) puts it

Insofar as probability theory is the study of mathematical models of random phenomena, it cannot give rules for the construction of sample description spaces. Rather, the sample description space of a random phenomenon is one of the undefined concepts with which the mathematical theory begins. The considerations by which one chooses the correct sample description space to describe a random phenomenon are a part of the art of applying the mathematical theory of probability to the study of the real world.

Whether the questions we ask are reasonable or not is therefore a geographical rather than a mathematical judgement. But if the calculus of probability is to be used we must first ask a question that allows us to construct some sample space. Similarly, the mathematical theory does not tell us *how* a set function can be assigned to an event although it does tell us the conditions which must be met if the calculus is to be employed. The particular method chosen to estimate probability is therefore an empirical rather than a mathematical problem.

From the axioms of probability theory a large number of theorems is derivable. These theorems relate entirely to the abstraction contained in the primitive definition of the sample space and the events contained in that sample space. It is possible to specify of the points contained in the sample space that each point (elementary event) has an equal and independent chance of occurring on any particular trial. The mathematical properties of a variety of probability functions describing events may then be deduced. Hence

probability distributions may be defined and their properties specified. The most important distributions defined from a sample space
in which equal and independent chances are found, are the binomial, the normal, and the Poisson. These play a very important
role in the application of probability models to actual situations.

By relaxing the condition of an equal chance it is possible to derive
a whole series of compound or generalised distributions such as the
negative binomial or the Neyman Type A, by generalising or compounding the simple Poisson law. The criterion of independence may
also be relaxed and theorems deduced applicable to general stochastic
processes. Random walks, markov chains, queueing and renewal
theory, birth–death processes, emigration–immigration processes, and
so on, may all be deductively explored starting from the basic axioms
of mathematical probability.

The whole complex of propositions and theorems contained in the
mathematical theory of probability will not be reviewed here. Good
accounts are given by Bartlett (1955), Feller (*1957*) and Lindley
(1965, I). There are two propositions which will be mentioned
briefly, however, since they link the deductive and inductive
exploitation of probability theory and help demonstrate how probability concepts may be applied to certain empirical phenomena. The
central-limit theorem and its rather weaker relative, the *law of large
numbers*, are thus extremely important theorems. They are stated as
follows:

(i) The *central-limit theorem* states that 'the sum of a large number of
     independent identically distributed random variables with finite
     means and variances, normalised to have mean zero and vari-
     ance 1, is approximately normally distributed.' (Parzen, 1960,
     372.)
(ii) The *law of large numbers* can be stated a number of different
     ways. In the present context it is probably easiest to give the
     Bernouilli version which states that 'as the number n of trials
     tends to infinity the relative frequency of successes in n trials
     tends to the true probability p of success at each trial, in the
     probabilistic sense that any nonzero difference ε between f_n and
     p becomes less and less probable of observation as the number
     of trials is increased indefinitely.' (Parzen, 1960, 229.)

Both the central-limit theorem and the law of large numbers can
be deduced from the axioms of probability. They therefore provide
the theoretical justification for frequency interpretations which
envisage p converging on some value in the limit. Under frequency
interpretations these two laws are often assumed to be true. But these
laws also define the characteristics of samples drawn from the sample

space under a specified sampling procedure, and hence help to provide criteria for identifying randomness in actual data-sets. In many situations, however, it is impossible to show such randomness and here frequentists usually are content to assume it.

The central-limit theorem and the law of large numbers are of great importance to the relative frequency interpretation of probability. Repeated sampling from the mathematical sample space leads to the convergence of $r(A, R)$ to $p(A, R)$ and the characteristics of such a convergence may be used to specify a 'standard chance set-up' to which random sampling procedures may be anchored (above, p. 238). The application of the probability calculus to repetitive events depends to a considerable degree upon these two laws. Much of the classical theory of statistical inference also rests upon these laws.

The mathematical theory of probability provides an elaborate and powerful calculus for discussing simple and complex 'random' phenomena and uncertain situations. The empirical significance of this calculus rests on our ability to provide the calculus with an interpretation. The application of the calculus in geography will, however, be examined later.

B The probability calculus and non-deductive inference—statistical inference in particular

The problem of non-deductive inference—or the problem of induction as it is usually called—has already been referred to as one of the outstanding problems in the development of a logically sound method of rational scientific investigation. One of the applications of probability theory has been to the construction of languages suitable for making rational and objective non-deductive inferences. Carnap (1950, 207) distinguishes five forms of inductive inference:

(i) The *direct inference* from the *population* (defined as the class of all those individuals to which a given investigation refers) to the *sample* (a subset of the population which is defined by enumeration).

(ii) The *predictive inference* 'from one sample to another sample not overlapping the first'.

(iii) The *inference by analogy* 'from one individual to another on the basis of their known similarity'.

(iv) The *inverse inference* 'from a sample to the population'.

(v) The *universal inference* 'from a sample to a hypothesis of universal form'. In this instance it is important to differentiate between the population and the universe of which a particular population may be a part. We might, for example, regard all pebbles on a beach as being a population from which we are sampling, while the universe consists of all pebbles everywhere. Since *laws* in the strict sense are supposed to be universal statements, this last form of inference has been of major concern to philosophers of science, but Carnap suggests that it is in fact of less theoretical and practical importance than the *predictive inference*.

The problem, then, is to furnish a language suitable for making inductive inferences in all or some of the above situations. Probability theory here provides a syntax for such a language, but in this situation it is impossible to proceed without a semantic interpretation also, since the problem being tackled makes specific reference to empirically identifiable situations. In formulating a language suitable for making inductive inferences scientists must agree not only upon a common syntax but also upon a common interpretation. Thus Ackerman (1966, 36–7) puts the problem as follows:

> The intuitive idea is to proceed by expressing the statements involved in a fixed language, and then calculating the probabilities that the hypotheses or predictive statements are true when the statements contained in the evidence are assumed true. . . . If two people disagree about the appropriate language in which a non-deductive problem should be formulated, they may calculate different probabilities but their results will not be comparable in terms of these probabilities alone, since the probabilities will be functions of the language in which they are calculated. To apply probability notions to non-deductive problems, a preferred and definite language must be agreed upon in terms of which the problems are to be formulated.

There is no complete agreement among mathematicians, statisticians and philosophers as to what such a language should look like nor is there complete agreement as to what kinds of non-deductive inference a particular language may be applied to. These disagreements, not unnaturally, reflect differences in the interpretation put upon probability itself. Here, then, is a prime example of the difficulty of formulating an ethically neutral language of decision and a superb demonstration of how 'rule-conforming' behaviour is basic to our understanding of the nature of scientific knowledge (see above, pp. 39–40, 58). There are thus a number of different languages available for formulating non-deductive inferences, and which language is adopted depends upon the particular school of thought (or paradigm as T. Kuhn (1962) might call it) that a particular scientist subscribes to. The languages available range from the logical construct of Carnap (1950), through a series of languages tied to the frequency interpretation of probability, such as the theory of statistical inference developed by R. A. Fisher, the Neyman–Pearson theory, and Wald's sequential approach, to the frankly Bayesian (subjective) approach of statisticians such as Savage and

Lindley. It is impossible to examine all of these proposed languages here, and the reader is therefore referred to the general discussion provided by writers such as Ackerman (1966), Hacking (1965) and Plackett (1966). But there are a number of general points about the development and application of such non-deductive languages which require some examination.

There are two classic problems in non-deductive inference. The first involves estimating parameters and this helps in translating abstract theorems into empirically usable equations for predicting and explaining events. Estimating the parameters is essential if a model or theory is to be calibrated for empirical use (Lowry, 1965). The second problem involves the testing of hypotheses against sense-perception data. Hypothesis-testing is designed to show whether or not a particular theoretical formulation, when given a real-world interpretation (which may involve estimating parameters) is in some sense 'better than' some other theoretical formulation when tested against observation. This procedure poses difficulties and it is perhaps worth while examining some of these in greater detail.

(1) *Testing hypotheses—the general problem*

Non-deductive inference to choose the 'best' out of a set of alternative hypotheses follows the same general principles as everyday inference (Bross, 1953, 214). The aim of non-deductive inference, however, is to construct a language in which these intuitive steps are formalised into rules. These rules then ensure that anyone starting with the same set of alternative hypotheses and the same evidence will reach the same conclusion. The general procedure is to begin by specifying the universe, U, to be considered. This universe defines the domain of the investigation. U may then be partitioned into a finite number of sets, H_i, and these may be construed as hypotheses regarding U. A set of observations, E, may then be drawn from U according to some specified sampling procedure. For each H_i we may specify an expected and an unusual set of observations. The actual set of observations are then examined, and if they conform to the expected outcome under a particular hypothesis, then that hypothesis may be accepted as true. This general procedure can be made more specific with reference to Bayes' theorem which states that:

$$P(H_i \mid E) = \frac{P(H_i)\, P(E \mid H_i)}{(\sum_i P(H_i)\, P(E \mid H_i)} \qquad \text{(for any } i = 1, 2, \ldots n)$$

The denominator in any situation is constant and thus Bayes' theorem may also be written as:

$$P(H_i \mid E) \propto P(H_i)\, P(E \mid H_i),$$

which in words reads that the posterior probability is proportional to the product of the prior probability and the likelihood. The importance of this theorem is that it provides a rule of inference for estimating the truth of a given hypothesis provided both the prior probability and the likelihood can be discovered (Ackerman, 1966, 91; Plackett, 1966, 249).

(a) *The prior probability* requires estimation if Bayes' theorem is to be employed. It is, however, extremely difficult to provide an objective unbiased estimate of it except under special circumstances. The early statisticians thus *assumed* the prior distribution to have a particular form and the particular form chosen varied from person to person. For this reason Fisher, followed by most frequency statisticians, excluded the prior probability from statistical inference except where some firm estimate based on sample evidence was available. Recently, however, several statisticians (e.g. Savage, 1954; Lindley, 1965) have reintroduced the prior distribution and put Bayes' theorem back into the statistical decision problem. In doing so they have pointed out that there are many situations in which we possess a good deal of information about the truth or falsity of hypotheses before a particular set of evidence is collected. Thus the subjectivist statisticians are also prepared to allow 'degrees of belief' in the H_i to function as the prior probability. These degrees of belief are subject to the conditions of consistency and coherence as already outlined above (pp. 240–1). The difficulty is of course that prior probabilities established by way of introspection will vary from person to person and there is no way of challenging them objectively. Most subjectivists claim that this is preferable to throwing away all prior information and that in any case the orthodox frequency approach is both unnecessarily restrictive and beset by serious conceptual difficulties of its own. Under frequency assumptions 'probability theory can only be a useful device for abstracting the features of certain kind of experimental data', but subjectivists see no real reason why 'the "statistical experience" of the statistician should not be capable of scientific study, and consequently, subject to some useful formal abstraction' (Ackerman, 1966, 89).

(b) *The likelihood* of E given H_i also needs to be estimated. Here $P(E \mid H_i)$ is usually called the *likelihood function*, $L(\theta)$. It amounts to stating the probability of getting a particular set of data *if* a particular hypothesis is true. A good deal of work has been done on this, particularly by Fisher, and the likelihood function is important in both Bayesian and frequency languages (Anscombe, 1964, 167–9). But although the likelihood function is of general importance it is not free of some awkward assumptions. These will not, however, be considered here as they are of a somewhat technical nature (Anscombe, 1964; Hacking, 1965; Plackett, 1966).

(c) It should also be self-evident that the testing of alternative hypotheses is going to be affected by the choice of H_i. In some cases there is a restriction placed on the language in which the hypotheses themselves may be discussed. Frequency languages thus restrict hypotheses to statistical statements—i.e. statements about fairly large aggregates of events. There is thus an important formal requirement that the hypotheses 'should be expressed in the language of the basic probability theory' (Churchman, 1948, 26). If, therefore, we accept a particular language for non-deductive inference, we must be prepared to state our hypotheses in that language. This important formal condition provides a salutary warning. Tests of hypotheses about such events as shopping behaviour, migrations, boundary disputes, and the like, depend upon the prior conceptualisation of these events in terms of some basic probability language. It would thus be inconsistent to argue that the universe we are considering comprises 90 towns in Sweden and then go on to use statistical tests from the Neyman–Pearson theory which necessarily assume that the towns are a sample from a very large, if not infinite population (see below, pp. 274–86).

In testing among the H_i to find the 'best-supported', it should also be evident that which turns out to be best-supported depends upon the specification of the alternatives. It is one of the criticisms of the Bayesian approach that it requires the examination of a continuum of hypotheses 'almost all of which are incorrect' (Plackett, 1966, 255). The frequency approach gets round this problem by examining hypotheses in isolation—the classic procedure in frequency languages is to set up a *null hypothesis* (which proposes that there is no support for a given hypothesis from a set of data) and an *alternative hypothesis* (which proposes support) and then choosing between them on the basis of the evidence. Bayesians regard this procedure as being unnecessarily restrictive, for many hypotheses may be found for which sufficient support can be found to accept them. Accepting an alternative hypothesis does not justify accepting it as the *best* hypothesis (indeed many accepted hypotheses prove to be very wide of the mark) and acceptance certainly does not provide any automatic justification for acting upon that hypothesis.

(d) The notion of *terminal utility* also assumes considerable prominence in the Bayesian approach to statistical decision-making. This terminal utility refers to the value put upon making a mistaken inference. We can make various types of mistake. The classic way of looking at this problem is to consider it in relation to a null hypothesis test in which the conclusions can have the following four-way split:

	Null Hypothesis true	*Alternative Hypothesis true*
Reject the null hypothesis	Type-I Mistake	Correct Decision
Accept the null hypothesis	Correct Decision	Type-II Mistake

The question then arises as to what value we put on making these different kinds of mistake. It is generally agreed that a *Type-I* mistake, being a sin of commission, is more serious than a *Type-II* mistake. This, however, is in itself a value judgement made by Bayesians and frequentists alike. But it is also important to decide what probability of a mistake is going to be tolerated in a given situation. This demands on a willingness to act on the result of a given test and this cannot be determined independent of the consequences of a particular line of action and the value-system we bring to bear in judging those consequences. The Bayesian statisticians freely acknowledge the importance of this, and tend to allow varying subjective judgements to enter the problem. Bayesians, by specifically taking terminal utilities into account, are thus interested in the totality of the decision-making process and regard testing hypotheses as also involving a recommendation for action. Undoubtedly the subjectivists take a far broader view of hypothesis-testing, but they do so at the price of allowing evaluative judgement. The frequency approach subdues value judgements but turns out to be far more restrictive.

(2) *Testing hypotheses—the frequency approach*

The main aim of the frequency school of statistical inference was to develop a non-deductive language that was as objective as possible and at the same time produced acceptable results. Fisher (1956) thus rejected all subjective elements and sought to develop a non-deductive language that avoided the complications of the prior distribution and terminal utility. This is not to say that either he or the other statisticians influenced by him ever denied that prior information and terminal utilities were relevant. Thus Pearson (1962, 55) writes of the initial formulation of the Neyman–Pearson theory:

> We were certainly aware that inferences must make use of prior information and that decisions must also take account of utilities, but after some considerable thought and discussion round these points we came to the conclusion, rightly or wrongly, that it was so rarely possible to give sure numerical values to these entities that our line of approach must proceed otherwise. Thus we came down on the side of using only probability measures which could be related to relative frequency.

There are some far-reaching consequences that result from rejecting the use of the prior distribution and the Bayesian format for formulating non-deductive inference. Without the prior distribution it is impossible to assess the probability of making a mistaken inference. The concept of a *mistake* in the Bayesian non-deductive language has, therefore, to be replaced in the frequency language by the concept of *error*. Bross (1953, 22) suggests that the difference here is essentially that between compound and conditional probability. A *mistake* involves asserting a hypothesis which is not in fact true, and this is a compound probability of the form that event A *and* event B occur. Making an *error* involves asserting A is true *if* B is true and this is therefore a conditional probability. These two probabilities are rather different measures and they are related in Bayes' theorem so that the probability of making a *mistake* is the product of the prior probability and the probability of making an *error*. Since the prior probability is usually regarded as inadmissible in frequency languages, such languages can usually refer only to *errors* and not to *mistakes*. For this reason inferences made in a frequency language should not be regarded as providing any necessary justification for action.

The exclusion of the prior probability in frequency languages poses a problem. Two solutions have been put forward. One, suggested by Fisher, involves what is termed *fiducial probability*. This turns out to be rather different from the theory of confidence-intervals proposed in the Neyman–Pearson theory (Kendall and Stuart, *1967*, II, 134–58). In both cases, however, the aim is to state a non-deductive language that refers solely to the observation data in testing alternative hypotheses. Of these two solutions the *fiducial argument* poses the most difficulty, and many claim that these difficulties are so serious that the argument ought to be rejected (Plackett, 1966, 261–4; Hacking, 1965, 133–60). Kendall and Stuart (*1967*, II, 134–5) give a lucid explanation of the meaning of fiducial estimation and point out that the fiducial distribution as defined by Fisher

(1956, 51–7) is not a frequency distribution in the usual sense of the term. It really expresses 'the intensity of our belief in the various possible values of a parameter'. The distribution is thus rather similar to the idea of 'degrees of belief' and Savage (Kyburg and Smokler, (1964, 178) simply regards it as 'a bold attempt to make the Bayesian omelet without breaking the Bayesian eggs'. It seems better, therefore, to concentrate upon the Neyman–Pearson theory which undoubtedly forms the most influential frequency language for non-deductive inference.

The objective of the Neyman–Pearson theory is to provide a theory that will allow alternative statistical hypotheses to be tested. (Neyman, 1950). It is perhaps worth while examining the meaning of the term 'statistical hypothesis' in greater detail. Consider a random variable, X, which may take on values $x_1 < x_2 < \ldots < x_n$, and which may be defined on a sample space (that is, the observations are all drawn from the same sample space). A frequency function, $f(x)$, may be defined which describes the probability that the random variable will take on any particular value. A statistical hypothesis may then be defined as *an assumption about $f(x)$*. We may make assumptions about the form of $f(x)$—e.g. assume it to be normal, Poisson, negative binomial, etc.—or about the parameters of $f(x)$. There is a distinction to be drawn between a *simple* hypothesis in which all the parameters of the distribution are specified, and a *composite* hypothesis in which only a subset of the parameters governing the distribution are specified. The procedure for testing the latter kind of hypothesis is obviously more complex but it is not different in principle from testing simple hypotheses (Kendall and Stuart, *op. cit.*, chapters 22 and 23).

Testing a hypothesis amounts to defining a set of rules according to which we may reject or accept the hypothesis. These rules are, in effect, rules of inductive behaviour. The particular method used in the Neyman–Pearson theory is to divide the sample space (i.e. all possible sets of observations) into two regions. One region forms a region of acceptance and the other, called the *critical region*, forms the region for rejection. This critical region is usually put equal to some arbitrary value. This arbitrary value is usually called the *size* of the test. The real problem, however, is how to distinguish between those observations which favour and those which disfavour a given hypothesis. In other words we need to set up precise rules for determining the location of the critical region in the sample space. The Neyman–Pearson theory clearly recognises that this cannot be determined without knowing to what alternatives the particular statistical hypothesis is being compared. Thus 'an adequate theory of testing must consider not only the statistical hypothesis under test, but also rivals to it' (Hacking, 1965, 89). By using alternative hypotheses it is possible to construct a set of rules. These rules rely on the concept of *Type-I* and *Type-II* errors. The *Type-I* error is given by the chosen size of the test (sometimes called the level of significance). The *Type-II* error is, however, a function of the alternative hypothesis. The principle which is proposed is that among all tests possessing the same *size*, choose one for which the *Type-II* error is as small as possible. This leads directly to the concept of the *power* of a test. The power measures the ability of a test to discriminate and, further, it is a function of the alternative hypothesis. Statisticians thus talk of power functions and the like. The power function allows a best critical region to be defined, and a test based on this best critical region is called a most powerful test (Kendall and Stuart, *1967*, II, 165). This procedure can be demonstrated with the aid of a diagram (Figure 15.3) which is taken from Kendall and Stuart.

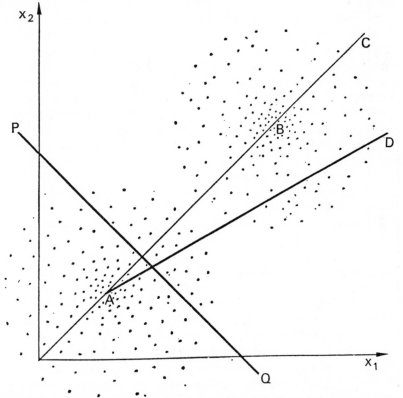

Fig. 15.3 A graphical representation of statistical testing between two hypotheses, A and B (see text; *from Kendall and Stuart, 1967*, II).

The figure shows two scatters of sample points in a bivariate situation. Let us assume that the scatter around A is that which would arise if H_0 (the null hypothesis) were true and the scatter around B is that which would arise if H_a (the alternative hypothesis) were true. Then controlling the *Type-I* error amounts to defining a region on the graph such that it contains 5% (or some other arbitrarily determined level) of the sample points in the cluster around A. We choose such an arbitrary low level in the general belief that it is important to minimise the possibility of making a sin of commission above all else. There are an infinite number of ways in which we could draw such a line. One region might thus be the area above the line PQ, another the sector CAD. It is clear from the diagram, however, that CAD contains a very much smaller proportion of the points expected if H_a were true than does the area above PQ. Thus the latter region will reject H_0 rightly when H_a is true in a higher proportion of cases than will the region CAD. It is therefore a more powerful discriminator.

It is possible to extend the Neyman–Pearson theory in a number of ways. From the diagram it should be evident that the whole sample space is being divided into two regions by the line PQ. This is not always a satisfactory procedure since it is quite possible to find a set of observations (located around D, say) which lead to the rejection of H_0 but can be construed only as very weak support for H_a. It is thus possible to develop mixed tests in which a third region is identified in which the outcome of the test is indeterminate and the decision on rejection depends on the outcome of some other unrelated experiment (Hacking, 1965, 93). These and other extensions of the Neyman Pearson theory need not detain us (see Neyman, 1950, 250–343; Hacking, 1965, chapter 7; Kendall and Stuart, *1967*, II; for general accounts).

The Neyman–Pearson theory of testing statistical hypotheses forms a standard non-deductive language. It is the basis of what might be called the orthodox approach to testing statistical hypotheses. It is intuitively appealing in that the insistence on small *size* and large *power* appears to conform to our intuition as to what a test should be like. The theory has, however, come in for a good deal of criticism, not least from Fisher (1956, 88–92) who accuses Neyman and Pearson of a 'wooden attitude' stemming 'only from their having committed themselves to an unrealistic formalism', and developing a theory that is 'liable to mislead those who follow them into much wasted effort and disappointment.' Fisher bases his criticism on what he calls an intrusive axiom which lays down 'what is not agreed or generally true, that the level of significance must be equal to the frequency with which the hypothesis is rejected in repeated sampling of any fixed population allowed by hypothesis.' The real criticism is the arbitrary determination of a size of a test before any test data are available, for this excludes all possibility of learning anything about 'significance' from the data themselves. Thus the domain of the Neyman–Pearson theory refers, as Hacking (1965, 97–106) points out, to pre-trial betting and not to post-trial evaluation of results. Tests based on the Neyman–Pearson theory are thus regarded, by many, as being misleading except under very special circumstances.

(3) Non-deductive languages for hypothesis testing

It thus appears as if there were at least three powerful non-deductive languages for testing statistical hypotheses:

(i) the *Bayesian* approach as developed by Savage (1954) and Lindley (1965, I and II);
(ii) the *Frequency* approach of Fisher (1956);
(iii) the *Frequency* approach of the Neyman–Pearson theory.

These three languages are not unrelated to one another, and it is surprising to note how many results and theorems these approaches, based on fundamentally different concepts, have in common. The exploration of the similarities and differences between the approaches thus forms a topic of research in statistics. It is fairly obvious, for example, that as the evidence increases the importance of the prior probability becomes less and less in determining the value of the posterior probability and eventually conclusions based on Bayes' theorem will not differ radically from conclusions based entirely upon the concept of frequency in the long run. Similarly, there are many tests which are admissible in all languages. The main difference appears to be in how the admissible evidence is to be controlled. Fisher (*1966*, 25) emphasises the logic of experimental design associated with testing procedures that allow learning by experience and the development of conclusions which are essentially 'progress reports, interpreting and embodying the evidence so far accrued' To do this Fisher was led to the controversial concept of fiducial probability. The Neyman–Pearson theory is much more rigid in its approach to statistical testing and again makes rather important assumptions about the nature of the data (long-run frequency) and about the expectation in the data, given particular hypotheses. The Bayesian theory is much broader but suffers from the problems associated with the subjective choice of the prior distribution.

In part, therefore, it can be concluded that 'where they differ, the basic reason is not that one or more are wrong, but that they are consciously or unconsciously either answering different questions or resting on different postulates' (Kendall and Stuart, *1967*, II, 152). One thing which has emerged from the recent vigorous debate on the foundations of statistical inference is that different languages are appropriate for different purposes and for different situations. Those concerned with the broader problem of decision-making and policy-making almost invariably resort to a Bayesian framework, and there is a natural extension from that concept of probability to more general statements of the decision-making problem, such as that developed by Luce and Raiffa (1957). On the other hand those concerned with conducting experiments on crop productivity under different fertiliser treatments will probably resort to the kind of procedure outlined by Fisher (*1966*). Thus although the fact of controversy among statisticians as to the nature of an appropriate non-deductive language may appear confusing and even discouraging to those seeking firm guidance, it does at the same time have the healthy effect of forcing us to evaluate both the *purpose* of a given investigation and the extent of our knowledge of the circumstances. Only when we are clear as to purpose shall we be able to (i) set up appropriate hypotheses and (ii) choose an appropriate method for testing them. It is worth while, therefore, remembering Churchman's (1948, 24) dictum that

we can best make explicit what we mean by a question, by formulating what we take to be the possible answers. In other words, a question remains ambiguous until one can state what the possible answers are like.

It is important therefore, that a good deal of hard analytic thought go into an investigation of the nature of the questions being asked and the purpose of the questions. Prior analysis, and in particular the elimination of ambiguity, can save an enormous amount of conceptual and interpretative difficulty later on. Without such prior analysis it will be difficult to avoid the question that may still arise after the employment of the most sophisticated non-deductive procedures—namely, what do the results of an investigation *really* mean? That we need consciously to choose one out of several non-deductive languages requires that the grounds for such a choice be made plain. Such a requirement can have only a healthy impact upon empirical research in geography. Such work, characterised as it so frequently is by a haziness of purpose and an ambiguity in formulation, can only benefit from a situation the logic of which requires deep hard analysis and thought. The danger is, however, that without such prior analysis and thought, ambiguous hypotheses will appear to receive a high degree of support from data-sets which have little or no relevance to the hypotheses. Inferential procedures always pose dangers. Unfortunately we cannot do without such procedures, since non-deductive inference lies at the heart of the theory of confirmation and validation. Under such circumstances we have no option but to use such non-deductive languages—but in doing so we should exercise supreme care.

III THE LANGUAGE OF PROBABILITY IN GEOGRAPHY

The term 'probable' occurs frequently in the geographic literature. For the most part, however, the term has been used in the broad sense usually implied in everyday speech. Sentences such as 'The yield of wheat in East Anglia can probably be explained in terms of the level of demand and the inherent fertility of the soil;' 'Teffont Magna was probably first settled towards the end of the twelfth century;' or 'We shall probably never know who first discovered America;' are examples of this kind of common usage. The use of the term 'probable' merely amounts to some kind of expression of doubt regarding the truth of the sentence. It would be interesting, however, to analyse such usage and attempt to discover if any deeper technical meaning of the term probability is implied. In a loose sense, most uses of the term amount to some kind of expression of the individual investigator's degree of belief in a statement, given the evidence available for it. There is rarely any attempt made to specify this degree of belief, although on occasion qualifying terms such as 'highly probable' or 'very improbable' do give an indication of the proximity of the degree of belief to 0 or 1. For the most part this everyday usage of the term remains ambiguous, unquantified, and unquantifiable.

There are two general exceptions to be made to this suggestion that the use of the term 'probable' in geography merely conforms to everyday usage. The first concerns the use of the term to define a particular philosophical position with respect to knowledge and understanding. Here the meaning, although not entirely free from ambiguity, is more specific and certainly of considerable interest in that it has had an important role to play in defining the objectives of geographic investigation. The second exception, which has become of great significance in the last decade in particular, concerns the use of the mathematical language of probability theory as a model-language suitable for discussing geographic problems. In part the deductive properties of this language have been used to model geographic phenomena, and in part non-deductive languages have been used in order to provide some quantitative test, some measure, of the truth of a given hypothesis in the face of a given set of data. The use of probability languages is now commonplace in geography. It is thus useful to enquire into the feasibility of mapping geographic problems into the calculus of probability or, inversely, to enquire into the possibility of providing the calculus with stimulating and reasonable geographical interpretations.

A The philosophical implications of probability arguments in geographic thought

There is some debate within science as to whether probability must necessarily provide the foundation for any adequate theory about reality or whether probability theory itself functions merely as a convenient model representation of real-world processes which can, if necessary, be explained by some theory that makes no reference to probability at all (Rescher, 1964). This debate has been most explicit in physics, where it has been held that Quantum theory—which makes explicit mention of probabilistic concepts—cannot be replaced by any more precise theory. Thus, the Heisenberg principle 'somehow indicates that an exact theory cannot exist' (Kemeny, 1959, 80). This has been disputed. There are, in fact, several different philosophical interpretations given to the employment of probabilistic concepts in science:

(i)   The world is governed by immutable chance processes.
(ii) Our ignorance is so great that we must resort to probabilistic arguments as a cover for our own ignorance (this was the view of Laplace, for example).
(iii) It is much more convenient to discuss aggregate events by way of probability theory than by way of any other theory.
(iv) The world is governed by precise laws but complex interactions can best be examined by way of probability theory.

Any number of variants of these positions may be found. It is impossible to prove, of course, whether the world is or is not governed by the laws of chance, but belief in a particular philosophical interpretation is important since that belief forms the starting point for the scientist's investigations. Neyman (1960) thus comments:

One may hazard the assertion that every serious contemporary study is a study of the chance mechanism behind some phenomena.

There can be no doubt that the approach to modern science has been deeply affected by the concept of probability and the principle of indeterminacy has bitten deep into the values held by scientists (Rescher, 1964). Given this climate of opinion, it is hardly surprising to find many geographers influenced. Phrases such as 'this probabilistic world' are not uncommon in modern geography. Thus Curry (1966A, 40) writes:

Today, in the mental atmosphere of the principle of indeterminacy it is not surprising that some geographers are seeking to use the probability calculus as well as rely on its philosophical connotations. Whether this trend is only the following of a current fad which seeks to hide our ignorance and sloth or implies living boldly in one's generation, eschewing the logic of aggregation and imprecise measurement is for the future to decide. Certainly, it is easy to sympathise with the resistance to regarding the earth's surface as governed by the mechanics of a roulette wheel and its development as a floating crap game. The triumphs of nineteenth-century science with its mechanistic cause-effect modes of thinking cannot lightly be set aside.

Acceptance of a probabilistic viewpoint provides a different starting point for an investigation, a different framework for analysis from the more traditional deterministic viewpoints. Given the disillusionment generally felt in geography with environmental determinism—a disillusionment that rubbed off, rather unnecessarily, on the determinist approach as a whole—it is hardly surprising that rumblings of a philosophical approach to geographic problems by way of 'probabilism' have long been heard. Indeed, the notion of probability appears to have lain at the heart of the methodology developed by Vidal de la Blache. This methodological position appears to have been misinterpreted in the Anglo-American literature. Spate (1952, 419–20), however, in resurrecting the term 'probabilism' suggests that this, rather than 'possibilism' was really implied by the French geographers at the turn of the twentieth century. Lukermann (1965) in a deep and penetrating study of the French School of Geography between 1884 and 1927 has shown the importance of the concept of probability to French geographical thinking:

French geographic thought possesses more internal coherence than other schools and impresses us with its apparent unity. This oneness of outlook hinges on a calculus which states that the picture we have of the world of experience, in its range and variety, can be described and explained *in toto* by statements of probability only, not finality—by conditions of dependence not necessity.

In adopting a probabilistic framework the French geographers did not abandon cause-effect modes of thinking. Rather, they cast the cause-effect mode in a different format. Lukermann (1965, 134) summarises their position as follows:

In saying that all events are caused but not determined and, therefore, not lawful, we leave the explanation of individual events and places to empirical description, not to deductive science. The latter is not causal and explanatory of individual events, but is rather descriptive and lawful of aggregate populations or categories. Individual events, in their range and variety, are explainable only in the chance intersection of mutually independent causal series. But in the aggregate, individual events are probable and lawful as regularities of the observed frequencies of these same events.

This philosophical position does not appear to have been accompanied by any extensive use of the mathematical theory of probability in the analysis of geographic problems. It appears to have served, therefore, as a basic standpoint for geographic investigation—a standpoint that seems to have eluded Anglo-American geography until the rather belated statements of Spate (1952; 1957) and Jones (1956) which merited only passing mention in Hartshorne's (1959, 58–9) *Perspective on the Nature of Geography*. The only detailed discussion of this issue is provided by Sprout and Sprout (1965). By the 1950s however, probabilistic concepts were being developed explicitly in geography as the mathematical calculus of probability became more and more important in the technical development of research methods in geography. By 1953 Hägerstrand had extensively investigated probabilistic simulation procedures, and by 1956 a brief encounter in the *Geographical Review* between Reynolds (1956) and Garrison (1956) as regards statistical inference in geography presaged the flood of probabilistic studies to come. In such studies, however, the main problem is to clarify the exact interpretation put upon the probability calculus. It is worth reminding ourselves, however, that such an approach does have philosophical connotations and in a sense provides a radically different standpoint from traditional modes of geographic investigation. Curry (1966A, 40) points up one such contrast as follows:

> In a sense, the formulation of a random process is the reverse of a deterministic one. In the latter we specify some 'causes' of certain intensity and interaction and obtain a result which will differ from reality by an 'error' term. In the former we begin, at least metaphorically, with unconstrained independent random variables and, by introducing dependencies and constraints, achieve results of various likelihoods.

Whether or not geographers will adopt *en masse* the philosophical position of probabilism remains to be seen. In part this will depend on the ability of the probability calculus itself to function as a reasonable language for discussing traditional geographic problems and here Curry (1966A, 41) suggests that 'there are many areas of concern to geography for which probabilistic thinking appears irrelevant.' But the acceptability of the probabilistic viewpoint will also partly depend upon the way in which the calculus leads us to consider new and exciting problems. This depends, however, on the ability to find interesting geographical interpretations of the calculus of probability itself and to this we shall now turn.

B The deductive development of the probability calculus and the analysis of geographic phenomena

We are here concerned with the probability calculus as a model suitable for the representation and analysis of geographic phenomena. The adoption of a probabilistic model does not, therefore, necessarily imply a probabilistic philosophy on the part of the investigator.

The methodological problem of finding a geographical interpretation for deductive probability theory amounts to this: can we identify circumstances in geography in which it is reasonable to conceptualise phenomena in such a way that they can be represented in the language of formal probability theory? The answer to this question in part depends upon prior concept formation in geography, i.e. the degree to which traditional geographic concepts and ideas do show a certain isomorphism with the structure of probability theory. But the answer may also depend upon the extent to which we are willing to revise our concepts and ideas in a manner that does show some isomorphism. This latter method may seem like trying to stuff the subject-matter of geography into a rigidly predetermined mould. To a certain extent this is true, but there is no necessary danger provided that (i) the new concepts developed are reasonably realistic, and (ii) it is fully recognised that other moulds (or models) are available and perhaps more suitable for treating geographic problems that do not readily fit into this mould. This latter method also clearly involves using the mathematical theory of probability as an *a priori* model-language. In fact the basic methodological problem is not affected by whether we start with the geographical or mathematical concepts, since it simply involves finding a translation from terms such as *sample space, subset of the sample space, set function*, and the like, to terms such as 'shopping for bread', 'the probability of a particular shopping expedition's taking in a baker's shop', and the like. This process is represented by the diagram on p. 264. The utility of deductive probability as a model for geographic phenomena depends entirely upon the mapping (and reverse mapping) and the assumptions which are involved in such a mapping procedure (above, pp. 184–90). This can best be demonstrated by an examination of the frequency and subjective interpretations of probability in the context of geographic research.

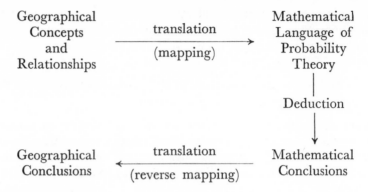

(1) *Geographic phenomena and frequency probability*

In the frequency interpretation of probability the sample space is most conveniently interpreted as containing all possible outcomes of an experiment. In order, therefore, to map geographic problems into such a framework we have to be prepared to treat phenomena *as-if* they are the outcome of some kind of experiment. There are many situations in geography where conceptualising phenomena in such a way does not do a great deal of violence to the phenomena in question. A journey to work, a journey to shop, passing on inform-ation to other people, a migration, the movement of soil down a hill-side, the impact of a raindrop on a soil surface, and many other phenomena, may in some senses be regarded as experiments and, even more important, as essentially repetitive experiments. In order for formal probability theory to be employed it is necessary to agree on a definition of the sample space. A sample space might thus be defined as all possible combinations of shops visited (irrespective of order) on a given shopping expedition. Defining a sample space in such a way does not seem an unreasonable way of examining shop-ping behaviour (this is not to deny that there are conceptual prob-lems—but there are conceptual problems in *any* investigation). It is then possible for us to talk of events (any subset of that space) such as visiting the butcher at the same time as we visit the baker, or visiting the baker on two consecutive occasions, and so on.

So far we have merely formalised the subject-matter of the investi-gation. It is now necessary to provide some way of assigning prob-abilities. Under frequency assumptions the probability is an observed ratio which is assumed to stabilise at the limit. In order to estimate probabilities, therefore, we need to assume that the observations which we possess relate to some hypothetical infinite population (or universe). This has operational implications since it is necessary, in order to go some way towards meeting this condition, that the population be specified and the sampling method conform to some chance set-up (e.g. a random sample).

Conceptualising the geographical problem in such a way involves making some important prior assumptions as well as involving some restrictions on the method of collecting the data. Having agreed to such a procedure it is then possible to (i) use the deductive theorems of probability theory as a model of the phenomena being investigated, and thus manipulate the data in a rigorous but flexible manner, (ii) state statistical hypotheses explicitly, and (iii) confront these hypotheses with the given data-set to estimate parameters and test the hypotheses. Once we have succeeded in mapping geographical phenomena into the probability calculus, considerable benefits (in terms of rigour and ease of manipulation) accrue. But the price of reaping these benefits is that we conceptualise geographic phenomena, generally speaking, as repetitive, recurrent, and independent events (or if there is lack of independence, we can specify its nature).

Consider the following list of events:

(i) Shopping expeditions in a large city;
(ii) Development of a whole system of settlements;
(iii) Conflicts among nations;
(iv) Distribution of capital cities in Europe;
(v) The foundation of London.

These are arranged in order according to the feasibility of treating them by way of the probability calculus. It seems perfectly reasonable to treat (i) and (ii) as situations where the probability calculus can be employed. On the other hand, (v) does not qualify at all under the frequency interpretation unless we are prepared to make some major adjustments in our thinking (e.g. treat London as one out of an infinite number of Londons that could have been founded). We possess in geography a continuum of situations ranging from those in which we lose very little information in mapping the problem into the probability calculus, to those where so much information is lost that the exercise becomes meaningless. Somewhere between these two extremes there presumably lies a point where the disadvantage of losing information balances the advantages to be gained from rigorous manipulation. Precisely where that point lies is, however, a matter of opinion.

In the last decade or so geographers have realised that a frequency interpretation of the probability calculus had great potential in the study of geographic problems. The early work of Hägerstrand (1953) on the diffusion of innovations, in which he sought to model the process by a *Monte Carlo* simulation procedure (which involves repeated sampling from a probability distribution), is worth mentioning, since it brought to the attention of geographers the possibilities that lay before them. It is also worth while mentioning the outstanding work of Curry (1962A; 1962B; 1962C; 1964; 1967) and Dacey (1962; 1964A; 1966A; 1966B; 1967) in developing probabilistic models in geography. General reviews of this work may be found in Curry (1966A) and Harvey (1967A), while Berry and Marble (1968) and Garrison and Marble (1967) provide collections of papers and essays many of which explore the connection between probabilistic concepts and geographic phenomena. Leopold, Wolman and Miller (1964) and Leopold and Langbein (1962) also examine probabilistic models in the context of the physical development of landscape. Most of this work has been done in the period since 1950. Geographers have apparently at long last discovered that the general theory of stochastic processes (see, e.g. Bartlett, 1955; and Feller, *1957*) provides a whole series of convenient model formulations (e.g. markov processes, queueing processes, waiting-time processes, diffusion laws, and so on) which can readily be used, provided that geographic phenomena can be translated or mapped into a language system that forms the basis for all such models, viz. the calculus of probability.

It is not intended to review here the numerous applications of frequency probability to geographic problems. There is however one important consideration which is worth airing. This concerns whether the probability language functions as a model-language for describing the processes or for describing the data (or both). Suppose that a set of observations are collected and that these can then be ordered in an array such that $x_1 < x_2 < \ldots < x_n$. It may then be possible to define a function, $f(x)$, that gives an almost complete specification of the x_i values. It may be possible to find a probability density function that effectively summarises the data-set. Let us assume, for the moment, that the data-set turn out to be normally distributed and therefore their form may be described by the following distribution:

$$f(x) = \frac{1}{2\pi\sigma^2} e^{-(x-\mu)^2 2\sigma^2}$$

where μ and σ^2 are two parameters that need to be estimated from the data (the usual method is to equate μ with the mean of the data and σ^2 with its variance). This function then provides us with a convenient *model description* of the data-set. We can now use the function for manipulation instead of the raw data. It so happens that the normal distribution is an extraordinarily useful descriptive device for data-sets since many rules of non-deductive inference presuppose that the data-set has this particular form. The point about this particular probability distribution (and many of its closely related brethren such as the log-normal) is, however, that it acts *merely* as a descriptive device and there is no attempt to model any process which would naturally give rise to the distribution of the x_i values. Attempts have been made to give process-type interpretations to distributions such as the normal, the log-normal, and so on, but by and large these have not been very impressive. What is impressive, however, is that there are a large number of situations in which data-sets do turn out to have a normal or log-normal distribution (Aitchison and Brown, 1957), under conditions of random sampling. In particular, of course, the normal distribution provides the basis for a theory of measurement error which has proved applicable to many empirical problems.

It is possible, however, to define functions with respect to processes rather than to the data. We may thus guess at some process mechanism and, having specified that mechanism, generate a probability density function whose form reflects the process in question. Here we are using the mathematical derivation of the probability distribution as a model for the geographic process. There are a large number of probability distributions which can be used in this way. Undoubtedly the most important family is that derived from the Poisson distribution. This frequency distribution is, as Coleman (1964, 291) points out, peculiarly appropriate for modelling naturally occurring processes, since it deals with numbers of events occurring continuously over time or space. Such a distribution has extraordinary value in the study of spatial distributions and opens up the prospect of developing 'geometric probability' as a basic language for discussing geographic form. The large family of theoretical probability distributions derived from the simple Poisson distribution are of great interest. Distributions such as the negative binomial (considered above, pp. 165–7), the Neyman Type *A*, the Polya-Aeppli, the beta-Pascal, and the like, all have interesting geographical interpretations. Consider the following real-life situation envisaged by Neyman (1939, 36) in his derivation of the Neyman Type *A*:

> Larvae are hatched from eggs which are being laid in so-called 'masses'. After being hatched they begin to travel in search of food. Their movements are slow and therefore, whenever in a given plot we find a larva, this means the mass of eggs, from which it must have been hatched, must have been laid somewhere near,

and this in turn means that we are likely to find in the same plot some more larvae from the same litter.

Or consider the account given by Anscombe (1950, 366) of the process governing the Polya–Aeppli distribution:

> If progenitors (e.g. plant seeds) are released randomly over an area at one time and their progeny (freely increasing by vegetative reproduction) are observed at a later time, we shall expect the number of individuals per quadrat to follow the Polya–Aeppli distribution.

These kinds of processes are clearly relevant to geography (Harvey, 1966B). But it is also possible to hypothesise geographic processes and derive probability laws which represent such processes. Much of Dacey's work is concerned with doing precisely this. In an article entitled a 'Modified Poisson probability law for point pattern more regular than random' Dacey (1964A) developed a probability model suitable for examining a probabilistic version of the spatial pattern hypothesised in central-place theory. In another article Dacey (1964B) examines 'A family of density functions for Lösch's measurements on town distribution'. In yet another article Dacey (1966B) examines an historical process of settlement formation and derives the probability distribution appropriate to it. In a series of articles Curry (1962A; 1964; 1967) has also developed theoretical arguments based on the Poisson probability distribution as a basic model of temporal and spatial processes.

From the point of view of the construction of geographic theory there can be no doubt that the language of probability provides considerable opportunities, and within that language there can be no doubt that the study of Poisson processes provides one of the most appropriate formats for discussing geographic problems. This conclusion depends in part upon the theoretical development already achieved by Dacey and Curry in geography, but it is also reinforced by the generally acknowledged fact that it is relatively easy to map empirical problems into a probabilistic theory that rests on Poisson probability (Coleman, 1964; Haight, 1967).

This theoretical use of Poisson probability depends, of course, on the prior conceptualisation of geographic processes as essentially repetitive, recurrent, and independent. As we have seen, there are many situations in which such a conceptualisation is reasonable. In particular it has been shown quite regularly in the social sciences that aggregate human behaviour can be represented as-if it is of this form (see Harvey, 1967B for a review of this question). Having made the conceptual leap thus far, however, we can extend the manipulation of both theory and data in all kinds of interesting ways. It is of some interest to examine here some of the implications of the law of large numbers and the central-limit theorem for theory-development by way of the ergodic principle.

The ergodic principle is implied by the law of large numbers and the central-limit theorem, both of which are tacitly accepted when we map a geographic problem into the calculus of probability under a strict frequency interpretation. Suppose we make repeated measurements, x_i, on events, a_i, in a sample space, S. Lindley (1965, I, 157) then states:

the mean can be found by taking each sample point or elementary event, a, finding the value of x_1 there, and averaging over-all a, so obtaining μ: or one can take a single sample point and find x_1, x_2, x_3 . . ., for this a, and the average of these will be plim $\bar{x}$. Provided the x_i's obey certain conditions the two means will be equal. Ergodic theorems are concerned with conditions for this to be so and the law here is a very simple case. The average over the sample space is often called a *spatial mean*: that over $\{x_n\}$ is called a *temporal* mean since n can be thought of as a measure of time.

In a stationary stochastic process both spatial mean and temporal mean are equivalent, and other relationships may also be regarded as equivalent. Now at first sight Curry's (1967; and above, pp. 128–34) assumption of ergodicity, with all its interesting theoretical results, may appear unrealistic. But in fact we have already tacitly assumed that the process is stochastic by choosing to represent the problem in the calculus of probability, and the only additional constraint is an assumption of stationarity. The reasonableness of this assumption may be judged on its own merits. Put in this way, Curry's assumptions seem much more palatable, even though they do form strong constraints on our thinking about settlement processes. Another interesting case of this kind of analysis is provided by Dacey's toroidal mapping of the Iowa settlement pattern. Dacey (1966A, 562) admits that the geographer may regard 'the joining of opposite boundaries of the study region' as 'a first order geographic sin', but goes on to point out that 'if the underlying process is assumed to be stationary, then any two subregions may be treated as identical with respect to this process and the joining of opposite edges is legitimate.' In both cases the underlying process is hypothesised as being stationary and stochastic. Such an assumption is not unreasonable in certain situations although it has to be admitted that geographic series are often not stationary and in addition they often contain major discontinuities (e.g. as a result of political or physical barriers). But once it is accepted there is no reason why both theory-construction (in Curry's case) and data-collection (in Dacey's case) should not take full advantage of all its implications.

Perhaps the lesson we should learn from this is that we should be willing to explore the implications of our assumptions to 'the limit'.

Frequency probability provides a convenient model for representing data-sets and provides also a whole series of stationary and dynamic models appropriate for examining certain geographic problems. Discussion of these models may lead to new hypotheses and the hypothetico-deductive unification of hypotheses which previously remained isolated in our system of knowledge. It should be recognised that these hypotheses are statistical in form and that formulating geographic hypotheses in this language does involve making important assumptions. How far geographic events may be treated in such a framework is therefore a question that can be answered only by a careful evaluation of the balance between the benefits to be obtained and the amount of information lost in the process. All investigations involve making assumptions of some kind. It is clear that there are many circumstances in geography where the language of probability provides a very efficient mode of discussion.

(2) *Geographic phenomena and subjective probability*

There has been very little investigation of subjective probability in a geographic context. Given Anscombe's comment (above, p. 242) that the Bayesian approach flourishes best in a well-developed field of study, this absence of a Bayesian approach to geographic problems is perhaps not so surprising. For the most part, therefore, this section will be concerned with the potential of such an approach rather than with a survey of achievements.

The Bayesian interpretation of probability theory provides us with a normative decision theory for conditions of uncertainty. It also provides us, Savage (1954, 20) points out, with a 'handy empirical psychological theory', which we may apply to actual decision-making situations to gain considerable insight. If it is true that spatial patterns in human geography are the result of human decisions and that the complexity of interacting forces producing a human spatial pattern can effectively be summarised by an analysis of decision-acts, then such a normative framework for examining decision-making will be of great use (Harvey, 1967B). The broader aspects of decision-making under uncertainty, as developed in the theory of games and decisions (Luce and Raiffa, 1957) have increasingly filtered into the geographical literature. But the explicit development of Bayesian probability theory has tended to lag behind. By far the most interesting formulation is that by Curry (1966B) who has pointed out that 'the statistics appropriate to decision-making in the resources field are Bayesian.' Curry develops an example of a farmer seeking to maximise his output in the face of an uncertain environment. In this case an objective prior distribution can be obtained by an examination of the climatic record or, less efficiently, from intuitive assessment of previous experience. The farmer may then assign utilities to, say, rainfall occurring at different times of year. These utilities will vary from farmer to farmer. The general question which Curry then considers is how, given these prior probabilities and terminal utilities, the farmer ought to change his production system as new evidence accumulates. Curry is particularly concerned with weather-modification schemes, but it is possible to generalise his argument to include natural environmental changes. A Bayesian analysis of this kind of problem helps us to understand the logic of the situation, and also provides a prescriptive theory of rational decision-making in the face of uncertain and changing circumstances. The normative treatment of problems in resource-use under uncertainty, such as those examined by Kates (1962) and Saarinen (1966), could undoubtedly make use of Bayesian probability. In addition, such an approach provides us with an idealised description of the farmer's decision-making process, and although it may be doubted if farmers actually proceed in such a manner, it may be that Bayesian models will provide us with a generalised decision-model which has empirical as well as normative utility.

It might also be useful to use the Bayesian analytic framework to examine our own decision problems. As geographers we develop a 'degree of belief' in a particular hypothesis, and we also tend to develop subjective terminal utilities regarding the value of making a mistake. New information is also accruing to us all the time. How, then, should *we* proceed to change our views and beliefs in a rational and efficient manner? If we are seeking to apply Bayesian forms of analysis to others, then why not analyse our own decision-making in the same way?

Perhaps a classic case where such an analysis would be appropriate is in regional decision-making. When faced with the problem of dividing up some area into a system of regions, different geographers usually identify different regions. These different regional divisions—well portrayed by Sinnhuber's study of the *Mitteleuropa* concept (Figure 15.4) and Lewis's (1966) study of changing concepts of the Great Plains region in the U.S.A.—are, presumably, the end-result of a whole chain of evaluative decisions. We see the end-product of this process and tend to discard the information about the way in which that decision was reached. Given that we only see the end-product, it is also very difficult to reconcile different views. If, however, we cast this decision procedure in a Bayesian framework we

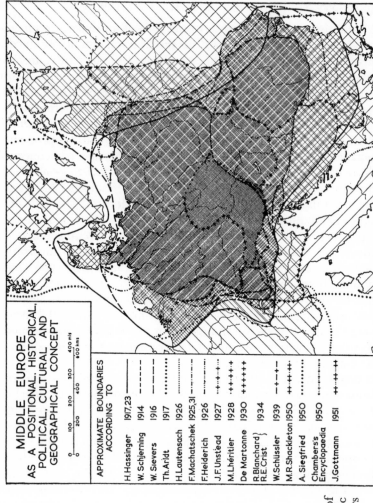

Fig. 15.4 The delimitation of Middle Europe as a geographic region—sixteen different views (*from Sinnhuber*, 1954).

might be able to derive a more consistent decision-making procedure and, at the same time, be able to reconcile divergent views more easily.

The first point to be made about it is that a particular system of regional divisions exists with only a certain degree of probability, given the evidence available. Choynowski (1959) has shown how a map-pattern containing information on percentages may be converted to a map-pattern based on probabilities, and this form of analysis may be used to provide us with a set of prior probabilities for data-collection units (preferably grid squares or something similar) which show how they are similar or different from one another, or significantly different from the average (Figure 15.5). New variables may be introduced and the new information may then be used to convert these prior probabilities into posterior odds. In this manner the most probable regional division, given the variables examined and the data available, may be devised. This procedure may well model the intuitive procedures used by geographers to reach decisions regarding regional division, but, more important, it may also provide us with a prescriptive model which will help to standardise geographic working on a topic that has great affinities with classification procedures. It may also be possible to develop a behavioural approach to regional decision-making. Iwanicka–Lyra (1967) provides a neat example of this in which the opinions of various 'experts' are used and reconciled to provide a definition of the boundaries of the Warsaw urban region. Such a procedure again appears to be in a field which could be approached by way of Bayesian statistics. Lastly, it is worth noting that we live in a society where changes occur fairly rapidly and, as location patterns change, so we must expect regional divisions to change. It always seems somewhat contradictory to speak of 'growth poles', 'spread effects', 'urban sprawl', and the like, and at the same time seek to establish regional divisions which will last for evermore. Again, it may be possible to develop a dynamic approach to regionalisation by using Bayesian analysis as a method for monitoring changes in evidence that must surely have an impact on what we regard as the most probable regional division.

This section has been largely speculative, since the Bayesian approach to decision-making has not been well developed in geography. But there appears to be considerable potential for such an approach, both as a means of analysing inherently geographic problems and as a means for understanding our own decision processes.

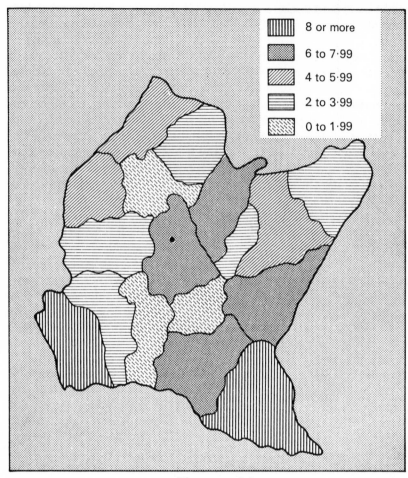

Fig. 15.5 (A)

Fig. 15.5 Maps based on probabilities. The occurrence of brain tumors in administrative units in Poland showing: **A:** Raw data expressed as a ratio per 10,000 head of population.

C Probabilistic inference in geography

Geographers have long used non-deductive procedures in the course of research, but only recently have formal non-deductive languages been used by geographers to provide a rational method for making non-deductive inferences. At the present time *statistical inference* provides the major method employed by geographers, and we shall therefore confine attention to this one aspect of probabilistic inference in geography.

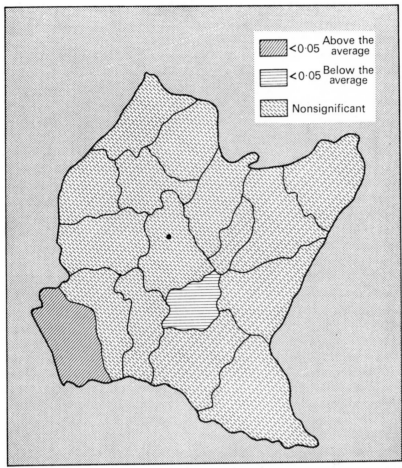

Fig. 15.5 (B)

B: Adjusted data to express *significant* spatial variation, given a probability model of significance (*from Choynowski, 1959*).

In 1956 Reynolds (1956, 129–32) pointed out that formal procedures of statistical inference could be employed in geographical research, but in doing so made a number of reservations that provoked a reaction from Garrison (1956). Two statements challenged by Garrison were that 'only a limited number of the phenomena that have a geographic interest are distributed in sufficient profusion to lend themselves to sampling', and that 'it is not surprising that at least one geographer has called for "new statistical systems of basic theoretical design to match and solve our own specific problems".' Garrison took particular exception to the last statement on the grounds that the 'logical methods of science are universal.' Since these warning shots were fired, geographers appear to have maintained a conspiracy of silence regarding the thorny problems attendant upon using formal procedures of statistical inference in geography. Such criticism as there has been has usually been an emotive reaction against the use of *any* quantitative techniques, rather than a balanced attempt to examine some of the real difficulties of statistical inference. We shall therefore attempt some examination of this problem here.

The application of formal non-deductive languages to inference in geography involves making assumptions. Some of these are general and stem from the need to map geographic information into the basic calculus of, say, the Neyman–Pearson theory, and some are specific to particular types or even individual versions of inferential tests. The geographer is therefore faced with choosing between a number of different basic languages, and, within each language, choosing 'the best' set of inferential rules out of a whole battery which are available. It is by no means easy to make such a choice, and to do so requires quite sophisticated understanding on the part of the geographer. Further, it is clear that more and more tests are being developed, many of them with particular kinds of circumstance in mind. An analysis of his inferential problem may lead a geographer to conclude that although appropriate tests do exist, none of these is very effective; hence a demand for specialised tests which deal specifically with geographic problems. But before such a demand can be made, a good deal of prior analysis is required. In general the appropriateness of a particular inferential procedure in a given situation depends on (i) whether the substantive geographical problem can be put in a format suitable for that procedure, and (ii) whether the assumptions necessary to that procedure are fulfilled. For the most part the techniques of statistical inference used in geography are those derived from the 'orthodox' frequency-minded statisticians. We shall therefore concentrate on the problems of applying conventional statistical techniques in geography. A number of steps are involved and we shall consider these separately.

(a) *Specifying geographical hypotheses* in a manner suitable for formal non-deductive inference involves translating the substantive geographic statement into an assumption about the frequency-function of some variable. It is required that the statistical hypothesis be expressed in the basic language of probability theory. Substantive geographical hypotheses may, however, be stated in a variety of languages. Many of the hypotheses derived from location theory, for example, are stated as deterministic propositions. These cannot, as we have already seen (above, pp. 136-40) be given a direct test. Thus Garrison (1957) converts the deterministic hypotheses of agricultural location theory, as developed from von Thünen, into statistical hypotheses in order to use regression procedures. Dacey (1966A) similarly converts a deterministic central-place model into a probabilistic model. We need to be sure that the translation of an initial statement into a probability statement is both reasonable and as accurate as possible (above, pp. 263-5).

(b) *Specifying the geographical population* is important in two respects. Such a specification, developed with respect to a particular hypothesis set, defines the *domain* of the hypothesis. What we are assuming, when we specify a population for testing a given hypothesis, is that the population is governed by the hypothesis. Three possible situations can occur:

(i) The hypothesis governs the population completely.
(ii) The hypothesis does not govern the population.
(iii) The hypothesis only partly governs, or governs only a part of, the specified population.

Now it is very difficult to specify the domain of a hypothesis in advance (above, pp. 91-6). The third situation is thus probably the most characteristic of scientific investigation, especially in the early stages, and it is the one that poses the greatest conceptual difficulty. What we are doing, therefore, is translating a substantive geographical hypothesis into a hypothesis (expressed as a frequency-function) about a geographical population. The second important point about geographical populations is, of course, that they provide the basis for sampling and the sample-population inference is one important category that comes under non-deductive inference.

Now it is extremely important that we should be clear as to what is meant by specifying a geographical population. To the statistician the population merely consists of abstract sets. But to the geographer the population 'comprises a class of objects, events, or numbers that are of direct interest' (Krumbein and Graybill, 1965, 61). Specification of the geographical population and of the measurements that can be made on it, is essentially the geographer's concern; here there is a clear need for unambiguous concept-formation in geography as an essential prerequisite to formal mathematical analysis (see above, pp. 184-90). Yet geographers have spent little time analysing the nature and mode of definition of the populations they deal with. Without such an analysis the application of formal non-deductive languages may be meaningless. The following account derives a great deal from Krumbein and Graybill (1965, chapter 4) and Duncan *et al.* (1961), who provide treatments of this problem of great utility to the geographer.

The definition of the population is a conceptual problem. We may choose to treat all farms in the world as the population. All farms in East Anglia form a sub-population of that population but for a given investigation we may choose to call all East Anglian farms the population. The definition of the population depends on the nature of the study. Krumbein and Graybill then state:

> Specification of this population involves at least three considerations: (1) a clear expression of what constitutes an individual or element in the population of objects or events, (2) specification of the kinds of numerical measurements to be made, and (3) specification of the limits of the population.

Thus the specification of the geographical population requires the prior definition of the *geographical individual* and this is a notion that geographers have been notoriously hazy about (above, pp. 216–7). In general we may divide geographical populations into two types based on (i) location individuals and (ii) property individuals. In certain cases we may treat (i) as an aggregated version of (ii), but in doing so we may involve ourselves in complicated inferential problems. A typical example is given in Krumbein and Graybill (1965, 76) where they suggest that the population of objects making up a foreshore may be defined as (i) cartons of grains of sand or (ii) individual sand grains. In sociology the same problem emerges in the form of (i) unit data versus (ii) data about individuals. Duncan *et al.* (1961, 43) examine in detail some of the consequences of specifying populations in these different ways and point out that 'one obtains a set of data for areal units, or unit data, only some of whose properties resemble those of the data on individual items in the population'. Consequently inferences made from areal data are inferences about similar populations of areas (similar in size, composition, etc.) and cannot be construed as inferences about individuals. The classic case of this is, of course, the difference between *ecological* and *individual* correlation. Duncan *et al.* (1961, 9) write:

> Can inferences about relationships drawn from areal data be applied to units other than areas? For example, if a statistician computes the proportion of the population which is of foreign birth in each census tract of a city and correlates it with the census tract proportion of home owners, finding a positive relationship, can he conclude that the foreign born are more likely than the natives to own their own homes? Much early work . . . appeared to assume that such inferences were justified, although no mathematical rationale for them was available. It was pointed out by W. S. Robinson in 1950 that they are not mathematically justified, and that individual relationships inferred from areal correlations may be seriously biased as to magnitude and even erroneous as to sign.

This particular point is of considerable interest since most use of correlation procedures in geography is based on areal data and draws inferences about individuals; and although Goodman (1959) has since shown that such inferences are justified under certain conditions, very little of this work actually takes account of this problem. In part the problem depends on whether the population is conceived of as being made up of aggregate elements (such as counties or cartons of sand) or of individual elements (such as farms and pebbles) and in part it depends upon whether locations or events are being referred to. Confusion on these issues underlies a good deal of the confusion over inferential procedures in geography. Inferences made about one population cannot, without assumptions, be extended to other populations and if inferential procedures are to be used then we must also assume that we possess some homogenous criteria for identifying the elements or at least some adequate weighting criteria to get round inhomogeneity (e.g. weighting data in correlation analysis as in Thomas and Anderson, 1965; also see chapter 19).

Assuming that an adequate definition of the *object* population can be stated, it is then required that we specify the measurements to be made on that population. If the population is specified as all farms in the world, several attributes of each farm, e.g., income, size, intensity of production, and so on, may be measured. Some of the methodological problems involved in measurement will be considered in chapter 17, but for the moment we need only record that the measurement process effectively translates a population of *objects* into a population of *numbers* about (or attributable to) substantive objects. It is these populations of numbers which are treated in statistical inference. In certain cases the population of objects will consist of only one member, but repeated measurements on that member may generate a population of numbers of considerable size —e.g. repeated measurements of the same base line in surveying and triangulation. The population is conceptualised in this last case in order to reduce measurement error to acceptable levels.

It is also important to specify the limits on the population. If the population is limited to all farms in East Anglia over half an acre in size, then our conclusions can relate only to that object-population and the measurements made upon it. Any extension of the conclusion beyond the specified population is strictly not warranted. In general, therefore, it may be concluded that the more the trouble taken over specifying the population, the less the problems will be when it comes to understanding the inferences drawn.

(c) *Specifying the sampling procedure* also forms a vital part of the procedure for making non-deductive inferences. The sampling problem will be dealt with in detail later (chapter 19) but it is important to note that statistical inference is essentially concerned with making inferences from samples; and if these samples are to be used in relation to substantive hypothesis, they must be collected in such a manner that they are representative of the population of objects and numbers being examined and relevant to the particular hypothesis being examined.

(d) *Specifying the appropriate test procedure* is one of the most difficult operations in non-deductive inference. Statistical theory provides us with a battery of models, each with specific assumptions, and the task we are required to undertake is to select one out of a whole battery of available tests on the basis of the fact that it is most compatible with the procedures involved in (i), (ii), and (iii), above. Most conventional statistical tests specify measurement made at least on the interval scale and specify also that the population of numbers obtained should be normally distributed (thus the sample distribution should be normal within the limits of sampling error) and that the sample size should be reasonably large (conventionally greater than 30). If these conditions are satisfied, then a whole range of so-called parametric tests may be employed. These are characteristically the tests presented by Fisher (*1936*; 1956; *1966*) and developed also in the Neyman–Pearson theory. It is clear that the use of such tests involves strong assumptions. Consider, for example, the assumptions underlying a 't' test:

(i)   Each observation must be independent of any other observation (i.e. obtaining any one observation must not prejudice the probability of obtaining any other observation).
(ii) The observations must be normally distributed.
(iii) The populations must possess the same variance (this is the condition of homoscedasticity).
(iv) The observations must be measured on the interval or ratio scale.

This particular test is very powerful (in the statistical sense of power), but the strong assumptions involved can rarely be fulfilled in geography. In some cases we shall be able to show that the necessary conditions are met in the data, but in other cases we are forced to assume their existence. Fortunately, the 't' test has proved to be fairly robust, working even when all the conditions are not strictly met. Nevertheless, we can conclude with Siegel (1956, 19) that:

All decisions arrived at by the use of any statistical test must carry with them this qualification: 'If the model used was correct, and if the measurement required was satisfied, then. . . .'

It is obvious that the fewer or weaker are the assumptions that define a particular model, the less qualifying we need to do about our decision arrived at by the statistical test associated with that model. That is, the fewer or weaker are the assumptions, the more general are the conclusions.

Thus new types of test are being evolved which, although they may be less powerful, are probably more applicable to geographical situations. These tests are usually called non-parametric and distribution-free, although as Galtung (1967, 341) points out, these terms are misnomers in the sense that the first really means 'non-interval-scale parametric', while the second really means 'non-normally distributed'. The greater applicability of these tests by virtue of their weaker assumptions is sometimes bought, however, by a general decline in power. Here the geographer needs to be a rational decision-maker, since he needs to select the most applicable and the most powerful test and these objects are not always compatible. In selecting a test, therefore, the geographer has to solve a particular case of the general problem that dogs the use of any mathematical calculus in empirical work, viz., to select the model that does least violence to the empirical situation, and at the same time is simple to handle and achieves the objectives that he has in mind. This is an evaluative procedure in which we have a real need of guidance from the statistician. But statisticians have not always been as helpful as they might. As the example of the 't' test demonstrates only too well, the theoretical assumptions necessary to derive a test are not always *relevant* assumptions for the application of that test to a particular data set. Statisticians have not always been willing to consider this and they have not therefore provided adequate 'rules of behaviour' for the practical application of the tests they derive. Tukey (1962) has suggested that this will become a major area of statistical analysis in the future. If so, the geographer is bound to benefit.

(e) *The drawing of inferences* is a fairly automatic procedure once a particular test is selected, since the rules are fairly strictly laid down. There are, of course, a number of evaluative decisions that can occur within the testing procedure, the most important being the significance level which is used, or the establishment of the critical region for the test. But in drawing inferences it is important to think back upon the general procedures so far followed, for these tell us precisely what our inferences are about. The formal inference procedure simply examines a particular *statistical* hypothesis in the light of a set of *numbers* drawn from a population by way of a specified sampling procedure. The inferences thus refer to the population of *numbers* and not to the population of objects. We may therefore say that the level of income on farms in East Anglia is significantly different from the level of income on farms in Wales, but we cannot go on to assert that the farms are significantly different. This latter inference is one made by extension from the formal inference. There is thus a distinction to be drawn between what Galtung (1967, 341) calls inferences about the *substantive* hypothesis (which is about geographic reality) and inferences about the *generalisation* hypothesis (which is about the data). Confusion between these hypotheses amounts to a confusion between the *statistical* significance of a test and its *theoretical* significance. As we have already seen (above, p. 256) there are many situations in which the rejection of a null hypothesis provides only very weak support for the alternative hypothesis. Even evidence of quite strong statistical support for the alternative hypothesis does not remove the need for a theoretical interpretation. Yet geographers frequently operate as if rejection of the null hypothesis provides automatic confirmation for some theory. The plain fact is that it does not. At this point, therefore, we have to face the difficulty of a *reverse mapping* from statistical (mathematical) conclusion to substantive geographical conclusions. This part of the inferential problem lies outside the formal non-deductive language.

One further aspect of non-deductive inference in geography may be dealt with here. It has been pointed out that formal inferences refer to sampling conducted with reference to a fully specified population. It should be clear that inferences drawn refer *only* to that population and they cannot be extended, without assumptions, to any other population. There are some difficult inferential problems associated with this condition, and these are particularly important in geography. These difficulties have been dealt with extensively by Galtung (1967, 361–70), and what follows is based on his account.

The basic proposition that Galtung begins with is that statistical tests are out of order if we do not possess a sample. Now there are a number of situations in geography where it is difficult to meet this condition without some fairly radical assumptions. A number of possible cases can be identified.

Consider, for example, the situation where we possess data on *per capita* incomes in the 50 states of the U.S.A. These states form the total population and the measure of *per capita* incomes provides us with a population of numbers about states. It would appear that techniques of statistical inference are not usable here without further assumptions. It may be appropriate to assume that in each state there is a measurement error and that the data we actually have are one sample out of an infinite number of possible measures which we could have made. Thus we are conceptualising the data as if they were a sample drawn from some hypothetical universe. The condition for using techniques of statistical inference here, therefore, is that the data 'may be conceived of as a sample generated from a universe, by a more or less specified model'.

A rather different situation emerges when we attempt to generalise results over space and time from populations which have specific time or space locations. Consider, for example, Dacey's tests on probabilistic central-place models which use the settlement pattern of Iowa (in most cases) as the sample data-set. In this case the true population is made up of Iowa towns and Dacey uses the 99 largest. Suppose there were only 99 towns in Iowa, then these 99 make up the total population. Therefore formal procedures of statistical inference are not justified without some further assumptions. One such assumption might be that Iowa is somehow representative of all town systems everywhere. Iowa thus forms one individual out of a collection of areas which, presumably, cover the whole world. If we seek to extend the population in such a manner, however, we are effectively generalising from a sub-population of towns found in one unit to all towns everywhere. The last part of this inference involves generalising from a sample of one, which, under any circumstances, is regarded as unforgivable. The same criticism must be made of samples collected at particular points in time, while samples collected in restricted areas of space and restricted areas of time pose a double conceptual difficulty.

One other answer to this is to construct a hypothetical universe. If, therefore, the sample is the same as the population, we may assume that a population exists and call it a hypothetical universe. In the case of Iowa we might hypothesise that the 99 largest towns now existing are a sample from 99 towns that could have developed in almost an infinite number of different ways in any spatial or temporal context. In some respects this is similar to the measurement-error argument, but it is more general and has intuitive appeal in that it allows us to assert the test of 'universal' hypotheses on the basis of very slender data. Such an assumption has enormous implications. Galtung (1967, 367–8) writes:

Since an 'hypothetical universe' can always be constructed in the mind of a researcher with a minimum of imagination this means that all data, in principle, should be subject to statistical testing.

It is also possible to construct all kinds of imaginable worlds and to regard our sample as only one unit from this higher-level population.

As a result we shall find it impossible to test hypotheses, since we have only one sample unit (e.g. the world) out of a population of imaginable worlds. This difficulty leads Galtung to conclude

that it is necessary to fix an upper limit and say that 'this is the level at which I accept my data' and then see to it that it really is possible to generalize to that level from the data (and that the samples do not cluster in space or time . . .).

In the same way that 'unlike other sciences geography can find great use for principles that are applicable to relatively small portions of the earth's surface' (above, p. 110), so geography is forced to make use of highly localised populations and highly localised samples which, in many cases, are precisely the same as the population itself. In these circumstances a number of different interpretations may be put upon statistical inference in geography. Consider the following five interpretations that might be put upon a statistical study of the spacing of the 99 largest towns in Iowa:

(i)   The population is the set of all Iowan towns and the 99 largest towns are a sample. Inferences are thus made from this sample (which is biased and almost comprises the population anyhow) to the population.

(ii)   The population is a set of all towns which are found in similar environments to that of Iowa. In which case the sample consists of the sub-population of Iowan towns and the assumption is that Iowa is somehow representative of situations to be found in other countries or states.

(iii)   The population is the set of all possible configurations of Iowan towns. In this case the 99 largest towns provide us with a sample of the infinite number of ways in which the towns could have been arranged.

(iv)   The population is an imagined hypothetical set of all towns everywhere and Iowa provides a reasonable sample to represent this hypothetical universe of towns.

(v)   The population is the set of all measures which could be made on the distances between settlements, given a specified law governing the distribution but a disturbance due to measurement error.

This last appears to be the construction that Dacey (1966A) puts on his own studies of the pattern of town spacing in Iowa. He assumes the pattern is generated by the central-place model, but that the actual locations deviate from theoretical optimality by some error term which may be conceived of as a kind of measurement error. Given the central-place model, the total number of towns, and the total area of Iowa, then the theoretical distance from a settlement to any other settlement can be calculated. The actual measures made on distances between Iowan towns then amount to a repeated measure made on this distance with a measurement-error term involved. The actual measures should be approximately normally distributed around the theoretical distance value if the model is correct. We have already noted Dacey's conclusions regarding the adequacy of the model using a test of this form (above, p. 138).

It is possible to construct further variants on the interpretations given above, but the main point is to demonstrate how confusions might arise if the population is not specified in advance, and if the

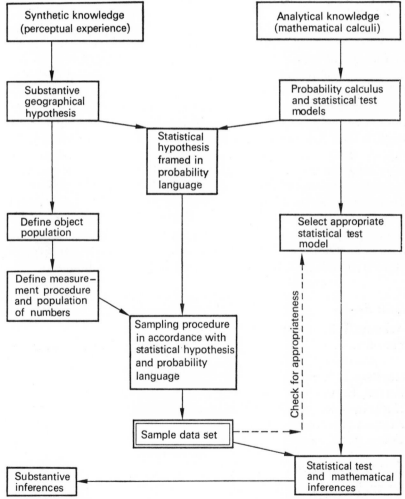

Fig. 15.6 The steps involved in applying statistical inference in geography.

statistical inference (for which rigid rules apply) is not kept apart from the substantive inference.

In general, the problems of statistical inference (summarised in Figure 15.6) in geography can effectively be resolved only by a careful evaluation of the test procedures available for a given set of circumstances, together with a full specification of the assumptions made in collecting and manipulating data. The various non-deductive languages available and the variety of test procedures available are, of course, internally consistent and may rigorously be applied. But the prior mapping of a substantive problem into the calculus appropriate for testing and the reverse mapping of the mathematical conclusions into substantive conclusions can be conducted only if assumptions are made. Here, as in most other situations, errors and difficulties may arise. But the major mistake comes not so much from making assumptions as from ignoring their implications. Certainly it is time that the conspiracy of silence that has reigned over inferential procedures in geography was broken. An explicit debate on the fundamental issues involved in inference in geography can do nothing but good. Such a debate will doubtless reveal the sense in which Reynolds' original reservations need to be heeded and the sense in which we may truly believe, as Garrison (1956, 429) suggests we should, that 'the logical methods of science are universal'. The *logic* may be the same (and it is a fundamental notion in this book that it is necessarily so), but the problems of applying that logic vary radically from situation to situation.

Basic Reading

Ackerman, R. (1966).
Fishburn, P. C. (1964), chapter 4 contains an excellent summary of probability theory.
Hacking, I. (1965).
Parzen, E. (1960).
Plackett, R. L. (1966).
Rescher, N. (1964).
Tukey, J. W. (1962).

Reading in Geography

Curry, L. (1964).
Curry, L. (1966A).
Curry, L. (1966B).

Dacey, M. F. (1964A).
Dacey, M. F. (1966A).
Leopold, L. B., Wolman, M. G., and Miller, J. F. (1964).
Lukerman, F. (1965).

PART FIVE
Models for Description in Geography

Chapter 16
Observation

The emphasis in this book has, so far, been heavily biased towards the *analytic* and *a priori* aspects of geographic understanding. We have thus dwelt upon the complexities of axiomatic systems, theoretical and model constructs, mathematical languages, and the like. Such an approach has not, probably, been to the taste of many geographers accustomed to the more traditional mode of approach to unravelling the mysteries that surround them. Indeed it may horrify those who seek understanding by poring over maps, by wandering, thoughtful and observant, down streets and over fields, by perusing records buried in dusty archives, by rummaging among old newspaper reports, by digging soil-pits and diving deep into limestone caverns, by watching and waiting, by sharpening their own perceptual experience, by training themselves to appreciate. Carl Sauer (*1963*, 400) once wrote of the joys of field work:

> Underlying what I am trying to say is the conviction that geography is first of all knowledge gained by observation, that one orders by reflection and reinspection the things one has been looking at, and that from what one has experienced by intimate sight come comparison and synthesis. . . . The important thing is . . . to recognize kind and variation, position and extent, presence and absence, function and derivation; in short to cultivate the sense of morphology.

To Sauer, the education of a geographer was essentially an education in how to experience; an education that relied upon interaction between student and teacher,

> engaging in a peripatetic form of Socratic dialogue about qualities of and in the landscape. Locomotion should be slow, the slower the better, and be often interrupted by leisurely halts to sit on vantage points and stop at question marks. Being afoot, sleeping out, sitting about camp in the evening, seeing the land in all its seasons are proper ways to intensify the experience, of developing impression into larger appreciation and judgement.

Such an approach to geographical understanding is deeply embedded in the geographical tradition. 'We must ask the earth itself for its laws,' wrote Ritter (Hartshorne, 1939, 55)—a precept that has governed geographical investigation in the minds of a long line of distinguished geographers. The classic work of the early geologists/ geomorphologists such as Gilbert and Powell in the American Far

West (Chorley, *et al.*, 1964) the keen observation of the French regional geographers, the open admiration of S. W. Wooldridge (1956) for the 'natural-history' approach, the aesthetic approach to landscape form of Lowenthal and Prince (1964), are but a few examples of the strength of this tradition. It may seem a far cry from this tradition to the harsh logical world of syntactical form and mathematical system.

But there is another tradition. One that rests less upon perceptual experience and more upon man's own imagination. John K. Wright (1966, 88) thus speaks of the various *terrae incognitae* left to the geographer to explore and concludes that 'perhaps, the most fascinating *terrae incognitae* of all are those that lie within the minds and hearts of men.' Lowenthal (1961) uses this quotation as a starting point for a brilliant exposition of the interaction between 'geography, experience, and imagination'. Lowenthal (1961, 260) himself concludes:

Every image and idea about the world is compounded, then, of personal experience, learning, imagination, and memory. The places that we live in, those we visit and travel through, the worlds we read about and see in works of art, and the realms of imagination and fantasy each contribute to our images of nature and man. All types of experience, from those most closely linked with our everyday world to those which seem furthest removed, come together to make up our individual picture of reality. The surface of the earth is shaped for each person by refraction through cultural and personal lenses of custom and fancy.

In part this world of image is the legitimate concern of the geographer, for it is itself a most significant variable determining the pattern and form of human occupancy. In part, also, this world of the image is the geographer's world, for, being human, the geographer himself develops images and concepts which are used to interpret and schematise the complexities of the real world that surrounds him. At this point we enter the world of the *a priori*, the world that is not susceptible of empirical verification except by indirect test, the world of hypothesis and artificial construct, the world in which it is difficult to differentiate between science fiction and science theory. Here, we enter the world of theory, the world in which imagination and invention reign supreme. We are back, therefore, at the point where theory begins, the point at which the free creativity of the human mind can be employed to its greatest effect. Science does not form the antithesis of art by denying the imagination. It merely seeks to control the imagination, to differentiate between mere science fiction and adequate scientific theory. Discovery, as almost every historian of science has pointed out, is something personal, an act that can be explained, if at all, only by the perceptive psychologist. It has little or nothing to do with the scientific method and with logic as such (Hanson, 1965).

The geographer can hope to handle the complex world with which he deals only by formalising shifting personal images into hard concepts. By developing concepts with generally agreed meanings the geographer may communicate and enter into discourse. 'Without basic concurrence as to the nature of things,' writes Lowenthal (1961, 242), 'there would be neither science nor common sense, agreement nor argument.' Some images are private and incommunicable, some become hardened into concepts, are written down as words with connotative meanings attached to them. But the meaning of words can shift and change. As Eliot has it:

> Words strain,
> Crack and sometimes break, under the burden,
> Under the tension, slip, slide, perish,
> Decay with imprecision, will not stay in place,
> Will not stay still . . .

Agreed meanings suddenly become contentious and, after a period of conflict, new agreed meanings emerge. Terms such as 'environment', 'region', and 'location', have interesting histories in geography. They illustrate how ideas, definitions, meanings, and, presumably, images, change as a subject expands and develops to embrace new worlds of perceptual experience. Glacken (1967) has thus traced the changing attitudes of man to his environment from the classical period to the end of the eighteenth century, showing how concepts change, how the consensus of opinion shifts, among those with a geographer's bent. Yet geographical understanding cannot develop without adequate concepts. Somehow these concepts have to be given a firm meaning, have to be made to stay still, however temporarily, for the purposes of communication and analysis. Substantive hypotheses can be developed only if adequate concepts are available—concepts which, by consensus, are given a well-defined meaning. Here, from the point of view of scientific investigation, we can proceed only in an *a priori* manner, build, in fact, *a priori* models of what the world is about and how the geographical reality we are seeking to understand is shaped and how it functions. *A priori* models are merely formalised images of the world, images which, with analytic tools, we have fashioned to be coherent and consistent.

The plain fact is that there can be no geographic understanding without geographic concepts and there can be no concepts without images. The image, as Boulding (1956) points out, is central to our construal of the nature of all understanding and scientific knowledge is no exception. As John K. Wright (1966, 75) has it, 'much of the world's accumulated wisdom has thus been acquired, not from rigorous application of scientific research, but through the skilful intuitive imagining—or insight—of philosophers, prophets, statesmen, artists, and scientists.' The analytic systems—the axiomatically developed theory, the symbolic system—which we have so far emphasised, act as 'juice-extractors' from postulate and concept statement (above, pp. 181–4). But the amount of juice that flows from such a procedure depends ultimately upon the inherent fruitfulness of the concepts and postulates set up.

It is difficult to write on, or give instruction in, concept-formation in geography. There is no set path by which new concepts can automatically be discovered, any more than it is possible to produce a great novel from a short course in creative writing. Some geographers are perceptive and imaginative, are extraordinarily sensitive to the physical, social, and intellectual environment that surrounds them, and have little difficulty in applying a fertile imagination to geographical problems. Some learn by hard graft how to think and perceive adequately in a limited environment. Some will just never learn. Imaginative geographers are probably born not made. It is difficult to legislate positively for the imagination although it is certainly possible to nurture it by imaginative teaching. Perhaps all that can be said is that a policy that ensures a broad range of possible perceptual experience is best—but even though you may take the horse to water you cannot make him drink. On the other hand it is possible to legislate negatively by stressing the danger of being mesmerised by a particular conceptual representation. In this way we may prevent ourselves from truncating our own experience, from running unobservant on deeply worn tracks, of becoming, in short, addicts 'of a particular way of perceiving' (above, p. 163). Ideas change, words change their meaning, new problems emerge, old ones fade away, and hence every position we take up is a temporary one. There is little or no sense defending temporary positions under the banner of orthodoxy. Perhaps the one major methodological lesson we have to learn from our own recent history, is that it does not pay to commit one's resources entirely to the defence of some rigidly defined position. Given the current rate of change in social and natural science, the best strategy appears to be one of maximum flexibility. And here the model concept, properly understood, provides us with a flexible approach to firm, if temporary, understanding of the geographic phenomena around us. Thus the formal representation of *a priori* images by way of *a priori* models perhaps provides the key to a vigorous and flexible geographic methodology.

At some stage, therefore, the geographer needs to come to grips with his own imagination, has to wrestle with the incoherent and partial images that come to him, has to learn to structure them. Here, Sauer's dictum that 'locomotion should be slow' can be applied to the geographer's perambulations in his own imagination, turning over fantasies, searching for that flash of inspiration, that new conceptual representation, which somehow helps to clarify what was previously obscure. The deep leather arm-chair and the decanter provide at least one strategy. In their anxiety to avoid an arm-chair image, geographers have on occasion taken to naïve empiricism. Prior thought and analysis can save a lot of time and trouble in the field. It is convenient, as Chamberlin (1897) pointed out long ago, to enter the field with a set of multiple working hypotheses. These hypotheses are artificial constructs, developed in the imagination in the light of what information is already available and in the light of what we suspect is reasonable.

But at some stage arm-chair philosophising must be confronted with experience. Hypotheses must be tested and evaluated with respect to perceptual experience. In this the geographer relies upon adequate modes of description. Description can be impressionistic. Sauer (*1963*, 403) writes:

> Beyond all that can be communicated by instruction and mastered by techniques lies a realm of individual perception and interpretation, the art of geography. Really good regional geography is finely representational art, and creative art is not circumscribed by pattern or method.

Coleridge's famous lines:

> In Xanadu did Kubla Khan
> A stately pleasure-dome decree:
> Where Alph, the sacred river, ran
> Through caverns measureless to man
> Down to a sunless sea . . .

turn out, according to Wright (1966, 119–23), to be an imaginative re-creation of the scene that greeted a Philadelphia botanist, William Bartram, when, in 1774, he visited Florida and came across the Blue Sink (otherwise known as Alligator Hole), the Manatee Spring, and Salt Springs Run. Nobody can doubt the quality of the poetry, the quality of the art, but it barely passes as good geography. This is not to say that we must necessarily give way to what Wright (1966, 76) calls 'a deep-seated distrust of our artistic and poetic impulses'. It may be, as Joseph Wimmer is said to have remarked, that 'the descriptive geographer is nothing other than a landscape painter and map drawer in words' (Darby, 1962, 4), but somewhere on the continuum from a dull prosaic list of 'what is there' to complete and unbridled poetic licence, we must draw a line between what we regard as good *geographic* description and what we regard as *geographically* inappropriate. Opinion has fluctuated as to where this line should be drawn. Some, deeply metaphysical and aesthetic, have sought to convey something of the feeling they experience. Others, obsessed with the notion of scientific objectivity, have sought to exclude all except what can be seen. Thus the controversy over the concept of landscape—or *landschaft*—contains a wide range of views extending from those who seek to convey the 'soul'—*Seele*—of a landscape by striking phrases, images, and metaphors, and those, on the other hand, who, being 'purists', would restrict discussion entirely to an 'objective' account of those objects in the landscape which can be seen (Hartshorne, 1939, 149–74). The writing of regional geography has similarly varied from the plain enumeration of fact upon fact, to a kind of *gestalt* mystical reincarnation of perceptual experience through the written word. The question of where we draw the line is partly a question of balance and partly a question of avoiding misrepresentation. In much of the literature, of course, this general problem relates to a discussion, of little worth but of great emotive significance, over whether geography be an art or a science. Darby (1962, 6) suggests a balance:

> Geography is a science in the sense that what facts we perceive must be examined, and perhaps measured, with care and accuracy. It is an art in that any presentation (let alone any perception) of those facts must be selective and so involve choice, and taste, and judgement.

The art in any scientific enterprise consists in evaluating, choosing, and selecting, and there seems little difference here between the regional geographer and the applied statistician searching for an appropriate test for a given substantive hypothesis. Geographic description, it is clear, cannot eschew selectivity or value judgements. But their incorporation does not give the geographer licence to do just as he pleases. The classic works of the French regional geographers thus strike a balance between the presentation of factual information and skilfully constructed literary accounts which succeed in evoking an image of the 'personality'—*personnalité*—of a region. At the same time it is possible to distinguish between substantial fact and intuitive impression. It is in a clear separation between fact and impression that the geographer's techniques of description require most careful application. In confronting armchair images with perceptual experience, it is necessary that the latter be reasonably independent of the former. To confront *a priori* image with intuitive impression smacks uncomfortably of tautology. Herein lies the greatest danger in admitting intuitive impression as a legitimate part of geographic description. Even the best geographer may have secret predilections for seeing what he wants to see—a kind of comforting *déjà vu* that short-circuits real understanding and appreciation.

In searching out adequate descriptions, therefore, the geographer requires control—control over the collection and selection of information, control over its manipulation. In this the geographer, like all other academics, proceeds by way of defining, measuring, and classifying the phenomena he is dealing with. He also develops characteristic ways of representing them—perhaps the most vital representational technique possessed by the geographer is that of mapping. In following these procedures, the geographer is participating in the same processes that go on in all areas of academic research. It is hardly surprising, therefore, that these procedures have been analysed, have been dissected, in order to ensure rigour and efficiency. It is thus possible to speak of a set of models available for collecting, defining, measuring, classifying, and representing, geographic data. Such models do not exclude value judgements, choice, and selectivity. They must be construed as aids to selection and choice, aids which ensure consistency and coherence in our geographic description. In the same way that we should regard it as inexcusable if a map were internally inconsistent (say, the symbols meant different things at different points), so we require standards of consistency and coherence in defining, measuring, collecting, and classifying. The next chapters will be concerned with the models that have been developed for such purposes.

Basic Reading

Darby, H. C. (1962).
Lowenthal, D. (1961).
Sauer, C. O. (*1963*), chapters 16–19.
Wright, J. K. (1966), chapter 5.

Chapter 17
Observation Models—Definition and Measurement

Reality presents the observer with a vast inflow of information. It is the function of observation techniques to select and order this information in a way that makes it manipulable and comprehensible. This process is a kind of search procedure, in which the signals received from reality are scanned for messages that seem to have some degree of regularity or coherence. Clearly, the way in which reality is searched, the particular set of filters (some might call them blinkers) which we use, has an enormous influence upon the kinds of question we ask and the kinds of answer we are able to give. The human mind has a limited capacity to receive, store, and process information—in the jargon of information theory it has a limited channel capacity. Given this basic elemental fact, how do we decide what information to include and what information to throw away? How can we adopt strategies that broaden, rather than restrict, our range of perceptual experience? How, in short, can we open our minds without becoming so saturated with information that our minds suffer stress and ultimately break down under the strain?

We characteristically make use of a variety of observation models in the process of selecting and codifying perceptual experience. The particular models we are concerned with in this and the next chapter are those of definition, measurement, and classification. Each model provides a set of filters for unscrambling the complex real world messages which we receive. These procedures are flexible—that is perhaps why we may best construe them as models—and it is clear that the particular design of the procedure may vary from situation to situation according to empirical circumstance and according to objectives. In other words, we may alter the design of the filter according to our needs. In general, our attitude towards these filters depends upon whether we are searching reality for hypotheses or whether we are structuring the information we receive in a manner appropriate for testing a specific hypothesis. In the former situation we tend to develop a flexible approach in which we alter the filters, direct them differently, and manipulate them, in order to search out regularities, patterns, natural breaks in complex systems, and the like, about which hypotheses may be generated. In testing hypotheses, however, the nature of the filters is more rigorously determined by the nature of the hypothesis itself. The filters here amount to an experimental design procedure appropriate for the hypothesis. It is clearly important that we design the filters in such a way that the information we finally gather in is entirely relevant to the given hypothesis and allows us to judge whether or not there is any degree of empirical support for that hypothesis. The adequacy of a particular test of a hypothesis is thus highly dependent upon the nature of the filters we construct.

These two rather different situations—*searching* reality for hypotheses and *structuring* reality to test hypotheses—have, therefore, a considerable impact on the way in which we use the various observation models that are open to us. It is also one of the more troublesome aspects of method that failure to distinguish between searching and structuring can lead to false inference and tautology. It is not appropriate, for example, having generated a set of hypotheses by imposing a set of filters, to use the same filters to test the adequacy of the hypothesis. Such a procedure is either redundant or downright spurious. A minor, but troublesome, difficulty, is that information collected with no specific purpose in mind is often difficult to use for the test of specific hypotheses. The perennial frustration of geographers seeking to test hypotheses using census data is adequate testimony of this—the data nearly always seem to be collected in inappropriate units and inappropriately measured and classified, for it to be useful for testing the specific hypothesis set up. This difference between searching and structuring relates, of course, to the particular strategy adopted in an investigation. Searching provides one route to scientific understanding, structuring to test *a priori* hypotheses provides the alternative (above, pp. 32–6).

The filters provided by the three procedures of definition, measurement, and classification are not entirely independent of each other, nor are they entirely free from awkward circularities. Definition, for example, presupposes a class of objects to which such definitions may be applied, and although, according to Caws (1959, 7), 'definition enjoys a priority, both historical and logical, over measurement', and 'science grew out of a situation in which the former was freely practised without any thought of the latter', it is difficult to envisage sophisticated operational definitions which do not relate somehow to measurement. Definition certainly presupposes classification, while classification and measurement likewise presuppose definition. These circularities are not the only ones, however, for it can be maintained that all definition is circular—either that or it is infinitely regressive. Ackoff (1962, 170) comments on this circularity:

> In a sense this is true, but defining takes place in three dimensions not two. When a full circle has been made, we are *above* our starting point and have brought to it a richer and more precise meaning than it had when we started. In effect, when we define one concept, this definition illuminates the concepts on which it depends as well as the concepts that depend on it.

In a similar way the three procedures considered here tend to bolster one another. Better classification procedure depends on better definition and better definition may depend on better measurement. Our understanding and our method may be circular, but it is also cumulative. Similarly, it is difficult to keep these observation models separate from other aspects of geographic investigation. The definition of a theoretical concept, for example, cannot proceed without reference to the theory of which it is a part. Ackoff (1962, 145) writes:

> Scientific definitions, like laws and facts, are not isolated islands floating in the sea of science, but are bits of ground firmly anchored to the land mass of scientific theory, laws, and facts, and hence are not less subject to change than any of these.

Coombs (1964, 5) similarly writes on measurement that

> The basic point is that our conclusions, even at the level of measurement and scaling (which seems such a firm foundation for theory building), are already a consequence of theory. A measurement or scaling model is actually a theory . . . admittedly on a miniature level, but nevertheless theory.

Classification systems similarly express primitive laws about structure (above, p. 33) while elaborate classifications may also be derived direct from theory.

It should not be concluded, therefore, that because the filters provided by definition, measurement and classification, are here examined in isolation, that they can or ought to be considered as independent methods in research strategy. In spite of these inter-dependencies, however, it is worth while considering the characteristics of such filters, if only to demonstrate the general rules which govern their use and their relative efficiency in translating the complex information-flow from the real world into comprehensible and manipulable information. Such a treatment may involve a good deal of *as-if* thinking, but it is highly revealing.

I DEFINITION

It is perhaps appropriate to begin by defining 'definition' in the broadest possible manner as 'any procedure for specifying meaning'. Clearly, such procedures are absolutely vital to the development of understanding and the communication of information. Yet Ackoff (1962, 174) remarks that 'defining is an aspect of the research process which all too few scientists take very seriously.' Many a controversy has arisen in geography, as in other sciences, from employing similar terms and concepts but using different methods for assigning meaning to them. Such semantic confusions are embarrassing, if not pernicious. The trouble is that there is no unique method for assigning meaning to a term. We thus possess a variety of methods and each is liable to yield us a rather different result. The problem is to choose, somehow or other, the 'best' way of giving meaning to a term. This choice, it turns out, depends upon the particular 'school of philosophy' a person belongs to, the purpose of a given investigation, the nature of the phenomena being investigated, and the amount of prior information available. There is, therefore, no quick, easy, and immediate, answer to the question of how we should specify the meaning of the terms we use.

The most elementary way of defining consists of poniting to an object and uttering a word. Such a form of definition is called *ostensive* and is basic to the evolution of language itself. Caws (1965, 44) suggests that this form of definition is the lowest form in science and even suggests that it is prescientific. This is not to say that it is unimportant, for ultimately the way in which perceptual experience can be transformed into terms is by such an act of pointing. Ostensive definitions are important in geography in many respects—the identification of rock types, of landscape forms, and the like, is often carried out ostensively. One of the basic functions of field work in geographic training is, after all, ostensive definition.

In ostensive definition the meaning of a term is conveyed by relating it to a percept. It is also possible to define a term by other terms. We may thus construct a dictionary of geographical terms which will provide a set of synonyms. It is characteristic of these dictionary-type—sometimes termed *lexical*—definitions that they allow the replacement of the term to be defined (the *definiendum*) by the defining terms (the *definiens*) without the truth or falsity of the statements containing the original term being in any way affected. A term may thus be a convenient shorthand for some much longer term. A number of dictionaries of geographical terms exist (Moore, 1967; Stamp, 1961; Monkhouse, 1965). Consider the following two definitions taken from Monkhouse (1965, 278, 326):

settlement: (i) Any form of human habitation, usually implying more than one house, though some would include a single isolated building; (settlement) may be rural or urban. (ii) The opening up, colonizing and 'settling' of a hitherto un-populated or thinly populated land, esp. of immigrants to a new country.

water-gap: A low level valley across a ridge, through which flows a river; e.g. the Wey (gap) near Guildford, the Mole (gap) near Dorking, through the N. Downs. . . .

This last definition is in part ostensive since it points to examples, but the basic idea is to describe the meaning of a term by way of other terms. The *definiens* thus describes a set of necessary and sufficient conditions for the application of the *definiendum*. A dictionary approach to definition amounts, therefore, to a description of the complex interrelationships between different terms and such a description ensures consistency of use and serves to eliminate ambiguity. But dictionary definitions are limited in a number of respects. They assume, first, that explicit definitions can be provided (i.e. reduction of the *definiendum* to the *definiens* can occur without any loss of information content) and, second, that definitions can be provided *internally* within a language system. In the first case we may find it necessary to provide *implicit* definitions and these are particularly important in the context of theoretical terms which cannot easily (if at all) be reduced to other terms. Terms such as 'water-gap' can be explicitly defined, but there are other terms, such as 'culture' and 'regional consciousness', which cannot be defined in the same way. These terms are the theoretical terms, the ideal types, the abstract constructs, and the like, which are extremely important in theory-construction. Here the role of definition changes from the provision of synonyms to the provision of the correspondence rules or *text* for a given theory (above, pp. 91–6). The problem of defining theoretical terms and idealisations is one that has taxed philosophers of science for some time. Nagel (1961, 98) writes:

no infrangibly conclusive proof is available, and perhaps no such proof is possible, that the theoretical notions employed in current science cannot be explicitly defined in terms of experimental ideas. . . . It is pertinent to observe, however, that no one has yet successfully constructed such definitions. Moreover there are good reasons for believing that the rules of correspondence in actual use do not constitute explicit definitions for theoretical notions in terms of experimental concepts.

Braithwaite (*1960*, 77) has similarly concluded that the theoretical terms of a scientific theory can only be implicitly defined by reference to the postulates which form the basis of the theory itself. Indeed

Braithwaite goes on to suggest that such theoretical terms 'cannot be explicitly defined by means of the interpretations of the terms in . . . derived formulae without the theory thereby becoming incapable of growth.' The meaning of this is simply that derived theoretical statements require translation into empirical statements, but this translation does not in any way constitute an explicit definition. Now the general importance of definition in this context has already been examined (above, pp. 88–9), and it will be remembered that there is a key difference between theoretical terms which can be related back to theoretical postulates, and many of the idealisations of social science which can be defined only intuitively (i.e. their definition relates to images or conceptual idealisations and is thus external to the scientific language system itself). We shall not consider this problem in any further detail here, except to refer to one important approach to definition which has attempted to solve these complex problems by making certain assumptions. This approach is termed *operationalism*, and it provides one solution to the difficult philosophical problem of specifying meaning.

A theory or model takes on meaning only 'when the symbols and things they represent are defined' (Ackoff, 1962, 141), when adequate correspondence rules are established (Nagel, 1961, 90–105), and when the difficult task of somehow translating abstract theoretical statements into tangible conclusions is accomplished (Braithwaite, *1960*, 53). Operational definition tackles all these problems by assuming that definition 'consists simply in referring any concept for its definition to the concrete operations by which knowledge of the thing in question is had' (Stevens, 1935, 323). The advantage of this rule of definition is that the difference between implicit and explicit definition disappears, and a rigid procedure is prescribed for ascertaining the meaning of the term. Thus Ackoff (1962, 175) writes:

> The formal requirements for (operationally) defining properties are that we specify what is to be observed, under what (changing and unchanging) conditions the observations are to be made, what operations are to be performed, what instruments and measures are to be used, and how the observations are to be made and treated.

Kaplan (1964, 40) likewise writes:

> To each concept there corresponds a set of operations involved in its scientific use. To know these operations is to understand the concept as fully as science requires; without knowing them, we do not know what the scientific meaning of the concept is, nor even whether it has a scientific meaning. Thus operationalism provides, not just a criterion of meaningfulness, but a way of discovering or declaring *what* meaning a particular concept has: we need only specify the operations that determine its application. Intelligence, in the famous dictum, is what is measured by intelligence tests.

Opinions vary as to the value of such an operational approach to meaning. Ackoff (1962), Rapoport (1953) are strongly in favour, Kaplan (1964) is sceptical. Certainly, there is a great deal to be said in favour of making definitions operational wherever possible. We may thus replace Monkhouse's (1965, 257) dictionary-type definition of a 'region' as 'a unit-area of the earth's surface differentiated by its specific characteristics' by an operational definition which specifies the procedure whereby regions may be identified (a good example would be Berry's (1967B) operational approach to regionalisation by way of multivariate analysis and grouping algorithms). Haggett (1965A, 188–90), for example, discusses the problem of giving an operational definition of an 'urban settlement' so as to ensure standardisation and comparability on a world-wide scale. A more operational approach to definition in geography, while in itself desirable, is unlikely to solve all problems however. Kaplan (1964, 40–2) discusses some of the objections to operationalism, the most important of which is the assumption that the conditions can be fully specified and that relevant conditions and measures can be distinguished from irrelevant. If we can believe Wallis's (1965) experiences with computer programmes for factor analysis, for example, we may anticipate that the outcome of a Berry-type approach to regionalisation will depend very much on which kind of computer programme is used. Operational definitions appear much more consistent, but in practice it is difficult to ensure absolute consistency.

The question of consistency, rigour, and precision of definition, however, is also of some interest. On the one hand the mathematisation of geographical theory requires precisely defined unambiguous concepts (above, pp. 184–90). On the other hand there are benefits to be had from retaining a certain 'openness', as Kaplan (1964, 62–71) calls it, in the use and meaning of a term. Thus:

> The demand for exactness of meaning and for precise definition of terms can easily have a pernicious effect, as I believe it often has had in behavioral science. It results in what has been aptly named the premature closure of our ideas.

This openness of meaning can take on a variety of forms. Terms which are defined with reference to the theory that contains them are likely to be changed as the theory is developed and extended. Some terms have a considerable degree of fuzziness—either with respect to border-line cases which we cannot easily assign to a defined class, or with respect to the 'norm' of the term which cannot easily be distinguished (e.g. the real meaning of 'rational man' and 'monopolistic competition' in economic location theory). It is also natural that as operational methods change, as new developments take place, so terms will change their meaning. Kuhn (1962) gives many examples of this in science in general, while terms such as 'environment' and 'region' in geography have shown in amazing variation in the way they are used and interpreted. There is therefore a genuine need for flexibility and mobility in the process of assigning meaning. This need is not necessarily incompatible with the need to provide firm unambiguous terms prior to mathematisation. There is no reason why, even in the early stages of development, precise definitions should not be set up to facilitate mathematisation. But in the early stages when we 'do not know just what we mean by our terms, much as we do not know just what to think about our subject-matter . . . we cannot choose wisely' (Kaplan, 1964, 77). What we have to recognise, therefore, is that we are constantly building temporary bridges from theoretical construct to observation terms and from theoretical construct to mathematical language. Such bridge-building requires precision in definition, but *real* precision is rarely, if ever, possible. It is perhaps worth reminding ourselves that the bridges are *temporary*, and that for every new situation we find ourselves in, for every new problem that arises, we should be prepared to build new bridges or at least satisfy ourselves that the old one is still substantial enough to carry us on our way without disastrous consequences. Kaplan (1964, 66) puts it this way:

> The vagueness of our terms does not consist in the fact that we are continually confronted with the problem of 'where to draw the line', but in the fact that we cannot solve this problem beforehand and once for all. The point is that lines are drawn and not given; that they are drawn always for a purpose, with reference to which the problem is solved in each particular case; that our purpose is never perfectly served by any decision; and above all, that no decision can anticipate the needs of all future purposes. Every term directs a beam of light onto the screen of experience, but whatever it is we wish to illuminate, something else must be left in shadow. Vagueness has never been better characterized than as the penumbra of meaning.

Definition, like all the other observation filters which we use, amounts to a temporary codification of experience according to certain rules. The important thing is to follow the rules in each set of circumstances, and not to be mesmerised by any one particular system of definitions set up with rather special circumstances in mind. To start from Base One each time may be rather tedious and appear rather inefficient, but methodologically it is the soundest procedure. This is not to advocate a plethora of definition systems as an end in itself, but rather to suggest that it will pay handsomely to take a close look at the properties of the definition system we are using. It is, fortunately, one of the characteristics of the history of science, that as our knowledge of our subject-matter increases and as our theories become more sophisticated and more reliable, so appropriate definition systems become easier to construct. Thus 'the closure that strict definition consists in is not a precondition of scientific enquiry but its culmination' (Kaplan, 1964, 77).

II MEASUREMENT

Measurement has been defined in various ways. Stevens (1959, 19) calls it 'the assignment of numerals to objects or events according to rule'; Ackoff (1962, 177) calls it 'the procedure by which we obtain symbols which can be used to represent the concept defined'; and Nunnally (1967, 2) calls it 'rules for assigning numbers to objects to represent quantities of attributes'. All these, and other, definitions have in common the idea of rules to be followed, and all concede that a variety of rules may be constructed, but that the rules should be explicitly formulated. There are, however, some important disagreements regarding the way in which such rules should be stated, the way in which different measurement rules should be applied, and the procedure for judging whether or not a given set of rules has been validly applied.

Measurement is not in itself desirable or undesirable; it is only useful or not useful. It has, however, a number of important functions in science and, as with other forms of observation, measurement can be judged only by asking how well it performs the particular function required of it. These functions vary from simple standardisation —by which we may compare objects and attributes according to some rule of measurement—through a search for a greater degree of precision in statements—by which relatively imprecise statements, such as 'hot', may be replaced by relatively precise statements, such as a temperature reading of 84°F—to the co-ordination of a particular set of events or attributes to a symbolic system—here the measurement system functions as a mapping rule from observation into abstract symbolic statement. The rules of measurement may, or may not, be useful in performing such functions. Standardisation, for example, is likely to be useful only if we can be certain that the measures we take genuinely reflect some property of the object in question. If we are interested in the way in which intelligence differs from person to person, it will be largely irrelevant to take a set of measurements on the length of the fore-arm. Similarly, the quest for a greater degree of precision is reasonable only if the aim of a particular study requires a particular level of precision. There is no point, for example, in using highly refined measures of length to construct a map at the one-in-a-million scale. There is a *real* increment in precision, also, only if we can have confidence in the operations which we use to determine a quantity and if we can be certain that the quantity actually refers to the quality we are considering. It may seem, for example, a considerable advance in precision to go from a statement like 'The economy is growing at a satisfactory rate' to a statement like 'The gross national product rose by 4·3% last year.' But if our ability to measure G.N.P. is in any case within 15% of the true value, then little advantage has been gained in the process of quantification. In the physical sciences many of these problems have been sorted out, but in the social sciences there is much greater room for improvement. Morgenstern (*1965*, 8), in a classic work that ought to be compulsory reading for all human geographers, thus writes:

it ought to be clear *a priori* that most economic statistics should not be stated in the manner in which they are commonly reported, pretending an accuracy that may be completely out of reach and for the most part is not demanded.

Pretending to a level of precision that is not attainable has pernicious rather than illuminating results. This is as true in academic research as it is for the speculator who hangs his decisions upon the monthly trade figures or the monthly indices of industrial output. Similarly, there is absolutely no advantage in translating a statement such as 'Soil moisture-content determines wheat yield' into '$Y = f(X)$' if the relevant measures of Y and X cannot be estimated (above, pp. 186–8). Co-ordinating empirical observation with abstract symbolic representations by way of the rules of measurement is realistic only if it can be shown that the rules of measurement can be validly applied to the empirical events in question.

These reservations regarding our ability to measure should not be regarded as an implicit condemnation of all attempts to quantify. The reservations serve merely to point out that quantification is not easy and that it is important to understand how and when quantitative techniques can successfully, as opposed to spuriously, be applied. An understanding of the methodology of measurement is here the key, for there can be no doubt that without an adequate way of measuring there can be little hope of advancing understanding, of creating sophisticated theory, and of rolling back the frontiers of our knowledge at an effective rate. This simple proposition may appear anathema to those who fear that somehow quantity is to replace quality, that, as Spate (1960) fears, the spirit of Lord Kelvin[1] is

[1] Lord Kelvin, the nineteenth-century physicist, once wrote that 'when you can measure what you are speaking about, and express it in numbers, you know

about to ride roughshod in geography (Burton, 1963, reviews many of the arguments against quantification in geography). In science quantity is not the polar opposite term to quality—it should better be regarded as a superior form of quality. In principle it ought to be possible, in all spheres of our understanding, to improve the quality of our understanding by some form of quantification. It is true that there are many areas of our understanding which prove intractable to quantification, but in such situations it appears that the intractability comes from our basic lack of understanding, our failure to conceptualise the situation reasonably, rather than from anything inherent in the situation itself. Thus Kaplan (1964, 167) writes:

whether we can measure something depends, not on that thing, but on how we have conceptualized it, on our knowledge of it, above all on the skill and ingenuity which we can bring to bear on the process of measurement which our enquiry can put to use. I believe that Nagel is right in saying of measurement that, from a larger point of view, it can be regarded as 'the delimitation and fixation of our ideas of things.' . . . To say of something that it is incapable of being measured is like saying of it that it is knowable only up to a point, that our ideas of it must inevitably remain indeterminate.

Measurement often appears an easy way to structure observation. There are, it is true, many situations in which detailed formulations of the rules for measurement are superfluous simply because the rules and the method of procedure are intuitively obvious. Measuring the length of a road or the weight of output hardly requires the specification of rules, since the conventions in such situations are well known and generally accepted. But such situations are exceptions, particularly in the social sciences, where the rules for measuring attributes such as 'utility', 'motivation', 'concentration', 'localisation', and the like, are far from being obvious. For this reason we shall go on to consider some of the technical problems involved in measurement. We shall (i) consider the nature of the various rules we can use, (ii) consider the problem of applying measurement schemes so devised to actual situations, and (iii) examine how we can assess the validity and the degree of error involved in such a measurement procedure.

A Measurement models

Until recently, measurement models could be grouped into four distinct categories (with one or two hybrid variants) all of which provided simple scalar systems for measuring objects or their at-

something about it; but when you cannot measure it, when you cannot express it in numbers, your knowledge is of a meagre and unsatisfactory kind.'

tributes. More recently, complex measurement models—usually called multidimensional scaling models—have been developed. But it is simplest to begin with the four simple scaling systems, which can be identified from the characteristics of real-number systems— i.e., they can be identified by their *mathematical* characteristics rather than by their empirical applications. Stevens (1959, 24) thus describes four kinds of scales called nominal, ordinal, interval, and ratio.

(1) *Nominal scaling*

It should be made clear from the start that there is some controversy over whether nominal scaling is a measurement system in its own right or whether it is simply a classification procedure. Torgerson (1958, 17) and Nunnally (1967, 11) dismiss nominal scaling from true measurement on the grounds that it simply provides a device for labelling or classifying objects rather than measuring the attributes of objects. Ackoff (1962, 180) and Stevens (1959, 24) include it as the simplest form of scaling. Nominal scaling really amounts to classifying objects or events and numbering them. Putting numbers on the jerseys of football players, or labelling different regions as 1, 2, 3, ... are examples. With nominal scaling there is no intention of performing *any* mathematical manipulation (we would not say that region 8 minus region 3 equals region 5, for example). Nominal scaling has, therefore, no mathematical power apart from that of identification. Confusion can arise, however, from mixing up a particular derivative of nominal scaling—enumeration or counting —with nominal scaling itself. Counting, as Ackoff (1962, 182) points out, amounts 'to assigning "1" to each class member and adding the 1's which have been assigned'. Counting the number of objects in different classificatory boxes is, however, rather a different procedure from giving the boxes a number, and it is clear that simple arithmetic operations can be performed upon the numbers derived in such a way. But here the scale refers to an attribute of the class of objects rather than to an attribute of the object itself and counting by enumeration may be regarded as a discrete form of ratio scaling. It is very difficult, however, to separate nominal scaling from classification and we shall not, therefore, consider it further here.

(2) *Ordinal scaling*

An ordinal scale is one in which a set of events or objects can be ordered from 'most' to 'least' but in which there is no information regarding the amount of the measured attribute that separates the objects or events. Ordinal scaling thus amounts to ranking objects or events in order of magnitude. Such a ranking provides us with rather meagre information, and ordinal scaling is thus regarded as the most primitive way of devising a true measurement scale. There are, however, a number of versions of ordinal scaling. A good account of these is provided by Ackoff (1962, 184–9), while other accounts can be found in Torgerson (1958, 25–31) and Fishburn (1964, chapter 4). It is important to distinguish between these different types of ordinal scaling because confusion between their properties can have disastrous results. Some of the different modes of ordering are worth considering therefore.

Torgerson (1958, 25–31), for example, distinguishes between ordinal scales with and without a natural origin. In some cases it is possible to argue that ranking is positive or negative around some unique natural origin. In ranking preferences, for example, people may be able to identify a point of indifference (sometimes termed the neutral point in the scale) on one side of which things are ranked according to positive criteria (e.g. active liking) and on the other side of which things are ranked according to negative criteria (e.g. active dislike). Identification of such a unique natural origin provides us with a good deal more information, but it limits the valid mathematical transforms that may be conducted on the data. In general, ordinal scales can be subjected to any increasing monotonic transform of the numbers used (this amounts to saying that any order preserving transformation can be conducted on the original data) without any information loss. But given a unique origin such transformations are restricted to those that preserve the unique origin.

The rules implied in ordinal scaling vary. Consider the following three rather different ways of ordering phenomena according to their attribute measure, x_i (Ackoff, 1962, 184–9):

(a) *Complete ordering* is the strongest form of ordinal scaling. It involves ordering the x_i measures in the manner $x_1 > x_2 > \ldots > x_n$. No two objects are allowed to occur at the same point on the scale. In general we may state that the scale is *irreflexive* (no element can equal any other element in its measure), *asymmetric* (a relationship between x_1 and x_2 implies a complementary relationship between x_2 and x_1, e.g. $x_1 > x_2$ implies that $x_2 \not> x_1$), *transitive* (the relationship $x_1 > x_2$ and $x_2 > x_3$ implies $x_1 > x_3$), and *connected* (all the elements $x_1, x_2 \ldots x_n$, can be placed on the scale). Strong ordering thus involves a good deal of presupposition regarding the measures made, but at the same time contains a good deal of information.

(b) *Weak ordering* involves a measurement system of the form $x_1 \geqslant x_2 \geqslant \ldots \geqslant x_n$. The differences between weak ordering and complete ordering are that *reflexive* relationships are allowed and the condition of asymmetry has to be replaced by a condition of *anti-symmetry* (which amounts to specifying x_1 and x_2 as identical only when $x_1 \leqslant x_2$ and $x_2 \leqslant x_1$). A weak ordering does not specify, necessarily, a unique ordering among the x_i and, characteristically, equivalence classes are involved. For example, ranking people according to socio-economic class will probably involve subsets of the x_i measures which are indeterminate with respect to order. Such a form of ordinal scaling does not involve so many presuppositions about the nature of the data, but it does convey less information.

(c) *Partial ordering* is basically the same as weak ordering except that it is non-connected. We may weakly order a population according to socio-economic class, but there may be a group in the population for whom we do not possess any information and who, consequently, must lie off the scale altogether. Similarly, in asking people to order preferences for places or objects, there may be some places or objects which they have no information about and these must, therefore, lie off the scale we are devising. We cannot expect a complete or weak ordering of taste preferences among oranges, bananas, pineapples, and ugli fruit if the people we are asking have never savoured the ugli fruit.

These three examples of the various ways in which we can devise ordinal scales (and many other variants may be devised) demonstrate how rather different variants on the rules of ordinal scaling produce rather different measurement systems.

Basically, however, ordinal scales possess similar mathematical properties and may be subjected to similar forms of mathematical manipulation. The scale is unique up to any increasing monotonic transformation, which means that we can assign any number we please to magnitudes on the scale, provided they do not violate the order. Thus, if we have three objects $A > B > C$ as measured on some attribute, we can put $A = 100$, $B = 50$, $C = 39$, or $A = 5$, $B = 4$, $C = 0$, and so on, without in any way altering the basic relationship. This mathematical characteristic has important consequences when it comes to considering the kinds of mathematical calculations that may be allowable, given ordinal scaling. This important issue has been considered in detail by Siegel (1956, 23–6), who lists the appropriate statistical measures that may be used, given ordinal data. These are summarised in Figure 17.1. This information is important, for it shows how statistical manipulations are related to the form of scaling used and hence indicates the key role of

SCALE SYSTEM	DEFINING RELATION	POSSIBLE TRANS-FORMATIONS	CENTRAL TENDENCY	DISPERSION	TESTS
					APPROPRIATE STATISTICS
					non-parametric ⟵	*parametric and non-parametric* ⟶
NOMINAL	Equivalence	$x' = f(x)$ where $f(x)$ means any one-to-one substitution	mode	% in the mode	chi-square contingency coefficient Goodman-Kruskal's Lambda phi-coefficient
ORDINAL	Equivalence Greater than	$x' = f(x)$ where $f(x)$ means any increasing monotonic transformation	median	percentiles	Spearman's rho Kendall's tau Kolmogorov-Smirnov Goodman-Kruskal's Gamma phi-coefficient
INTERVAL	Equivalence Greater than Known ratio of any two intervals	any linear transformation: $x' = ax + b$ $(a > 0)$	mean	standard deviation	*t*-test *F*-test (analysis of variance) Pearson's *r* point biserial (one variable dichotomous) etc.
RATIO	Equivalence Greater than Known ratio of any two intervals Known ratio of any two scale values	$x' = cx$ $(c > 0)$	geometric mean	coefficient of variation	

Fig. 17.1 The Measurement Scale Systems and Their Associated Methods (Mathematical and Statistical) *(after Galtung, 1967; Siegel 1956; and Stevens, 1959).*

measurement in relating observation to abstract mathematical statistics.

(3) *Interval and ratio scaling*

It is usual to consider these two types of scale separately but the reason for this appears to be largely historical. In both interval and ratio scaling we can rank-order objects with respect to a measured attribute, but in addition we are able to specify how far apart the magnitudes are from each other. In other words we can measure the distance between two objects on the scale. With interval scales, however, no natural origin can be identified, and any origin is arbitrary. A classic example is provided by the various scales for measuring temperature—e.g. Fahrenheit and centigrade—which can be used to specify the distance between readings, but which do not yield an absolute measure. This restricts possible mathematical use to transformations which are linear (e.g. we can convert centigrade readings into Fahrenheit readings by using an equation of the form $F = 32 + 1·8 \ C$). Any linear tranformation of this form preserves the information in the original data. More complex transformations distort relationships however. In more simple terms, it does not make sense to suggest that 20 °C is 'twice as hot' as 10 °C (this means that we must also assert that 68°F is 'twice as hot' as 50°F!).

Ratio-scale data, however, possess a natural origin and absolute magnitudes can be determined. Weight, mass, length, and the like, can all be measured on the ratio scale. Some would also say that temperature measured on the Kelvin scale is a ratio measure. These measures can be subjected to more complex transformations that involve multiplication. It thus makes sense to say that 10 miles is twice as far as five miles.

A wide range of statistical measures and procedures may be used given interval and ratio data. Most classical methods in statistics thus presuppose interval measurement at the very least. Ability to measure on an interval or ratio scale is thus an important prerequisite for translating observation problems into variables that can be treated by way of the Neyman-Pearson theory of statistical inference (above, pp. 280–1). Some of the mathematical manipulations possible, given interval and ratio scaling, are listed in Figure 17.1.

(4) *Multidimensional scaling*

The scales we have examined so far have implicitly assumed the notion of measurement on a single, unidimensional, underlying continuum. For many problems in measurement this assumption appears reasonable. But there are some attributes which are plainly multidimensional. We have already considered one such attribute —spatial position—in great detail (chapter 14). There are other attributes—colour for example—which are clearly multidimensional and others that we strongly or weakly suspect of being multidimensional (intelligence, motivation, utility, and so on). Most of the initiative regarding multidimensional scaling has come from psychology, where almost all attributes being measured appear to have a degree of multidimensionality, or to use an easier term, the attribute is *complex*. In such a situation, Torgerson (1958, 248) writes,

> Instead of considering the stimuli to be represented by points along a single dimension (i.e., a unidimensional space) the stimuli are represented by points in a space of several dimensions. Instead of assigning a single number (scale value) to represent the position of the point along the dimension, as many numbers are assigned to each stimulus as there are independent dimensions in the relevant multidimensional space. Each number corresponds to the projection (scale value) of the point on one of the axes (dimensions) of the space.

If the properties of the space are known it is possible to estimate the distance between objects provided certain assumptions can be made. Suppose, for example, that we are seeking to measure a complex attribute such as economic development and let us assume that there are two dimensions to this complex attribute (say, power consumption *per capita* and rate of population growth). If these two dimensions are independent we can represent them by orthogonal axes (Figure 18.4) and plot countries in this two-dimensional space. The distance between two countries, p and q, on the complex attribute, D_{pq}, can then be calculated by the pythagorean theorem.

This simple example demonstrates the principle of multidimensional scaling. It can be extended to n dimensions and there is, therefore, a problem of determining how many dimensions the underlying complex attribute has. This problem will be considered later in connection with numerical methods of classification. There is similarly a tremendous difficulty in determining the interrelationships between the dimensions. Here it is as well to remember Russell's dictum (above, p. 195) that 'measurement presupposes geometry', for the measurement of distance on a complex multidimensional attribute depends entirely upon an ability to specify the geometric characteristics of the n-dimensional space formed by those n dimensions. There is no need for the axes to be orthogonal and no inherent reason why we should use Euclidean geometry to estimate relationships (apart from the fact that it is simple and well known). There is similarly no reason why all the axes should possess the same scale characteristics. We may thus find ordinal scales (of various strengths) mixed with interval and ratio scales.

The problem in multidimensional scaling, therefore, is how to measure the distance between objects located in this n-dimensional space. Thus Torgerson (1958, 251) writes:

> Any of a large number of different geometric spaces could conceivably be used as the basic spatial model for a multidimensional-scaling procedure. However . . . the Euclidean model is the only one that has been considered at all seriously.

Since 1958, however, a vast amount of work has been done on multi-dimensional-scaling procedures, particularly in psychology (Cattell, 1966; Kruskal, 1964), and the applicability of some of these to geographic problems has also been considered (Downs, 1967). It is clear, however, that the technical problems involved (particularly on the geometric side) are precisely the same as those involved in measuring *geographic* distance and that the same basic methodology applies. Transformations, projections, and complex distance measures (above, pp. 219–23) are just as applicable in the attempt to measure distance on some complex attribute as they are in the attempt to measure geographic distance. The generalisation of distance measure from the special case of Euclidean geometry taken from Adler (1966, 282–3; above, pp. 223–4) is thus a form of a general equation for relating distances to projections given by Torgerson (1958, 293):

$$d_{jk} = [\overset{r}{\underset{m}{\Sigma}}(|\, a_{km} - a_{jm}\,|)^c]^{1/c}$$

in which the distance between two stimuli j and k in an r-dimensional space consisting of $m = 1, 2, \ldots, r$, orthogonal axes is given by the formula with c taking on a variable value. When $c = 2$, the simple Euclidean measure of distance results; when $c = 1$, a rather special distance measure devised by Attneave (1950), variously known as the 'city-block' or 'Manhattan metric', results. Much larger values of c can be used and there is no reason for c to be an integer (Kruskal, 1964).

There is not space to go into multidimensional scaling methods in any detail here. Those interested should refer to the psychological literature (particularly Torgerson, 1958; Kruskal, 1964). But it is interesting to note the basic similarity between the measurement problem as it occurs in scaling complex attributes by way of various stimuli, and a parallel situation in a more traditional geographic context. Thus (Tobler 1966A) has used the psychological techniques of multidimensional scaling in constructing map projections, while Gower (1967) has used an understanding of map-projection techniques to transform complex data sets to more manageable dimensions.

B The application of measurement models in observation

From the preceding discussion it should be clear that measurement consists in assigning numbers to objects by *mapping* those objects into an abstract space of some specified and determinate structure. It should also be clear that

the order of a set of objects is something that we impose upon them . . . the order is not given by or found in the objects themselves (Kaplan, 1964, 180–1).

Further, each measurement model possesses specific mathematical characteristics which define the valid mathematical operations that can be conducted. How, then, can we choose measurement models appropriate for given problems and how can we construct them?

The answer to this question depends on some assessment of measurement as an empirical observation device. This is in part a complex philosophical issue, and in part a purely practical one. But, as usual, philosophical and practical considerations overlap. As usual, also, there is anything but common agreement among scientists and philosophers of science.

The inherent measurability of phenomena has thus been the subject of some debate. Campbell (1928), for example, divided all measurement into *fundamental* measurement (the magnitude of which does not depend on any other magnitude) and *derived* measurement (formed from compounding fundamental measures), and suggested that science should be concerned with fundamental measures as far as possible. Such a view is still highly controversial (Ackoff, 1962, 195–201; Ellis, 1966, chapter 5). Most writers agree, however, that certain properties are much easier to measure than others, and that the measurement of some complex attribute can often successfully be approached by breaking down complex dimensions into simple one-dimensional attributes which can then be compounded to provide the complex measure. This, after all, is the approach inherent in multidimensional scaling. But there are many situations in which it is easier to measure some complex attribute directly than to measure its components. This appears to be particularly true in the behavioural sciences, where something like demand may be estimated more easily than component measures. In general, complex theoretical concepts, such as utility, welfare, stress, and the like, prove extraordinarily difficult to measure simply because we do not know enough about the actual meaning of the concepts themselves. Theoretical understanding thus makes measurement much easier, partly because we then understand precisely what it is we are measuring, and partly because we possess laws of behaviour to help us in devising adequate measurement systems. Measurement as an observational device, therefore, depends upon a satisfactory theory and upon the statement of adequate correspondence rules to operationalise theoretical concepts. Without adequate theory we need to resort, once more, to *a priori* models.

An adequate theory will presumably tell us something about the structure of observed events. Given that we know this structure, it should then be possible to derive some way of measuring those events. It is then clear that there is no point in trying to use interval-scale measures, when the underlying structure is discontinuous. It would be possible to debate this issue with reference to all kinds of problem, but it is perhaps useful to consider that of scaling preferences. Suppose, for example, that we are concerned with deriving some measure of place utility and we did so by attempting to measure the preferences of various people for various places. To start with we know very little about 'place utility', and it is thus a very ill-defined concept. Certainly we do not possess any adequate theory. How then should we measure it? We could seek information at the nominal scale—a simple 'Would you or would you not live there?' kind of question. The answers would not yield us very much information, nor would they yield magnitudes capable of any sophisticated manipulation. On the other hand they presuppose least as regards the nature of the magnitude being measured. Suppose we then asked people to rank-order places according to their preferences, and thus mapped them on to a completely ordered ordinal scale. Such a measure is more sophisticated and yields a good deal more information and, further, can be manipulated. Yet it presupposes a good deal. For example, it assumes *transitivity* in the scale—can we really be sure that people's views on place are transitive? Even if we can assume transitivity at any one instant of time, can we assume transitivity over a period of time? Suppose we now sought to measure place utility on an interval scale. The information content would be great, the ability to manipulate extremely good, and we could bring to bear on the data so obtained sophisticated mathematical techniques. Yet the measurement presupposes an enormous amount. In all probability the data reflect more regarding our assumptions than they do of the place utility a person holds. We thus find ourselves in the situation described by Fishburn (1964, 78), in that

as we pass from measures which are most imprecise . . . to the measure which is most exact (the interval measure), the assumptions of the measurement procedures become more demanding and less likely to hold in practice.

This example demonstrates an important general principle. Our method of measuring requires that we make assumptions (these assumptions are basically the same in form in mapping any object into an abstract mathematical space). We need to be sure that these assumptions can reasonably be met in the empirical situation. In general, therefore, the choice of a particular measurement model amounts to choosing that measurement system that yields us the greatest detail of information (and is capable of the most sophisticated manipulation), while presupposing least about the nature of the underlying structure. This problem of presupposition is particularly important in the social sciences where the observer often interferes with performance of an individual being observed (Webb, *et al.*, 1966). This leads, therefore, to the general problem of how we should construct and operationalise methods of measurement, methods which are as meaningful and objective as possible.

One method of constructing measures is to compound simpler measures in some specified way. This leads to an interesting framework for thinking of measurement systems known as *dimensional analysis* (Bridgman, 1922; Langhaar, 1951; Ellis, 1966, chapter 9). This provides us with 'a code for telling us how the numerical value of a quantity changes when the basic units of measurement are subjected to prescribed changes' (Langhaar, 1951, 5). It also provides us with a means for comparing different measurement systems and evolving a hierarchy of such systems. We may thus begin with simple measures of basic properties and construct more sophisticated measures. Consider, for example, a one-dimensional measure of length $[L]$. We may square it and obtain a two-dimensional measure of area $[L^2]$, and cube it to obtain a three-dimensional measure of volume $[L^3]$. Density of population thus has the dimensions $[P/L^2]$, or $[P\,L^{-2}]$. We can extend such a form of analysis to consider complex structures of measurement. Curry (1967, 224) has thus used dimensional analysis to look at the basic mathematical properties (dimensions) of purchasing in a social system. The point about such analysis is that it allows us to develop a systematic way of evolving new measures and, perhaps, deriving new indices which are of interest. Most indices of use in human geography, such as the coefficient of localisation, various indices of economic development, indices of relative growth, and so on, have been derived in a nonsystematic fashion. Dimensional analysis provides a neat way of evolving measures, comparing measures, and exploring structure theoretically.

Operationalising measurement models amounts to devising a procedure for mapping objects on to some predetermined scale. An infinite number of procedures could be developed and an infinite number of predetermined scales devised. The methodological problem, therefore, is to find that procedure and that scale that does maximum justice to the situation in the light of a given objective. Again, this is not meant to encourage a plethora of scales and procedures. Indeed there is much to be gained from consistency; there are many areas of geographic research where particular measurement models have been adopted and consistently used for many years without any ill effects. But the danger is that a measurement system will be taken for granted. When a discipline moves into new areas of research, an unawareness of the basic problems inherent in measurement 'can lead to extremely ineffective research' (Ackoff, 1962, 215). It is ineffective because measurement models are merely filters through which we monitor complex messages. Over the years these filters become more refined and better adjusted to our needs. But as we turn to new areas, so we must anticipate that new measurement models must be used. It is then that we are forced to recognise that measurement is a useful procedure only when we know very well what we are doing and why we are doing it.

C The validation of measurement models and the assessment of measurement error

A valid measure is one that measures what it is supposed to measure. Even if it does measure what it is supposed to, it may do so rather badly; and therefore we require some understanding o. the errors involved in using a particular measure. Validation and assessment of error thus provide us with a means for checking on a particular measurement model and provide us with information on how well that particular model performs the job it is supposed to.

It ought to go without saying that it is impossible to assess the validity of a given measure without considering its purpose. Validation thus amounts to some empirical assessment of how well a given measurement model performs in a given situation. Nunnally (1967, chapter 3) suggests three ways of examining validity, each being appropriate in rather different circumstances. *Predictive* validity is involved when the purpose of the measure is to estimate something else. A typical example would be an entrance examination which is designed to measure future performance (IQ testing for streaming in schools, etc.). The validation of such a measure really depends upon the correlation between the measure made and some measure of future performance. *Content* validity is much more difficult to assess for the measure cannot be tested by correlating it with something else, since the measure is conceived of as a direct measure on some attribute. We cannot validate a measure by examining its results in such a situation. But we can validate it by examining the mathematical form and operational procedures involved in relation to what we know about the nature of the objects being measured. This amounts to a thorough check on the design of the measurement procedure—a kind of mini-experimental-design procedure as a precursor to measurement. The aim of such a validation procedure is to ensure that the measure relates only to the attribute being measured (i.e. there are no other interfering forces) and that the attribute is measured in an unbiased manner. *Construct* validity is even more difficult to assess, for here we are basically concerned with the problem of measuring abstract constructs—utility, satisfaction, motivation, and the like. Such theoretical constructs (idealisations) refer to a given domain of behaviour. Utility, for example, is a notion employed to cover situations where people choose one out of several alternatives, and, perhaps, act on such a choice. In measuring such constructs we tend to select one aspect of the domain the construct is thought to govern, and to perform some experiment on it, the outcome of which is regarded as a measure of the construct itself. In utility, for example, the classic procedure is to attempt to analyse it by way of betting behaviour in a gaming situation (von Neumann and Morgenstern, *1964*; Fishburn, 1964). There is an enormous inductive assumption here, of course, for we assume that we are measuring utility and not just betting behaviour in a gaming situation. The more we know regarding the construct itself (i.e. the better we are able to define its domain) the easier construct-validation will become, for then we can select a wide range of alternative situations in that domain and relate the measures to one another, presumably emerging with one or two procedures which provide us with the information we need. Construct-validation is, however, a complicated business (Nunnally, 1967, 83–99). Problems of validation are thus numerous, but without some effort in this direction we are likely to find ourselves misinterpreting the meaning of the measures we make. Good accounts of the problem may be found in the references by Nunnally (1967), Coombs (1964), Ghiselli (1964), and Torgerson (1958).

All types of measurement are subject to error, and the magnitude of that error affects the usefulness of the measure in a given situation. The ideal situation, of course, is one in which the magnitude of the error involved is so small that it can be ignored for a given purpose. The significance of measurement error is thus not independent of purpose. Ackoff (1962, 205–14) lists four sources of measurement error:

(a) *Observer error* results from the inability of the observer to abstract himself entirely from the measurement process. In the physical sciences observer error is largely due to the inability of the senses to discriminate finely enough; hence the use of the normal curve of error to estimate the true measure. In the social sciences, however, all kinds of problems develop, associated with the inability of the investigator, when dealing with other people, to abstract himself from the questions being asked. The observer frequently projects himself into the measurement situations, and the measure may thus contain a bias according to the observer (Webb, *et al.*, 1966). The error in such a situation is not reasonably conceived of as random, it is, rather, systematic.

(b) *Instrumental error* results from biases in the instruments being used. A thermometer may not be calibrated properly, a balance may be slightly out, and so on. No instrument is perfect, and it is reasonable, therefore, to ask for information on the sensitivity of the instrument and level of error involved.

(c) *Error due to the environment* results when conditions in the environment change to affect the observer, the instrument, or the thing observed. A change in temperature may affect a measure of length, a change in credit restrictions may alter a person's views on place utility (by affecting his potential to move), a hot humid day may affect the stress felt by the observer and the observed. It is one of the functions of experimental design to control for such errors, but in the non-experimental sciences such control is extremely difficult. Again, it is necessary to have some estimate of the error involved.

(d) *Error due to the observed* may result either from the inherent variability of what is being observed (e.g. people change their views on place utility over time, and may even think differently in the evening from the way they felt in the morning) or from the behaviour of the observed being affected by the behaviour of the observer. The classic case of the latter is the *indeterminacy principle* in physics formulated by Heisenberg. In the social sciences the indeterminacy principle appears to be very general, especially in questionnaire work. It is almost impossible for an interviewer not to influence an interviewee, since observer and observed are placed in a highly reactive situation. Controls may be introduced but it is unlikely that error will be eliminated or that the error will be non-systematic. An enthusiastic interviewer is likely to get rather different results from a cynical interviewer, while respondents have a disturbing habit of giving answers they think the interviewer wants.

These four sources of error are inherent in any measurement procedure, and they may all be combined, to produce errors of considerable magnitude. It is possible to reduce errors by careful design and by a careful manipulation of the measures obtained (they can be screened for errors for example). But it is unlikely that error will be completely eliminated and this degree of error must affect the uses to which a given measurement procedure can be put. Morgenstern's (*1965*) critique *On the Accuracy of Economic Observations* has much to say about the errors involved in economic statistics and concludes that, for the most part, the degree of this error invalidates many of the uses to which they are put. It is vital, therefore, to pay attention to measurement error. As Ackoff (1962, 214) has it

the continuous reduction of error . . . is a major objective of science and is one of the principal measures of its progress.

D Measurement in geography

In the preceding sections we have concentrated on some of the methodological problems inherent in the measurement process. In many areas of geographic research these problems may appear largely irrelevant not because they are unimportant but because appropriate methods of dealing with them have been devised. There is nothing very controversial about the way in which measurement problems in surveying are overcome. In most traditional areas of geographic research, therefore, there is a well-developed methodology of measurement. In human geography, on the other hand, where there has been a very rapid change in emphasis over the past few years, no such conventional methodology has yet been evolved. Given the growing interest in the behavioural aspects of location—studies in spatial perception (Gould, 1966; Lowenthal, 1966; Downs, 1968), locational behaviour (Pred, 1967), environmental perception (Kates, 1962; Saarinen, 1966), and the like—it is important that the bases on which we measure be clearly understood. It cannot be denied that our methods of measurement in these areas of research leave a great deal to be desired. This is in part due to the inherent difficulty involved in measuring 'images', 'values', 'satisfaction', 'utility', and the like, but it also results from a failure to grapple with basic methodological problems. It is therefore appropriate to close this section on measurement by returning to the problem of measuring place utility (above, pp. 317–18).

A number of measurement procedures have been devised for measuring place utility. Rather than survey all of these, or devise new measures, however, it is convenient to set up just one model and examine the validity of it. Consider, therefore, the following model. We begin by choosing eight areas in a city which are relatively homogeneous with respect to their living conditions. Let us assume that such areas can be identified and that we can put names to them, e.g. 'Whiteoak', 'Clifton', 'Chelsea', and the like. We then choose a sample from the population living in the city by some specified method and ask them to rank-order the eight names they are given in order of their preference for living there. What information does such a model yield us?

In direct terms the model simply yields us a response to given stimuli. Torgerson (1958, 46) suggests three ways in which we can conceive of such a model:

(i)   *The subject-centred approach*, in which the variation in response is attributed to the individuals (i.e. we are measuring the utility of each individual with respect to places).

(ii) *The stimulus-centred or judgement approach*, in which the variation in response is attributed to variation in the stimuli (i.e. we are measuring the utility of places as seen by individuals).

(iii) *The response approach*, in which variation in response is attributed to both stimuli and individuals (i.e. we are measuring a mixture of the utility of places and the utility scales of individuals).

The response approach is probably the least satisfactory because it mixes two rather different things. But as our model stands, it clearly belongs to this category. Now it is possible that the sampling design can be used to transform the model into a judgement model. Suppose we know, for example, that the key variables with respect to the utility scale of individuals are education and income. By stratifying our sample to hold education and income constant we may control out the variation in the utility scale of individuals in large measure and hence obtain a measure of the utility of places. Let us now pursue this particular tack.

We are now certain that we are scaling the stimuli and not the people. The question then arises what the stimuli actually represent and how they are actually arranged by people. The stimuli consist of a series of names, and the response relates directly to these names. We can go on to make statements about place utility only provided that certain assumptions can reasonably be made (here we are involved in construct validation). Consider the following reasons why people might arrange the names in a given order:

(i)   Some names sound nicer than others—thus people might rate

'Clifton' persistently higher than 'Coalpit Heath' simply because it sounds nicer as a name.

(ii) People may have a variable amount of information about the places they are asked to rank. Variation in the rank-ordering may thus partly measure variation in the information which people have. In situations where people possess zero information, it may be inappropriate to use a completely ordered scale; a partially ordered scale would be better.

(iii) People may possess different 'images' of places (because of reputation, and other factors which are difficult to specify), and they may order them solely in terms of these images.

(iv) They may order places with respect to the actual living conditions in the areas named and their preference for such living conditions.

These four reasons are not the only ones that could affect the outcome, nor are they independent of each other. The net outcome, therefore, is that we do not possess any very good control over the stimuli we are presenting, nor are we sure what it really measures. If we compound with such difficulties those associated with variability in the population, we find ourselves in an extremely complex situation in which it is difficult to judge the validity of the measure proposed. The specification of error is likewise extraordinarily difficult. It is true that many of these difficulties may be partly overcome by a sound sampling procedure. Such procedures will, however, be considered in chapter 19.

It may be argued from this example that, rather than pursue the chimera of measurement, we should abandon the idea altogether. There is indeed some justification for adopting such advice. But the interesting thing is that in order to measure effectively we require deep analytic understanding and considerable thought regarding the controls necessary and the errors naturally incurred. Measurement may not be a satisfactory end in itself, but we can be sure that in pursuing this end we will turn up problems and difficulties, the solutions to which will provide major advances to our understanding. This, however, depends entirely on a sound understanding of the nature and principles of measurement. Without such an understanding we are plainly lost.

Basic Reading

Ackoff, R. L., *et al.* (1962), chapters 5 and 6.
Churchman, C. W., and Ratoosh, P., (eds.) (1959).
Coombs, C. H. (1964).
Kaplan, A. (1964), chapter 2.

Morgenstern, O. (1965). Siegel, S. (1956), pp. 21–34.
Torgerson, W. S. (1958).
Webb, *et al.* (1966).

Reading in Geography
Haggett, P. (1965A), chapter 8.

Chapter 18
Classification

Classification is, perhaps, *the* basic procedure by which we impose some sort of order and coherence upon the vast inflow of information from the real world. By grouping sense-perception data into classes or 'sets' we transform a mass of unwieldy information so that it may be more easily comprehended and more easily manipulated. Classification is, after all, basic to language since 'by every general name which we introduce we create a class' (Mill, *1950*, 90). If language were restricted to proper names only, communication would become impossible. Thus classification is one of the basic tools we use in dealing with the world around us.

To concentrate attention upon this aspect of classification would, however, largely miss the point of classification as a rational scientific device. We can represent classification as a set of rules for assigning data to their appropriate classificatory boxes. These rules can be conceptualised in abstract logical terms (set theory has a great deal to offer here). The application of these rules depends, however, on the objectives. Thus, as with both measurement and definition, classification may be regarded as a means for searching reality for hypotheses or for structuring reality to test hypotheses. It may also be regarded as a beginning point or the culmination of scientific investigation. We possess, therefore, no means of assessing the adequacy or efficiency of a given classification independently of the job it is designed to do. There have been many occasions in the past when it has seemed as if geographers had forgotten this elementary fact. Classifications have thus been produced without its ever being quite clear what purposes they are designed for. The geographic literature is replete with complex classifications of towns, land uses, climates, regions, morphometric features, and the like, which appear to have been devised with no particular purpose in mind. It is scarcely surprising that many of these classifications have never been used for anything. In recent years, however, we have been reminded very forcibly that purpose and classificatory form are inextricably bound up together. Berry's (1958; 1967B) extensive work on classification and grouping procedures, R. H. T. Smith's (1965A; 1965B) timely reminder that town classifications—a favourite for non-purposive classification apparently—have relevance only given a definite purpose, and Grigg's (1965; 1967) intensive analysis of the regional concept in the light of the principles of taxonomy, have rectified the situation to a considerable degree. Geographers have not been alone in their misconduct, and indeed their sins appear minor compared with those of sociologists and political scientists who have, according to Brown (1963, 168–71) a distinct taste for 'completely useless' classificatory schemas. Thus:

When someone produces a 'bulky system' he must also answer the implied question 'A system for what?' He cannot merely reply 'It organises the data.' *Any* criterion will organise data—will order items into classes—but only some systems of classification will be scientifically useful. The work of selecting the 'fundamental entities' of a field is united to the work of finding the answers to its basic problems. If we seek explanations only *after* we have produced a system, we have no reason to suppose that we shall discover them.

The particular filter that we call classification serves to order data and does so very comprehensively. It is most important, therefore, that we regard it as a flexible device, one to be altered to meet our needs in any given situation. The danger is that impressive classifications will ultimately ossify and inhibit research effort rather than encourage it, if they are regarded as sacrosanct. This is not to deny the great benefit that accrues from comprehensive classifications or even to deny the benefits to be had from the stability of such systems. But we should be prepared to change our classifications when it becomes clear that they have outlived their usefulness. Flexibility in application of the rules of classification does not, however, deny the logical rigour necessary in formulating the rules themselves.

I THE LOGIC OF CLASSIFICATION

The logical rules governing the development of classification systems are designed to ensure internal consistency and coherence. They also show how classification systems may be formed. These rules can be stated by way of usual logic (Stebbing, 1961, chapters 5 and 6), or we may use the theory of sets to demonstrate them (Tarski, *1965*; Stoll, 1961; Christian, *1965*). We shall take a set-theoretic approach.

If we conceptualise the information flow from the real world as consisting of elementary pieces of information—termed 'bits'—we can then attach to each element certain attributes or properties which it possesses. This is rather like attaching to each element a series of labels which describe various attributes and properties. We

can begin by denoting the elements by lower-case letters $a, b, c, \ldots$
We may then define a set, denoted by upper-case letters $A, B, C, \ldots$,
as a group of elements which possesses some attribute or property
in common. We may determine whether a given element, x, belongs
to a set X according to whether it possesses the requisite property
or not. If x is a member of the set X, we write $x \in X$. If x is not a
member of the set X, we write $x \notin X$. A particular element x may
belong to many different sets. Thus a firm may belong to a set of
firms producing iron and steel, A, a set of firms with more than 1000
employees, B, a set of firms whose shares are quoted on the Stock
Exchange, C, and so on. In such a case we write that $x \in A$, $x \in B$,
$x \in C$, etc.

Now a set can be described in two ways—*by enumeration* or by
definition. Describing it by enumeration amounts to listing all the
elements contained in it. Using the usual brace notation for sets dis-
cribed in this way we might thus write out the set $\{a, b, c\}$, which is
made up of the three elements a, b, and c. We could also write that
$b \in \{a, b, c\}$, but that $x \notin \{a, b, c\}$. The grouping $\{a, b, c\}$ may then be
examined to see what properties or attributes a, b, and c, have in
common. In practical terms this amounts to an intuitive grouping
procedure and then a search among the objects so grouped to find
out what properties and attributes they have in common. Describing
by *definition* amounts to specifying a particular property which must
be satisfied by the members of the set and by no others. Here we begin
with the property or attribute and group accordingly. Thus if we
were considering a set, A, defined on the property that it is a set of
firms producing iron and steel, we should write $x \in A$ (where x is a
firm producing iron and steel). Such a method for describing sets will
obviously be more appropriate when we have prior knowledge as
regards the relevant properties for classification in a given situation.

Having described sets in an appropriate manner, we can then
resort to the manipulations of set theory to discuss the relationships
among sets and the evolution of some system of classification. A
number of relationships among sets may be defined:

(i) 'If A and B are sets which are so related that every element of
A is also an element of B, then A is said to be a *subset* of B'
(Christian, 1955, 46). This condition will be denoted by the
relationship $A \subseteq B$. By this definition every set is a subset of
itself and to get round this problem any set which is a subset
other than the set itself is defined as a *proper subset*, written
$A \subset B$. If A is not a proper subset of B we write $A \not\subset B$. Thus
if A is the set of all settlements with more than one million
inhabitants and B the set of settlements with more than 1000

inhabitants, it is clear that set B includes set A and therefore $A \subset B$. The opposite relationship does not hold, however, and therefore $B \not\subset A$. This gives us the condition of *equivalence* between two sets, namely, $A \subset B$ and $B \subset A$.

(ii) We may combine sets in specific ways and these operations are useful in developing classification systems. The union of two sets A and B, written $A \cup B$, defines a new set that has elements which are members of A or B or both (Figure 18.1A). The intersection of two sets A and B, written $A \cap B$, creates a new set whose members are members of both A and B (Figure 18.1B). The *complement* denotes all those elements in a *universal set, U*

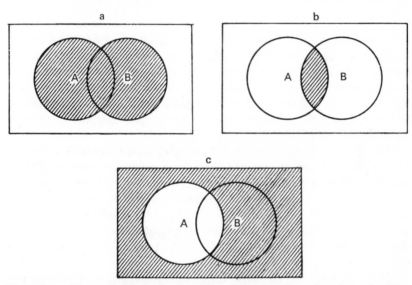

Fig. 18.1 Venn diagrams for set-theoretic operations showing : **A :** The union of two sets (shaded area); **B :** The intersection of two sets (shaded area), and **C :** The complement of set A (shaded area).

(defined for the purposes of a given discussion), which are not members of a specified set. The complement of set A is written A' (Figure 18.1C) and is defined by the relation $A' = \{x \in U : \sim (x \in A)\}$ which in words reads that the complement of set A comprises all the elements in the universal set U except those contained in A.

(iii) It is useful for purposes of manipulation to define an empty set ϕ which has no elements. It may be empty simply because we can find no elements to put into it (e.g. towns of more than 50 million inhabitants) or because the properties that define it are mutually exclusive (e.g. all towns that have less than one million and more than one million inhabitants).

The utility of a set-theoretic treatment of classification is that it allows us to formulate consistent rules for forming classes. Basically we can proceed in two ways—dividing the universal set, U, (say, all towns) according to a particular property, or by grouping elements into sets and these sets into larger sets, and so on (Kemeny, *et al.*, 1957, chapters 2 and 3). In the case of division we may divide according to a particular property to ensure that the subsets of U are disjoint and exhaustive. A group of sets A_1, A_2, . . . A_n, thus partition U, if, and only if,

(i) $A_i \cap A_j = \phi$, where i and j are any two sets with $i \neq j$.
(ii) $A_1 \cup A_2 \cup \ldots \cup A_n = U$.

It is then possible to cross-partition the A_i by using further criteria. Suppose we partition U into A_1, A_2, . . . A_n, and then develop another partition B_1, B_2, . . . B_n, such that both the A_i and the B_j are disjoint and exhaustive. It is then possible to form a new partition by considering all the subsets of U of the form $A_i \cap B_j$. We may continue to break down the universal set in this fashion, introducing a new criterion for partitioning at each stage.

Grouping procedures proceed in a different direction. The individual element x may be regarded as a set composed merely of itself and no other element, $\{x\}$. We may then group the sets $\{x_i\}$, $\{x_2\}$, . . ., $\{x_n\}$, according to certain criteria of similarity. Thus $\{x_1\} \cup \{x_2\}$ might equal A, $\{x_3\} \cup \{x_4\} \cup \{x_5\}$ might equal B, and so on. The sets so formed may then be united to form super-sets, and the procedure can continue until the last union of sets yields the universal set.

These procedures, or rules, for classification indicate certain important criteria for ensuring the consistency of such classification systems. Particular attention should thus be paid to the criterion of exhaustiveness (i.e. no object is left outside the classification, given the definition of the universal set, U), and mutual exclusiveness (i.e. that no object can be assigned to two different boxes at the same time). Such logical properties say nothing, however, about what substantive criteria are to be used, or how and in what order they are to be used, in devising a classification system. This is essentially an empirical problem, and we shall therefore turn to the problem of operationalising these abstract logical schemas. Before considering the procedures in detail, however, it would be useful to consider some of the purposes of classification and some of the problems inherent in selecting the properties or attributes to act as criteria in classification.

II THE PURPOSE OF CLASSIFICATION

It has already been made clear that the utility of a given system of classification cannot be assessed independently of its purpose. The many purposes of classification can, however, be grouped into two types.

(1) *General or 'natural' classifications*

These are intended for general use by all scientists (Sokal and Sneath, 1963, 11–20). The notion of a 'natural' classification was developed in detail by Mill (*1950*, 300–1):

> The ends of scientific classification are best answered when the objects are formed into groups respecting which a greater number of general propositions can be made, and those propositions more important, than could be made respecting any other groups into which the same things could be distributed . . . a classification system thus formed is properly scientific or philosophical and is commonly called a natural, in contradiction to a technical or artificial classification or arrangement.

The naturalness of the classification, as far as Mill was concerned, depended on the causal structure of explanation that could be invoked in distinguishing the various classes. As Mill envisaged it, therefore, the natural forms of classification represented the culmination of scientific enquiry and rested upon a sophisticated understanding of structure—an understanding which, it turns out, neither the nineteenth nor the twentieth century has possessed. Most modern views are more pragmatic and tend to regard general classifications as performing the function of a data-storage system as well as providing a comprehensive nomenclature for working on empirical problems. The appropriate criteria for such a classification thus become maximum efficiency in data storage (this is essentially an information-retrieval problem) and maximum efficiency in naming things. But in many cases these different functions are incompatible, and efficiency in performing one function is offset by inefficiency in performing another. Thus Sokal and Sneath (1963, 6) comment on the present system of taxonomy in biology that it

> attempts to fulfill too many functions and as a consequence does none of them well. It attempts (1) to classify, (2) to name, (3) to indicate degree of resemblance (affinity) and (4) to show relationship by descent—all at the same time. We shall show . . . that it is impossible not only in practice but also in theory for the current system to perform these tasks adequately.

A general classification may be designed to serve many purposes but it is unlikely to serve all possible purposes with any but a low level of efficiency. Sokal and Sneath (1963, 7) thus suggest, at the minimum, a separation of classifications based on *resemblance* (the objects show certain affinities), *homologous characters* (the objects have a common origin), and a *common line of descent* (the objects possess a similar evolutionary history).

In preserving information and ordering information it is often important to preserve the same classification to ensure comparability. If the standard industrial classification changed every six months many forms of academic research would be hopeless. But although stability is important, from the point of view of information yield, this should not imply no change at any price. There is only one thing worse than a classification system that is changed too frequently, and that is one that is never changed even though it clearly has no relevance to current conditions.

(2) *Specific or 'artificial' classifications*

These can be thought of as classifications devised for a special purpose. Such classifications are very much involved in experimental-design procedure, and special classifications may be developed to test specific hypotheses, or to deal with specific kinds of problem. The problem itself defines the criteria and method to be used in such cases. Clearly, however, there are many situations in which the specific purpose is approximately met by a general-purpose classification and it is therefore easier to use that information than to derive a new classification system from scratch. There are also many specific-purpose classifications which, embracing as they do a wide range of properties and objects, are difficult to differentiate from general classifications. It is probably best, therefore, to regard classification systems as being arranged on a continuum from the least to the most general.

III THE SELECTION OF PROPERTIES AND THE PROCEDURE FOR CLASSIFYING

A classification system involves, at some stage or other, a definition of the significant criteria to be used in classifying objects and may, further, involve arranging the criteria in some order of importance. This implies that we know what properties or attributes are important for differentiating objects and events in a given situation. Hence the strong link that binds classification and theory is forged. We can thus determine the relevant properties by reference to theory. If we state the properties without specific reference to theory we must necessarily presuppose the existence of some theory. This elementary fact of life in any scientific investigation needs to be carefully considered. The very power of classification rests upon this strong and irrevocable connection with theory. But therein also lie dangers, for it is all too easy to presuppose a theory without knowing it.

The problem thus arises how we determine the significant properties on which to classify and, once determined, how we then use them to assign objects into classes. The determination of the relevant properties is essentially a substantive issue, and no general rules can therefore be applied, except to say that, presumably, the most important properties should be given the greatest weight. But the interrelationship between classification and theory does provide one important methodological guide. It has been pointed out above (pp. 88–96) that a theory is related to a given domain of circumstances (objects and events) by its text. We may thus conceive of reality (a kind of super-universal set) as being partitioned into a number of domains, each ruled by a given theory. In reality the partitions are blurred, some domains are empty of theory (because it has yet to be developed) and some domains are inefficiently ruled. Consider now a universal set, U, defined for a given purpose (assume we are dealing with all towns). This particular universal set is a very small subset of the super-universal set of all objects and events everywhere. If we know clearly in which theoretical domain U lies, then it should be relatively easy to identify the relevant criteria for classifying the objects and events. But there are many situations in which U obviously lies in several different theoretical domains simultaneously. This may be because the theory is not well developed enough or it may be because we are trying to classify objects and events that really lie in radically different theoretical domains. Mill (*1950*, 92) long ago drew attention to the fact that objects and events of a radically different kind could not easily be brought into the same classification— a rule that Grigg (1967, 486) has recently restated in the geographic context. This problem boils down to one of effective definition. All towns may be an appropriate universal set, but stones and beetles, cabbages and kings, heads and sherry, appear not to be. But there are many situations in which it is difficult to decide upon the universal set to be considered. As theory develops so new relationships emerge and so it becomes possible to classify what previously seemed to be very disparate events and objects into some unified system. Even cabbages and kings have metabolism in common, and both contain a very high percentage of H_2O. Then, also, there is the problem of what to do in the relatively empty domains, the areas of our experience where we have little or no theory to guide us. Here we must of necessity travel hopefully, although in contrast to the popular adage, it would be much better to arrive at adequate theory in such domains.

The choice of relevant properties on which to classify thus depends upon the purpose of the classification and upon what we regard as the significant properties with respect to that purpose. In selecting the significant properties we require all the information we can muster regarding the events and objects to be classified.

Given a purpose, and given adequate information regarding the criteria to be used in classifying, how then should we proceed to assign objects and events to classes? The exact procedure will, of course, depend on the circumstances. Thus some writers emphasise a difference between classification proper—in which it is possible to classify elements into relatively homogeneous classes—and what they term *ordination*—which involves making divisions on a continuum (Greig-Smith, *1964*, 158). In the geographic literature this difference is not usually recognised, although it is clear that the classification of climates, for example, amounts to ordination rather than classification proper. Classifying on continuous phenomena, however, is not in principle different from classifying discrete objects, although it is much more difficult. We shall use 'classification' here in a very broad sense to include ordination as a special, but difficult, case.

The particular difficulty that ordination poses, however, is related, as Grigg (1965; 1967) has pointed out, to the problem of identifying the *geographical individual* (above, pp. 215–17). In assigning objects or events to their classification boxes we require at least that objects and events can be identified. Given variables that are continuously distributed in space (or time), it is impossible to identify individuals except by making certain assumptions. We may thus sample in a temperature surface at a number of points and treat the readings at the points as individuals and then use formal classification procedures. But this involves an important extra assumption. Perhaps the main point from a methodological point of view is not so much in actually making the assumption as in reminding ourselves that we have made it. Herein lies the utility of distinguishing carefully between classification proper and ordination.

There are basically two different kinds of classification procedure. We may thus speak of 'classification from above'—often termed 'logical division', or 'deductive classification'—and 'classification from below'—often termed 'grouping', 'inductive classification', and the like. In addition we may distinguish between monothetic classifications (which are inevitably associated with logical division) and polythetic classifications (usually associated with grouping pro-

cedures). We shall therefore consider these two rather different procedures in some detail.

(1) *Logical division or 'classification from above'*

Logical division simply involves partitioning a universal set, U,
according to the logical principles already set out for such a procedure. Stebbing (1961, 107–9) gives a detailed account of the logical
rules involved, while Grigg (1965; 1967) has explored the application
of logical division to the construction of regions.

The division of the universal set takes place in a series of steps, and
at each step one property or set of properties is used to differentiate
between classes (Figure 18.2). Rigid and successive logical divisions
define what Sokal and Sneath (1963, 13) call *monothetic classifications*

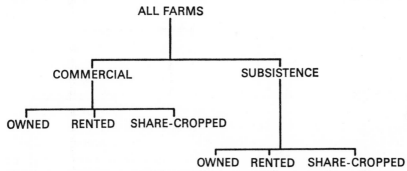

Fig. 18.2   Diagram to illustrate two stages in the logical division of a
universal set of farms into mutually exclusive classes.

which are characterised by the property that 'the possession of a
unique set of features is both sufficient and necessary for membership
in the group thus defined.' The kind of classification which emerges,
however, is very much affected by the criterion selected at each step
and the order in which the criteria are employed. In developing
classifications by this route, therefore, we need to place the criteria
in order of significance and this, of necessity, assumes that we know
a great deal about the phenomena being classified; in other words,
we possess an adequate theory about structure and we can employ
that theory deductively to identify the classes. Such an approach has
its dangers. Sokal and Sneath thus go on to point out that

> Any monothetic system will always carry the risk of serious misclassification if
> we wish to make natural phenetic groups. This is because an organism which
> happens to be aberrant in the feature used in the primary division will inevitably
> be removed to a category far from the required position, even if it is identical with
> its natural congeners in every other feature. . . . The advantage of monothetic
> groups is that keys and hierarchies are readily made.

The procedure involved in logical division is clear and seductively simple. The procedure is thus heavily relied upon. Mill (*1950*, 301) regarded it as the only effective and logical way of classifying phenomena, and much of the work done in the nineteenth and early twentieth century resorted to this procedure. Thus Grigg (1967, 482) has pointed out that world regional classifications are almost always devised by employing the principle of logical division. But the logical clarity of division is not always matched by its realism. It presupposes a fairly sophisticated understanding of the phenomena being investigated, else the classifications evolved may be totally unrealistic,

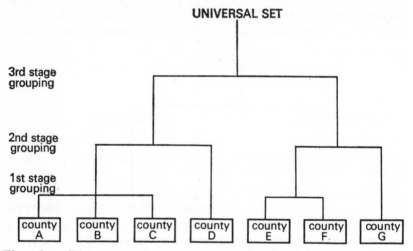

Fig. 18.3 Diagram to demonstrate grouping 'tree' for a set of seven counties.

nothing better than an inspired guess. It thus seems as if logical division were best suited as a method for classifying when the classification is viewed as the culmination of scientific theory. This is not to deny that it may be useful in other situations. But when it is used in situations where we know little the dangers inherent in its use need to be fully recognised. Classification by logical division in the absence of adequate theory amounts to stating an *a priori* model and the consequent methodological difficulties require clear recognition. Having imposed an *a priori* model in the form of some classification system we cannot legitimately infer that the model constitutes a theory (above, pp. 145–54).

This difficulty can be demonstrated by an example. Suppose we wish to find out if hierarchies exist in settlement patterns, and begin by dividing all   , hamlet, village, and town, on an *a priori* basis and then test to see whether these groups are characterised by different functions. We might be tempted to conclude from a positive result to such a test that a hierarchy does indeed exist and that it possesses three distinctive levels. This conclusion is not, however, valid without further evidence, for the conclusion is presupposed in the *a priori* classification model. Without further evidence the argument remains entirely circular. Suppose, on the other hand, that we possess a sophisticated well-validated settlement-location theory which states that, under a given set of conditions, a three-level hierarchy will exist. We may then use this theory to assign given settlements to their predicted level in the hierarchy and then, in order to test the applicability of the theory, test to see whether the settlements so assigned indeed do perform radically different functions. The argument in this case is not circular. In the absence of theory, however, we are in great danger of proving what we have already assumed to be true. This form of circular argument is not unknown in geography (Brush, 1954; Vining, 1955; Berry and Garrison, 1958; provide an interesting argument on this very issue).

(2) *Grouping or 'classification from below'*

The logical properties of grouping procedures should be clear from the set-theoretic presentation above (pp. 327–30). Operationalising grouping to yield classifications, however, is a much messier procedure than that involved in logical division. In situations where we face a good deal of uncertainty grouping probably yields much more realistic classifications. Hence grouping is often regarded as an inductive procedure by which the phenomena being examined are searched for regularities and for significant interrelationships. Grouping thus seems peculiarly appropriate to situations in which we do not know what the significant properties are. But there is no inherent reason why grouping should not proceed on the basis of theory. The major difference between grouping and logical division, from a philosophical point of view, lies in the specification of the universal set. In grouping procedures it is of necessity specified by enumeration, whereas in logical division it is specified by definition. In logical division we may obtain many classes with no members, but this is impossible given grouping procedures. Hence any general conclusions made from grouping must proceed by induction. A typical grouping pattern is demonstrated in Figure 18.3.

In grouping we begin by enumerating the universal set as containing x_n elements and for each of those elements we list a set of attributes or properties, each of which has potential for differentiating among the x_i elements, and all of which we regard as being relevant.

Now it should be clear from this that classification by grouping is not free from awkward *a priori* assumptions. In particular we need to preselect both the elements to be grouped (all towns in England, all towns in England and Wales, all towns in Western Europe, and so forth), and the variables which, as a group, we regard as being relevant (e.g. employment characteristics, socio-economic characteristics, and so on). But the presuppositions involved are far less binding than those involved in logical division, and in particular we need make no assumptions about the order or interrelationships among the variables used to differentiate classes. Classifications that emerge by grouping are usually *polythetic* (Sokal and Sneath, 1963, 13), which means that a particular class of elements so classified will share many features in common, but no element in the class needs to possess all the features used to identify that class. Such classifications are much more permissive and probably much more realistic. But there are difficulties in assigning the elements to classes (or forming them into groups) because such an assignment depends on *degree* of affinity. Cases may, and frequently do, arise in which one element could just as well be grouped with alternative and rather different groups.

Many of these procedural difficulties have led to the specification of rules for identifying similarities and assigning elements to groups. Recently, these rules have become more rigorously formulated with the aid of mathematical procedures. Quantitative methods of classification have thus become important in many disciplines. Geography has been no exception. Since the mid 1950s quantitative methods of classification have become an important part of the geographer's method. The initial work was largely developed by Berry (1958; 1960; 1961; 1967B) but at the present time these techniques are being extensively employed in all areas of geographic research and in cognate disciplines (such as sociology, psychology, geology, and soil science). Given the current importance of such quantitative methods, and the fact that many of the procedural problems involved in grouping are dealt with in this literature, it is appropriate to consider these techniques in some detail.

IV QUANTITATIVE TECHNIQUES IN CLASSIFICATION

Phenomena are classified by reference to their attributes. Measurement is something made on an attribute of an object. It therefore follows that classifications can be devised by reference to the measurement made on an attribute rather than by reference to the existence or non-existence of the attribute in question (which in itself may be regarded as a simple nominal measure). Measured attributes of phenomena contain a good deal of information so that, provided all the assumptions required in the measurement process are reasonably fulfilled, we might expect that classifications derived by way of measured attributes will contain a good deal more information (and thus be more realistic) than classifications that are derived by other means. In order to group phenomena on a quantitative basis we require:

(i) A set of objects or events, k_1, k_2 . . . k_n, to be grouped.
(ii) A set of relevant attributes or properties, p_1, p_2, . . . p_m.
(iii) A set of measures, x_{ij}, on the attributes of objects (the measures may be on the nominal, ordinal, interval, or ratio scale, or on a mixture of scales).

We can then define an m by n matrix, X, made up of the x_{ij}'s:

$$
\begin{array}{c}
\textit{Objects} \\
k, \ k_2 \ . \ . \ k_j \ . \ . \ k_n
\end{array}
$$

p_1	x_{11} x_{12} · · · · · x_{1n}	
p_2	x_{21} · · · · · ·	
p_i	· · · · x_{ij} · · ·	
p_m	x_{m1}, · · · · · x_{mn}	

Attributes or properties

Quantitative methods of classification involve searching this matrix of measures for appropriate groupings. In order to do this we require a rule for distinguishing groups. The commonest procedure is to minimise within-group variance on the measures and to maximise between-group variance (there are a number of variants on this rule, according to Berry, 1968). This explicit mathematical rule for grouping and classifying corresponds to a general intuitive notion as regards classification, namely, that the classes should be as distinct from one another as possible and internally as homogeneous as possible.

In order to apply this rule, however, we need to be able to estimate the distance (sometimes called the taxonomic distance) between two objects as they are measured on the m variables. Here we can refer to the principles of multidimensional scaling (above, pp. 313–15), for the m variables we are using to classify form an m-dimensional space in which each object or event is located. We require a measure of distance between objects as they are positioned in that m-dimensional space. Consider, for example, a two-dimensional orthogonal space (say X measures the percentage of the population in a town in service activity and Y the percentage of the population educated beyond school-leaving age). Suppose we locate six towns in this two-dimensional space (Figure 18.4). It is intuitively obvious that there are two groups, p and q, clearly separated from each other. Taking the mean of both groups we can calculate the square distance between groups and the mean squared distance within groups. In

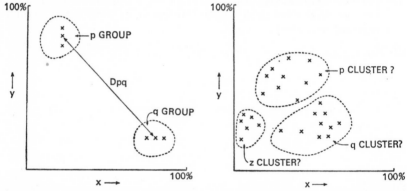

Fig. 18.4   The measure of distance for classifying in a two-dimensional orthogonal space, showing : **A** : A simple case for grouping six observations in the space made up of two variables; **B** : A complex case in which it is difficult to discern groupings in the space.

this case the grouping is obvious, but in cases where it is not (Figure 18.4B) it is possible to search among all combinations of the objects located in the space and select that particular combination that maximises the squared distance between groups and minimises the mean squared distance within groups.

In principle such a method may sound easy enough (even if rather tedious) but there are a number of difficulties. The most serious difficulty arises with respect to the geometry of the m-dimensional space. If it is Euclidean then no problems arise, but this amounts to saying that the correlation matrix between the attributes has no entry significantly different from zero, i.e. all the attributes are independent of each other. If the attributes are in any way correlated (and they nearly always are) then the space is non-Euclidean and we need to know the structure of the m-dimensional space in order to calculate the distance between objects. This implies that we also need to know what the interrelationships are among the attributes. Now there are two ways we can know this. Given a sophisticated theory about structure, we can predict what the interrelationships ought to be among the attributes and hence define the m-dimensional space theoretically. If we do not possess such a theory, we can seek to group the attributes according to the way in which they co-vary over the n objects. Such a procedure amounts to developing a general (if temporary) theory about the structural interrelations among the attributes and then using this information to define the structure of the m-dimensional space. Grouping of the objects is thus contingent upon some prior analysis of the interrelationships among attributes. The net result is to draw attention to both aspects of our basic data matrix—the attributes as well as the objects. The same modes of analysis may, in fact, be employed to examine both aspects. Following conventional usage (Cattell, 1965) we may differentiate between R-mode analysis (which examines the interrelationships among the m attributes or variables) and Q-mode analysis (which examines the interrelationships among the n objects or observations).

The procedural problem posed by the nature of the m-dimensional space should not be regarded as a minor technical difficulty, to be thrust aside at the first opportunity, to be assumed away without much analytic effort. The tremendous importance of classification in fact centres on this one basic problem, for its solution implies a considerable increment in our understanding of the phenomena in question. To develop understanding of the interrelationships among attributes is one of the basic aims of any investigation. The importance of classification as a search procedure rests on the fact that it poses this very basic problem. Indeed, it may be argued that the solution of this one procedural difficulty, implying as it does the construction of a theory of interrelationships, is far more important than the end-product, the classification system itself.

Given that we possess measures of some kind on the m attributes of the n objects, it is possible to estimate the nature of the interrelationships among attributes or objects by using the various measures of similarity, association, and correlation, that have been devised. There are a large number of such measures available, but which is appropriate depends on the nature of the measurement system used (nominal, ordinal, etc., see Figure 17.1), on the size of the sample (small-sample statistics usually being different from large-sample statistics), and on the form of the distribution (non-normally distributed data requiring non-parametric measures unless the data can be transformed to a normal distribution, in which case the parametric tests may be used). We may thus use a wide range of measures varying from chi-square contingency measures and phi coefficients, through Spearman's rank correlation and Kendall's tau, to the product-moment correlation coefficient, or we may devise special measures for examining association directly (Sokal and Sneath, 1963, chapter 6; Greig-Smith, *1964*, chapters 6 and 7; Miller and Kahn, 1962, chapters 12 and 13). Such measures compare each attribute with every other attribute in the R-mode and each object with every other object in the Q-mode. The problem then arises how we are to group attributes or objects, given this basic information. By way of example, we shall examine a number of approaches to this technical problem of identifying classes.

(1) *Mahalanobis' generalised distance* (D^2) *statistic*

The D^2 statistic was developed by Mahalanobis (1927; 1936) and is, therefore, one of the earlier measures to define the degree of affinity between different classes. It should be emphasised that the measure applies to already established classes and states how similar they are to one another. The measure may thus be used to group classes into more general classes. The statistic has since been developed by Rao (1948; 1952), while Miller and Kahn (1962, 258–73) provide a worked example in geology. Berry (1960; 1967B) and L. King (1966) provide examples of its use in geography, while Stone (1960) provides an example of a distance-measure analysis of regional similarity. The statistic is particularly useful for discriminating within a relatively homogeneous population. Rao (1948, 183) notes that the use of the statistic in its simple form depends upon the attributes (characters) measured being independent. In this case:

the formula for D^2 would be simply the sum of the squares of differences in mean values for the various characters. But when the characters are correlated . . . they may be replaced by a set of *transformed characters* which are linear functions of the observed characters and are mutually uncorrelated. When once these transformed characters are obtained, the calculation of D^2 is reduced to finding the simple sum of squares.

The transformation is quite complicated but the important thing to note about it in this context is that it functions to transform an m-dimensional non-Euclidean space into an m-dimensional Euclidean space which facilitates the measurement of distance. This way out of the technical difficulty posed by non-Euclidean relationships is characteristic of many quantitative methods of classification.

(2) *Principal-components and factor analysis*

Principal-components and factor analysis provide a dominant mode of approach to quantitative classification in the geographic literature. Its tremendous importance must undoubtedly be attributed to the pioneering work of Berry (1960; 1961; 1962; 1965; 1966; 1967B; 1968) which has been taken up by many other geographers (e.g. Ahmad, 1965; Carey, 1966; Henshall and King, 1966; L. King, 1966; R. H. T. Smith, 1965B). The method has been used by others in a geographical context (the early studies by Kendall, 1939; and Hagood, 1943; are particularly important) and has been extensively used in cognate disciplines such as psychology (Cattell, 1965), soil science (Bidwell and Hole, 1964), botany (Goodall, 1954), geology (Imbrie, 1963), etc. The studies of British towns by Moser and Scott (1961) and of American cities (Hadden and Borgatta, 1965) are major contributions of geographic interest.

Although in many respects similar, factor analysis and principal-components analysis should not be confused. Thus Cattell (1965, 411) has pointed out that

much confusion and disputation confounding means with aims would be avoided if the mathematical purpose of component analysis were semantically distinguished from the experimental aim of factor analysis.

This distinction is also emphasised by Kendall (1957, 37):

In component analysis we begin with the observations and look for the components in the hope that we may be able to reduce the dimensions of variation and also that our components may, in some cases, be given a physical meaning. In factor analysis we work the other way round; that is to say, we begin with a model and require to see whether it agrees with the data and, if so, to estimate its parameters.

From the point of view of classification this difference amounts to classifying without any theory (in the case of principal-components) or classifying with theory (in the case of factor analysis). The detailed technical differences between principal-components and factor analysis need not concern us, as adequate accounts can be found in the literature (Cattell, 1965; Cooley and Lohnes, 1964; Kendall, 1957). Both methods operate on a data matrix directly or on a correlation matrix of either attributes (R-mode) or objects (Q-mode). Consider a correlation matrix of association among the variables which will be an m by m matrix. Principal-components analysis amounts to a projection of the m-dimensional space on which m variables load on to an r-dimensional component space (usually $m = r$) which has the properties that each of the component dimensions is orthogonal to every other and that the first component is defined as that vector that extracts the maximum amount of variance from the basic correlation matrix. Successive components are orthogonal but extract, at each stage, the maximum amount of residual variance. It is usual with principal-component analysis to find that the first few components extract a very high proportion of the total variance. Suppose the first two components extract 70% of the variance from a 25 by 25 matrix. This amounts to saying that two components contain 70% of the information in the 25 attributes and, in addition, these two components are orthogonal to each other. The advantage of this from the point of view of classification is obvious. It is thus possible to calculate how the n objects 'score' on the components. The matrix of component scores forms an n by r matrix which locates objects in an r-dimensional Euclidean space. It is then possible to group objects on the basis of the information in this matrix. However, it is interesting to note that principal-components analysis also collects together the attributes, for by examining the component loadings the attributes themselves may be grouped, interrelationships better understood, and an underlying structure of the attributes identified. It is not technically necessary to interpret the interrelationships among the attributes before classifying. But the generality of the classification devised will depend on the stability (and presumably the generality) of the components used.

In factor analysis a number of basic underlying dimensions are hypothesised and searched for in the data matrix (or the matrix of correlations between either objects or attributes). In this case the same mathematical procedure is followed except that the resultant matrix will have only k factors (the number of factors is determined by hypothesis and will presumably be much less than either m attributes or n objects). In addition, the correlation of an attribute or object with itself is partitioned into a *general* portion (known as the communality) and a *unique* portion (conceived of as being irrelevant to the general study or a kind of random noise or both). It is not necessary for the factors to be orthogonal (they may be oblique). However, the original space is transformed into a much more parsimonious space whose structure is known. It is therefore possible to classify on this basis.

There is some question, however, as to the appropriateness of using factor-analytic techniques in geography when, for the most part, we have no clear idea of underlying structure of either attributes or objects. Berry (1966) in regionalising economic development uses factor analysis which involves, among other things, estimating the degree to which a variable is *generally* (as opposed to *uniquely*) correlated with itself. Berry thus estimates the communalities (the degree of general correlation) on the principal diagonal of the correlation matrix. L. King (1966, 206) dissents from this view and with respect to urban groupings suggests that 'a great deal more understanding of urban relationships would seem to be required before such communality estimates can be employed confidently'.

(3) Grouping procedures

The discussion of the D^2 statistic and principal-components and factor analysis has shown how attributes or objects lying in a complex-non-Euclidean space can be mapped into a Euclidean space (or a non-Euclidean space whose properties are known) and thus the distance that separates them be easily calculated. The next stage in the classification procedure is to use these distance measures to group the objects or attributes. Many studies have resorted to an intuitive grouping, usually by inspecting the pattern of component scores (which may be done graphically with two components). Various other methods, which may not require a Euclidean space, have also been used. Techniques such as cluster analysis, linkage analysis, profile analysis, and so on, can thus be used without a principal-components or factor analytic transformation. Sokal and Sneath (1963, chapter 7) and Miller and Kahn (1962, chapters 12 and 13) provide good reviews of such techniques.

Recently numerical techniques have also been developed for splitting or grouping populations in a multidimensional Euclidean space. Edwards and Cavalli-Sforza (1965) have thus developed a method for partitioning points in a multi-dimensional Euclidean space into two clusters on the criterion of maximising the between-clusters sum of squares (and consequently minimising the within-clusters sum of squares). Classification here takes place by successive splitting (and it has the properties of logical division in many respects). The procedure is an iterative one, and if there are n elements in the space the first split requires $2^{n-1} - 1$ iterations to identify the optimal split. Successive splitting in this fashion does not, of course, guarantee that the clusters established after several stages possess maximum sum-of-squares differences between them. In general, the more splitting that takes place the less accurate the technique will be.

Ward (1963) and Howard (1966) have devised a similar procedure, except that it starts with all the elements and groups them successively (i.e. they group from below). Starting with p groups, it is possible to examine all $p(p - 1)/2$ ways of uniting two of them to form $p - 1$ groups. Grouping upwards in this fashion produces a characteristic linkage tree (Figure 18.3). Grouping upwards does not necessarily yield optimal classification after several steps, however, and in general the more steps involved the less accurate the technique will be.

Clustering and grouping procedures have been used in the geographical literature to identify classes. Berry (1966; 1967B) provides a good account of these uses in geography.

(4) Discriminant analysis

In discriminant analysis we start out with a set of classes which have known attributes. But in polythetic classification there may well be difficulty in deciding which out of several possible classes an individual element belongs to. Discriminant analysis provides a set of rules for ensuring that we 'make as few mistakes as possible over a large number of similar occasions' in assigning individuals to predetermined classes (Kendall, 1957, 144). Good accounts of the method may be found in Miller and Kahn (1962, 276–83), Kendall (1957, chapter 9) and Cooley and Lohnes (1964, chapter 6) while Casetti (1964) has used it in a geographic context.

We may imagine our m-dimensional attribute space as being divided into two regions, R_1 and R_2 such that $R_1 \cup R_2$ exhausts the space. The problem is, then, to draw the boundaries of the two regions in a manner which is least likely to result in the misclassification of any individual placed in that region. The linear discriminant function provides one way of defining these regions, given two classes which are known to be different. The procedure for estimating the linear discriminant function need not concern us (Miller and Kahn, 1962, 276–83, provide an excellent account).

Casetti (1964) however, in extending the technique to geographic problems has pointed out that discriminant analysis can be used as a measure of the efficiency of a particular classification, and it is a short step from this to an iterative procedure by which an optimal classification may well be identified. We may thus change the linear discriminant function (i.e. alter the parameters of it) and search for that linear discriminant function which most efficiently discriminates between groups. This is not, however, the prime aim of discriminant analysis. Its main function is to provide a set of quantitative rules for assigning objects into predetermined classes.

The quantitative approach to classification thus possesses four aspects:

(i)   Quantitative analysis of the interrelations among the attributes or among the objects,
(ii) Transformation and reduction of the correlations to a geometric structure with known properties (usually Euclidean);
(iii) Grouping or clustering the objects or attributes on the basis of the distances measured in this transformed space.

(iv) Once classes have been firmly identified, the development of rules for assigning phenomena to classes.

All of these aspects of quantitative classification have been tried in geography and it is perhaps worth while closing this section with a brief comment on their utility and their limitations. Basically, the limitations of the techniques depend upon the nature of the data available and the assumptions that have to be made in order to justify the mathematical manipulation of that data. It is well known, for example, that there are problems inherent in the factor-analytic method (how to estimate communalities and the stability of the factors under sampling variation, for example). Thus Miller and Kahn (1962, 295) point out that in factor analysis 'due to the variation in mathematical systems and choice of technique within a system, synthesis of the results of several factor analyses are difficult if not practically impossible'. Studies by Wallis (1965) and Matalas and Reiher (1967) bear out this contention, although it is worth noting that factor analysis is rapidly changing as a technique (Cattell, 1965; 1966). The various methods of clustering, linking, and grouping, similarly face difficulties of application. The nature of the basic data also needs to be carefully considered. Here the choice of variables (attributes), the sampling design in collecting the data, the appropriate measure for the phenomena being investigated, all require careful consideration, for they must necessarily affect the legitimate manipulations that can be made on the data. Thus the choice of the product-moment correlation coefficient for regionalisation problems appears singularly inappropriate, since one of the technical requirements of this statistic is independence in the observations. Since the aim of such regionalisation is to produce contiguous regions which are internally relatively homogeneous, it seems almost certain that this condition of independence in the observations will be violated.

In short, the employment of quantitative techniques requires a very thorough evaluation of procedure and data as a preliminary to the analysis. Even given that we are reasonably satisfied (and we have to recognise that we can never be perfectly satisfied) regarding data, assumptions, and procedure, we still require a careful evaluation of the results. After all, we are concerned to find general classes and general relationships among attributes, we are concerned with finding comparable underlying structures in complex data matrices, and, above all, we are concerned with identifying a theory about structure which can command our confidence as an analytic, retrodictive or predictive device. Quantitative techniques for classification bode well as search procedures. They can lead us to new ideas, new frameworks for analysis, and so on. They have tremendous potential in geographic research. Whether that potential is realised depends on whether mindful evaluation occurs rather than mindless application by those using such techniques.

V CLASSIFICATION—A CONCLUDING COMMENT

There are a large number of classifications extant in geography. These vary in form from those that classify by similarity, by relationship, or by both (e.g. the formal, functional, or synthetic approach to regional classification), from those that take a morphological approach, a genetic approach, a functional approach, a causal approach, and so on, to those that are multivariate and incorporate several different approaches at the same time. Almost all the phenomena with which geographers deal have, at some time or other, been classified. Land uses, towns, climates, soils, coast lines, rivers, economies, and above all, perhaps, regions, have been variously approached. Yet classification is essentially a means to an end, a filter through which we transform sense-perception data for a given purpose. The objective and the classificatory form are not independent however, and if we fail to identify our purpose we must expect our purpose to be governed—perhaps insidiously—by the classification system we adopt. We can either adjust the means to a given end, or weakly allow the means to determine the end. It may seem easy to separate out means and ends, but classification is buried deep in the way we think, speak, and write, it forms a basic nomenclature with which we work, often largely unaware of its systematic presuppositions. How else, it might be argued, can we explain moraine than by its genesis? Yet moraine is itself a term based on genesis. The question thus amounts to asking how else can we explain phenomena classified by origin except by reference to origin? Classification is a powerful filter which we can either use or be used by. 'A classification,' wrote Cline (1949), 'can prejudice the future.' We currently possess in geography a sound basic methodology for classification. Let us ensure that we do not allow the classifications developed to prejudice our future and let us also ensure that we make an efficient use of them in pursuing a deeper understanding of the phenomena we are concerned with.

Basic Reading

Tarski, A. (*1965*), chapter 4.
Sokal, R. R., and Sneath, P. H. A. (1963).

Reading in Geography

Berry, B. J. L. (1958).
Berry, B. J. L. (1967B).
Grigg, D. B. (1965).
Goodall, D. W. (1954).
Smith, R. H. T. (1965A).

Chapter 19
Data Collection and Representation in Geography

Geographical data are obtained when the geographer records facts about some aspect of geographic reality or when he accepts or adapts facts recorded by others. Geographical facts may be regarded as some objective record of observation. Such a record is objective in the sense that it is *intersubjective*—which means that repeated observations of the same phenomena by different people will yield the same factual statement—and *reliable*—which means that an observer repeatedly recording the same phenomena will produce the same factual statement. These criteria are not totally realisable, of course, since all recording is subject to some kind of error (*cf.* above, pp. 319–22), but basically we may regard geographical facts as belonging to our collective perception of the world, and hence communicable, rather than to the private world of image and fancy (Lowenthal, 1961).

The procedural rules involved in defining, measuring, and classifying, go some way to ensuring this kind of objectivity in geographical reporting. In this chapter we shall pursue this approach and discuss some of the rules available for collecting, storing, and representing, geographical data.

I THE GEOGRAPHER'S DATA MATRIX

Data collection amounts to a set of rules for constructing and filling in some sort of data matrix. This data matrix refers to *individuals* (objects or events) and to various *observations* made on the *attributes* of those individuals. Geographical data may thus be represented by a matrix of individuals against attributes (above Figure 14.5). This matrix becomes a cube when we observe the attributes of individuals over time. This way of representing geographical data has been developed by a number of writers (Berry, 1964A; Chorley and Haggett, 1967), and has already been referred to above (pp. 212–17; 313–15; 338–45). It automatically involves us in the methodological problem of identifying geographical individuals and those meaningful attributes of them which we can measure and observe. The basic *geographical* individual is one identified by way of a space-time language. It may be a point (zero-dimensional), a line (one-dimensional), an area (two-dimensional), a volume (three-dimensional), a space-time volume (four-dimensional), and there is no logical reason why we should not go into even higher dimensions although the utility of doing so may be doubted in space-time languages. Similarly, the observations made may refer to unidimensional or multidimensional attributes. The data-matrix (or cube) approach effectively collapses a much more complex space into two or three dimensions for recording purposes. Some of the methodological problems involved in this procedure have already been examined. Attention will be confined here to some practical problems in constructing such data matrices and some of the implications which flow from their method of construction.

(a) *The individuals* being observed require precise and unambiguous definition. This is essential to the construction of a data matrix and, indeed, to all empirical work (above, pp. 212–17). It is not possible to consider here the many ways in which individuals may be specified, but some of the practical problems involved can be demonstrated with reference to the important class of two-dimensional geographical individuals, namely, *areal units*. Appropriate areal units for geographical study may be identified in a number of different ways, depending on the objectives of the study and the phenomena under investigation. It is possible to distinguish, for example, between *natural areal units* (based on discrete objects such as farms, countries, lakes, and so on) and *artificial areal units* (most important for dealing with continuous phenomena such as temperature, distance, and so on). In the natural case the boundary of the individual can be determined by reference to the phenomena in question, whereas in the artificial case the boundary has to be imposed. It is also important to distinguish between *singular* areal units (individual farms, for example) and *collective* areal units (a region made up of many farms, for example). This distinction is particularly important when it comes to inferences, for, as was pointed out above (pp. 277–9), inferences made at the collective level cannot be extended (without major assumptions) to the singular level.

The areal units which might be used can sometimes be arranged into some kind of hierarchical structure. A country contains a set of states, the states include a set of counties, and each county includes a set of farms. This hierarchical arrangement is often imperfect;

a farm may overlap into two or more parishes, a town may over-lap into two states, and so on. The hierarchy of areal units is much looser than that which we might construct by the logical division of some universal set (above, pp. 335–7). This loose hier-archical arrangement poses conceptual and inferential problems. The same areal unit may be conceived of as singular or collect-ive. Consider, for example, two counties, A and B, which we wish to compare with respect to agricultural activity. If we say that A has 60% of its land under arable cultivation compared with 24% in B, then the counties are being treated as singular areal units. If we say that 60% of the farms in A are arable farms compared with 24% of the farms in B, then the counties are being treated as col-lective areal units. This difference is important for the two state-ments are completely different from each other in import. Further, to say that 60% of the *land* in A is under arable compared with 24% of the *farms* in B being arable, says nothing of any significance. We are, in short, trying to compare completely different types of areal individual. To establish a relationship between climatic con-ditions in a set of counties and the percentage of land under arable, says nothing whatsoever about the relationship between climatic conditions and the percentage of arable farms.

Similar problems of comparability and inference exist when differ-ent levels in the hierarchy are simultaneously studied. These diffi-culties are usually referred to as the *scale problem*. We shall begin by considering the special case in which distinct steps in the hierarchy of areal units can be recognised, i.e., areal units at one level can be included in areal units at the next level. In such a 'nested' hierarchi-cal situation it should be observed that comparisons can only be made between similar individuals (i.e. individuals at the same level in the hierarchy) and that inferences made about relationships at one level cannot be extended, without making strong assumptions, to any other level (McCarty, *et al.*, 1956; Haggett, 1965B; Duncan, *et al.*, 1961; Harvey, 1968B). This is not to say that conditions at one level are irrelevant to conditions at another. It does indicate that the nature of the analysis is contingent upon whether the individuals being compared or analysed are at the same, or different, levels. Three kinds of situation can be identified:

(i) Same-level analysis means that the individuals can be directly compared since they are at the same level in the hierarchy.

(ii) High- to low-level analysis yields a contextual relationship (e.g. price policy and price support at the national level forms the context within which variation in farm production can be analysed).

(iii) Low- to high-level analysis yields an aggregative relationship (e.g. national output is made up of the output of individual firms).

All of these situations are of interest, but each requires its particular mode of thought and, incidentally, a particular data-collection procedure.

Such natural hierarchies in the phenomena being studied cannot always be identified. With continuously distributed phenomena the areal units are imposed rather than natural, and it is clear that areal units of any size can equally well be chosen. Such arbitrary areal units may be arranged into a hierarchy, but this design is imposed not given. How, in such a situation, can appropriately sized areal units be chosen? One possible method is to choose some arbitrary co-ordinate system and identify uniformly sized areal units within that co-ordinate system. We may collect data, for example, by way of grid squares (Hägerstrand, 1967). It is also possible to use areal individuals which are discrete in one respect as areal units for the collection of other data. It is very tempting in such situations to regard the discrete individual as being natural for the secondary purpose for which it may be used. This is, of course, fallacious. Geographers long ago discovered that national units were not appropriate for discussing climatic characteristics (Hartshorne, 1939, 44-7); and there is, similarly, nothing 'natural' about using counties as units for discussing spatial variation in farming types. Administrative units, which from the point of view of administrative structure may be regarded as singular areal individuals, are frequently used to collect information regarding either continuously distributed or small-scale discrete geographical phenomena. In such a situation there is nothing natural in the way the data are aggregated and it may well be that the administrative units are not 'similar' or comparable from the point of view of the data aggregated within them. Hence arises the necessity to adjust such areal data for the variable area of the administrative unit itself—an adjustment that has not always been made. Thus Chisholm (1960) criticises Dickinson's (1957) discussion of commuting patterns in W. Germany and Belgium on the grounds that Dickinson's analysis ignores the effects of varying sizes of administrative unit. Migration studies have been particularly concerned with this type of problem (Hägerstrand, 1957; Kulldorf, 1955), while Robinson (1956) and Thomas and Anderson (1965) discuss the necessity of weighting data according to areal size prior to correlation analysis. A general discussion on these and similar problems is provided by Duncan et al. (1961) and Haggett (1965B).

The construction of the basic data matrix for areal individuals thus faces a twofold difficulty. Firstly, the appropriate size of the data-collection unit has to be determined. Secondly, it has to be ensured that the units so devised are comparable with one another (or else some weighting system has to be devised for the data recorded in them). Both of these problems appear intractable without major assumptions being made. There are, however, some glimmers of hope for their solution emitted by the interaction between individuals and attributes.

(b) *The attributes* of individuals to be observed and recorded also require unambiguous definition (above, pp. 301–6). But, more important to the present problem, it is also required to select a finite number of *significant* attributes out of an infinite number of possible attributes which we could observe. The decision as to what attributes are significant depends upon objectives and purpose and, ultimately, upon theory. If, however, it is assumed that geographers are concerned with the variation of phenomena over space, then, clearly, we may restrict attention to those attributes that vary over space. This cannot be determined independently of the scale of analysis chosen. The significant attributes to be observed regarding spatial variation among countries are thus likely to be completely different from the attributes to be observed regarding spatial variation within farm units. In the former case variability in tariff policy may be significant whereas in the latter it is irrelevant.

This interaction between the scale of analysis and the significance of different attributes may be used creatively. Consider, for example, the analysis of a phenomenon such as population distribution. The spatial variability in population density portrayed in a set of data depends upon the size of unit used in the collection of population information. Suppose we can use units of any size. It may then be possible to identify that scale (size of unit) at which the spatial variation in the data is at a maximum. The criterion in this case might be the same as that employed in grouping procedures (above, pp. 337–8), namely, that the within-unit variance should be minimised and the between-unit variance maximised. This whole procedure amounts, in fact, to developing rules for drawing regional boundaries so that the resultant regions convey the maximum amount of information about spatial variation, while remaining comparable with one another for analytic purposes. Some technical aspects of such rules are developed later.

The problem of comparability between arbitrarily determined regional units is also a difficult one. Geographers have undoubtedly

suffered in the past because much of their data has been collected
(in official censuses and surveys) in administrative units which vary
in shape and size and which are not easily comparable. The reaction
to this has been a growing demand for such data to be collected in
uniform-size units (kilometre-square cells for example) to ensure
comparability. This demand rests upon a Euclidean concept of
space—a concept which is not really appropriate for the analysis
of many types of geographical phenomena (above, pp. 207–12).
The proposal that all geographical data should be collected in kilometre-square cells means that socio-economic activity, which varies
in intensity from place to place and often has a non-Euclidean
spatial form, is being recorded in a Euclidean framework—thus a
kilometre square in the centre of London is regarded as equivalent
to a kilometre square in the highlands of Scotland. This is not to
propose a non-Euclidean co-ordinate system for recording such data
(although such a system is not beyond our ability to construct), for
it may well be that the Euclidean framework is the most appropriate
for recording and manipulating observations on many different
types of phenomena simultaneously. But we should be warned that
if the distribution of some geographical phenomena *is* non-Euclidean
(and there is considerable evidence that much of it is), then imposing arbitrary Euclidean areal units for observing it is going to create
as many problems as it solves. It may seem reactionary to defend
the crazy patchwork of administrative units of different sizes which
have traditionally been used to record data, but such units have
tended to be adaptable in the face of temporal change and tend to be
much smaller in areas of great social and economic activity than
they are in areas of little economic activity (see, e.g. Haggett, 1965A,
169). In some cases it may be, therefore, more appropriate to use
data collected in old-style administrative units than data collected
in the new-style uniform Euclidean cell. Ultimately, of course, the
appropriate co-ordinate system for a given phenomena will depend
upon 'the intensity of acting forces' exerted by it. The nature of
the co-ordinate system to be used, together with the size of the most
appropriate areal unit embedded within that co-ordinate system is,
like the selection of any space-time language, an empirical problem.
Only with better knowledge of the phenomena we are investigating
will it be possible to modify totally arbitrary decisions on the procedure to be followed in constructing data matrices and in recording
supposedly objective geographical facts. Here, therefore, what may
have seemed obscure and highly abstract issues concerning space-
time languages (above, pp. 215–17) have considerable relevance
for practical decisions. The procedural rules to be followed in

constructing a data matrix need not necessarily remain arbitrary and intuitive for ever more. Some approximate rules will be developed at the end of this chapter (below, pp. 383–6).

II FILLING OUT THE DATA MATRIX—SAMPLING

Assuming that geographical individuals have been identified and that the attributes to be observed have been decided upon, the next step in recording geographical facts is to fill in the data matrix or data cube if observations are to be made over time. The simplest procedure here is, of course, to provide a measure on the attribute of the individual in each cell in the data matrix (the measure may be of any kind, including nominal). Here the rules of measurement are crucial (above, pp. 306–25). It may prove tedious and unnecessary to fill in every cell in this matrix, particularly when there are large numbers of geographical individuals to be observed, and the task becomes impossible, given an infinite number. Complete enumeration of the geographical population and complete observation on all individuals in that population is thus a rather rare occurrence. When this is not possible it is necessary to resort to some sampling design.

The aim of sampling is to form a small data matrix (or data cube) out of an enormous data matrix (or cube) in such a way that the small matrix (or cube) provides approximately the same amount of information needed for a given purpose as would the larger matrix (or cube). The sample data will not provide exactly the same information, but they provide estimates of the population data within specified limits of error. The rules of sampling are extremely important. Thus Stuart (1962, 9) writes:

> The sheer weight of tabular material which emerges from a sample survey of any size often tends to make the survey-user insensitive to the credentials of the sample itself. And yet, I shall argue, the credentials of a sample are not only important for the interpretation of the results, but are in principle the *only* information of value we can have in this respect.

All non-deductive inferences in geography are therefore contingent upon an appropriate sampling procedure developed with respect to a specified population. The problem of specifying the population and the individuals contained within it has already been considered (above, pp. 277–9; 351–4). Attention will here be focused on the problem of sampling. Without a specified sampling procedure non-deductive inferences are likely to be useless or downright misleading, while different sampling designs exist in profusion and hence require evaluation with respect to a given study. In general the selection of one out of many potential sampling designs depends on:

(i) The purpose of the investigation.
(ii) The nature of the phenomena being investigated.
(iii) The mode of inference envisaged (e.g. particular forms of statistical inference require particular kinds of sampling procedure).
(iv) The cost (in time, manpower, and money) incurred. The first three factors are important in assessing the validity of a given sampling procedure, while the fourth is an extremely important practical consideration governing the feasibility of a given sampling design.

The procedural rules that govern the application of sampling designs have been examined by numerous writers in great detail. Stuart (1962) provides a good elementary introduction, while Cochran (1953) discusses the theory of sampling at length. The practical problems involved in sampling are treated very thoroughly in Yates (*1960*)—the classic work on the subject—and Moser (1958) and Hansen *et al.* (1953) also provide good accounts. The special topic of spatial sampling, which is of particular interest to geographers, has been considered by writers such as Matérn (1960), Krumbein (1960), Krumbein and Graybill (1965, chapter 7), Greig-Smith (*1964*, chapter 2), and Kish (1965, chapter 9). Accounts in the geographical literature are provided by Berry (1962), Berry and Baker (1968), Haggett, (1965A, 191–200) and Holmes (1967). On account of this vast literature on sampling theory and sampling design, consideration will be restricted to a few issues of basic methodological and geographical interest.

Given a population of n individuals and a sample of N individuals drawn from that population there are $\binom{n}{N}$ possible samples of size N and if N is not fixed there are 2^n possible samples which can be drawn from that population (this includes the special cases where $N = 0$ and $N = n$). The problem in sampling is to determine the 'best' sample out of these 2^n possible samples. To help us in this situation a number of 'standard' sampling procedures have been designed. The general theoretical characteristics of these procedures have been examined in the literature in varying detail, and their accuracy, efficiency, and precision, have been evaluated for certain types of empirical situation (see e.g. Stuart, 1962; Cochran, 1953; Hansen *et al.*, 1953). We can thus talk of purposive or probabilistic sampling, sampling with or without replacement, simple random sampling, systematic sampling, stratified sampling, cluster sampling,

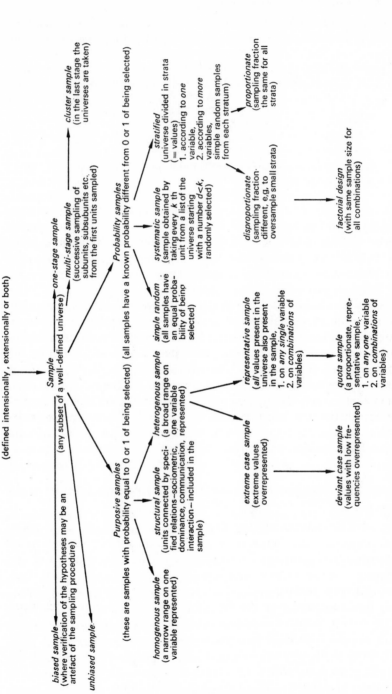

Fig. 19.1 A typology of sampling Designs *(from Galtung, 1967).*

multistage sampling, multiphase sampling, and so on. Galtung (1967, 56) and Haggett (1965A, 195) provide simplified typologies of these sampling designs. Galtung's typology (Figure 19.1) identifies twelve major sampling forms but, as he points out, with three-stage sampling even this simplification yields 12^3 (1, 728) possible sampling designs. By the time we add to this the use of various sample sizes and, with stratified sampling, the various sampling fractions that can be chosen within strata, we find that even 'standard' sampling designs provide us with a very wide range of choice.

Unfortunately, there are no standard rules which we can use to reduce this range to a relatively simple choice among a few alternatives. Thus 'the entire richness is essentially at one's disposal' (Galtung, 1967, 58). The selection of a particular sampling design, therefore, can be made only on the basis of a particular research objective. This may seem, as Galtung (1967, 49) suggests, a fairly trivial conclusion, but in fact it is 'a prescription which becomes less trivial in the light of the number of cases where standard recipes are followed just because they exist and are simple to follow'. The selection of a sampling procedure, like the selection of an appropriate geometry or the selection of any mathematical language, is an empirical problem. It depends upon an empirical assessment of the reasonableness, efficiency, and feasibility, of a given design in the light of a given objective. Most texts discuss the application of various criteria in selecting a sampling design (e.g. Hansen, *et al.*, 1953, 4-11; Kish, 1965, 23-6), but before these can be discussed an important distinction must be made between *purposive* or *judgement sampling* on the one hand (for which only intuitive criteria exist) and *probability sampling* on the other (for which the criteria are measurable).

(1) *Purposive or judgement sampling—the 'case-study' approach to geographic research*

Purposive or judgement sampling can take a number of forms, the most important of which is the selection by an 'expert' of 'typical' or 'representative' samples. The selection of the sample depends on the judgement of the 'expert'; and it is clear that different 'experts' may select different samples, and that there is no way of showing objectively how representative the sample is of the population. This is not to say that judgement sampling is worthless, for, particularly in the early stages of an investigation, it has an important role to play. But there is no means of knowing how and in what respects the sample chosen is biased. These limitations are discussed by Yates (*1960*, 9-10), Hansen *et al.* (1953, 5-9), Kish (1965, 19), etc.

Judgement sampling has been of great significance in geography, for, as Blaut (1959) and Haggett (1965A, 191) point out, the 'case-study' approach in geography amounts to the selection of a 'typical' or 'representative' individual from a population and a study in depth upon that individual. Most traditional studies are of this form, and until a few years ago most of our understanding of geographic phenomena rested entirely upon analytic insights gained from the intensive study of judgement samples. A typical city, a typical farm, a typical cyclonic disturbance, a typical river capture, and so on, tend to form the very basis of our understanding. These typical examples tend to correlate very strongly, of course, with the idealisations (or *ideal types*, see above, pp. 91–6) which express our conceptualisations of the reality we are dealing with. We tend to look for that particular individual in a population that has the characteristics of our idealisation. In other cases idealisations are developed from a particular individual that is thought to be representative of a population. Here the dangers of purely circular argument are considerable.

The general problem posed by the case-study approach is its lack of generality. Inferences made from the typical example selected are uncontrollable with respect to the total population. Such inferences are a matter of intuitive judgement and we cannot rely upon formal non-deductive languages in making them. We thus remain uncertain how objective and how reasonable such inferences are. Blaut (1959) thus advocates a combination of the case-study approach (which allows intensive study of structure and relationship, and hence hypothesis formation) with a probability-sampling approach (which allows the generality of the conclusions to be assessed and, perhaps, the hypotheses to be formally tested with respect to a given population). This line of approach has a great deal to recommend it, for, as we shall see, the design of a probability sample is easiest when we have a considerable understanding of the structure of the phenomena we are investigating and a clear statement of a set of hypotheses to be considered.

In practice much geographical investigation rests on judgement samples. Inferring general-population characteristics from a typical example amounts to inferring a generalisation on the basis of a sample of one (above, pp. 281–6). It would be interesting to assess, for example, how far our knowledge of internal city structure rests entirely on a judgement sample that contains the city of Chicago and nothing else. From Park and Burgess through Hoyt to Berry at the present time, we find that Chicago has tended to dominate our thinking about urban structure and form. Generalisations about urban form (concentric-zone 'theories', multiple-nuclei 'theories', and the like—which incidentally scarcely conform to the scientific notion of what a theory should be) from the structure of Chicago (with occasional corroborative evidence from elsewhere), are not necessarily false or unreasonable—they are merely not generally substantiated. Such substantiation can be provided, it is generally agreed, only by the objective procedure of probability sampling.

(2) *Probability sampling*

Probability sampling refers to 'a formal procedure for selecting one or more samples from the population in such a manner that each individual or sampling unit has a known chance of appearing in the sample' (Krumbein and Graybill, 1965, 148–9). Given that each individual in the population has a known and non-zero probability of being sampled, and given the formal procedure designed for collecting the sample, it is then possible to use formal procedures of non-deductive inference (e.g. statistical inference in particular) to evaluate the relationship between the sample and the population. It is then possible to evaluate several different sampling designs in relation to a given objective and measure which is the most efficient.

In probability sampling the main concern is with the conditions holding in the population. The sample provides an estimate of these conditions. The difference between the population condition and the sample estimate is known as the *accuracy* of the estimate. When it is possible to survey the complete population (and therefore calculate actual conditions) it is possible to evaluate the accuracy of a sampling procedure. An accurate sample-design will thus provide us with unbiased estimates of the population values. *Precision* refers to the spread of the sample estimates around the true population value. It is usually measured by the variance of the sample estimates. In evaluating sample designs it is obvious that accuracy and precision provide two important criteria. In most cases we are able to measure precision directly from the sample data, while accuracy is more difficult to assess.

A formal probability-sampling procedure allows us to map the sample-population relationship into the calculus of probability and consequently allows us to draw non-deductive inferences which are objective in the sense that they are free from systematic bias. By following the procedural rules laid down in probability sampling therefore, we can be sure that we have obtained a sample which 'will reproduce the characteristics of the population, especially those of immediate interest, as closely as possible' (Yates, *1960*, 9). This is not to say, however, that all probability-sampling procedures are equally feasible or equally efficient. It has thus been shown that in a population made up of heterogeneous groups it is far more efficient (in the sense that a given level of precision can be achieved with a far smaller sample) to use stratified sampling than simple random sampling, provided the strata identified are reasonably homogeneous within and heterogeneous between. Stratification thus amounts to a classification procedure which minimises within-group variance and maximises between-group variance; and it is often useful, therefore, to precede a sampling design by some grouping procedure. In other cases it proves impossible to conduct a given type of sampling procedure either because it is not operationally feasible (the selection of a pebble on a beach at random for example) or because it is too costly. In many cases, therefore, cluster sampling (sometimes known as batch sampling or 'grab' sampling) may prove the best procedure in spite of its many technical limitations. The selection of a particular sampling design thus depends upon the accuracy and precision required, the efficiency and feasibility of a particular procedure, and the cost involved.

The sampling design selected, however, also depends upon the substantive objectives of an investigation. In particular, it depends upon the structural or functional characteristics of the population which we are interested in. Clearly, many kinds of situation can arise here, but it will be useful to consider just two very general situations and examine how the sampling procedure is related to the objectives.

(a) *Sampling to estimate the characteristics of populations* turns out to be one of the commonest, and one of the simplest, situations in geography where probabilistic sampling procedures are employed. The aim here is to determine some specified characteristic(s) of the population by way of sampling procedure. Typically, we might be interested in estimating certain descriptive statistics (e.g. mean and variance) of the population by way of a sample. We might try to estimate mean annual income per household, average turnover per retail establishment, average size of pebble on a beach, and so on, by the use of sample estimates. A particular case of such a procedure which is of great interest to the geographer is estimating the proportion of land surface given over to a particular type of land use. Here the total population may be regarded as the infinite number of points that make up the area or the finite number of small areal units that divide it (e.g. fields).

Probabilistic spatial sampling is particularly important in geography (Berry, 1962; Berry and Baker, 1968; Haggett, 1965A,

191–200; Holmes, 1967) and provides one important way for generalising from a small number of observations to a much larger area (say a country or region). Such spatial sampling may be conducted using:

(i) Points.
(ii) Lines (traverses).
(iii) Areas (quadrats).

Each of these forms has its virtues and its difficulties. Sampling by way of points in the field may be very costly, and therefore traverses and quadrats may be more appropriate. Point sampling of point-type phenomena is also very inefficient (mainly because of operational difficulties, but also because in general it is true that sampling a characteristic that has a very low probability of occurring requires either a very large number of samples or an alternative sampling design which samples large units at one time, e.g. quadrats). Point sampling of areally distributed phenomena (e.g. land-use type) appears to be relatively efficient. Quadrat sampling has an advantage of operational simplicity and low cost in many field situations, and is also very useful for the study of point patterns, but it suffers from a number of difficulties. These are dealt with in great detail by Greig-Smith (*1964*, chapter 2) and we shall consider these later in this chapter.

Within each form of spatial sampling, however, it is possible to devise numerous probabilistic designs. Six ways of designing a point sample are thus illustrated in Figure 19.2. The problem then arises how to choose among these various potential sampling designs. Berry and Baker (1968, 94) point out that in the case of spatial sampling the 'choice of sampling procedure for any phenomenon depends on how that phenomenon is distributed.' They go on to suggest that if the phenomenon being studied is randomly distributed then most sampling forms are appropriate and the choice therefore amounts to selecting that which is easiest—which in the case of sampling on a map would be a systematic sampling design. If there is a linear trend in the distribution, stratified sampling will be more efficient than systematic sampling, and systematic sampling in general more efficient than simple random sampling. If there is serial correlation (spatial auto-correlation) in the distribution of the phenomenon (and there invariably is some amount of this), then it is difficult to devise simple rules, for the 'relative precision of the sampling procedures depends upon the shape of the serial-correlation function'. When this serial-correlation function is unknown, no optimal sampling design can be identified, but Berry and Baker suggest that a *stratified systematic unaligned sample* (Figure 19.2) is here

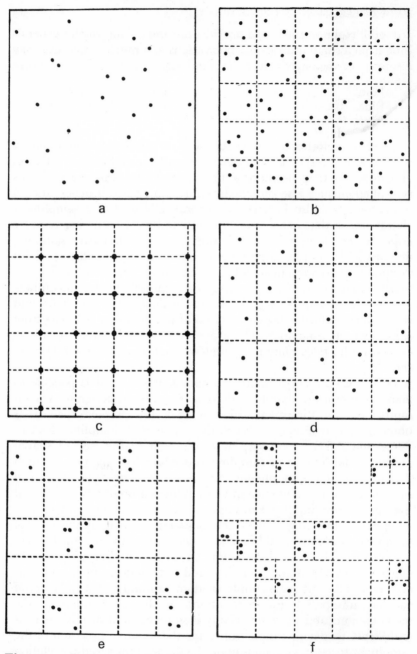

Fig. 19.2 Types of spatial point sampling Design : **A :** simple random sampling; **B:** areally stratified random sampling; **C:** systematic random sampling; **D:** systematic unaligned random sampling; **E** and **F:** two versions of nested random sampling (*after Berry and Baker,* 1968; *and Krumbein,* 1959).

an appropriate form. It appears, therefore, that an optimal sampling design can be identified only if we possess (or can assume) a good deal of information about the phenomenon being sampled.

(b) *Sampling to identify relationships among attributes in a population* is a substantive problem that may be tackled using probability sampling. It is, however, rather more complex than sampling to estimate descriptive statistics of the population, for the sampling procedure here operates as a kind of experimental-design procedure in an essentially non-experimental situation. In general geographic research faces the problem of dealing with multiple simultaneous interactions among a number of attributes within a population. In chapter 20 we shall examine complex causal interactions among collections of variables in some detail. The classic way out of such a situation in experimental science is to set up factorial experimental designs. Thus Fisher (*1966*, 94) writes:

> We are usually ignorant which, out of innumerable possible factors, may prove ultimately to be the most important, though we may have strong presuppositions that some few of them are particularly worthy of study. We have usually no knowledge that any one factor will exert its effects independently of all others that can be varied, or that its effects are particularly simply related to variations in these other factors.

This kind of situation poses difficulty at the best of times, but it is possible in the experimental sciences to set up experimental designs to determine significant relationships and physically to control our unwanted variables. In the kinds of non-experimental situation which geographers face it is not possible to impose physical controls (except very occasionally). It is here that the design of a survey and the sampling design itself become of crucial importance. Kish (1965, 394) suggests the following typology of sources of variance in any data set:

(i) The explanatory variables which are the object of research.
(ii) Variables which might interfere with the explanatory variables but can be controlled, e.g. by sampling procedure.
(iii) Variables which cannot be controlled and may thus be confused with the explanatory variables.
(iv) Variables which cannot be controlled but whose interference is random (or can be made random by a sampling procedure).

The variables in (iii) pose the greatest difficulty; and it is the aim of survey design and sampling design to relocate these variables into class (ii) or into class (iv). Efficient stratification can achieve the former, whereas the latter may be achieved by randomisation. Suppose, for example, we are interested in examining the relationship between terrain and farm size, and farm income. It is likely that terrain affects farm size and there is, therefore, a multicollinearity in the set of relationships. Blalock (*1964*, 89) writes:

> One way out of the difficulty posed by multicollinearity is to make two independent variables completely unrelated in a *sample*, even though they are highly correlated in the population. This can readily be achieved through stratification, which has a certain analogy with manipulations in experimental design.

It is thus possible to design sampling systems in such a way that a number of interrelationships are simultaneously studied. As Fisher (*1966*, 102–4) points out, such designs have considerable advantages in that they yield a good deal of information from one survey design. Haggett (1965A, 300; Figure 19.3) provides a geographical example

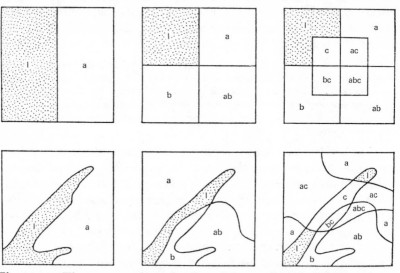

Fig. 19.3 The construction of a factorial sampling design for random sampling in populations affected by multiple interactions. At each stage the region is subdivided by a further factor, and sampling is then random within each factor-combination region (*from Haggett*, 1965A).

of such a design by isolating sixteen types of geographical area on the basis of the various factorial combinations of terrain (*a*), soils (*b*), farm size (*c*), and farm accessibility (*d*).

Multipurpose sampling designs, and designs to examine isolated relationships within complex relationships, are difficult to evaluate. But in the non-experimental sciences such procedures are clearly of great importance; and geographers are likely to draw very considerably on such kinds of design in the future. The more information we possess on the nature of the interactions in the population, however, the easier it will be to design adequate sampling procedures—and here the case-study approach has much to offer as an exploratory device.

(3) The sampling frame

In order to sample we require some sort of sampling frame which locates the individuals in the population. If, for example, we are using a random sample we need to locate the population in some way in order to apply the random sampling procedure. Typical sampling frames are registers of electors or ratepayers, a bundle of air photographs, a list of the individuals in the population, and, most important from the geographical point of view, the map. The importance of the sampling frame cannot be underestimated, for 'the whole structure of a sampling survey is to a considerable extent determined by the frame' (Yates, *1960*, 60). Yates and Moser (1958) consider the problems of providing a sampling frame, and they point out that a frame may be defective because it is inaccurate, incomplete, subject to duplication, inadequate, or out of date.

There are many situations in geography where the provision of a sampling frame is peculiarly difficult. Sampling mobile populations (cars, crowds, and the like) poses considerable difficulty, and sampling simultaneously over space and time (sampling in the data cube instead of in the matrix) is similarly quite complicated. More important, however, is the case where the total population is not available for study (either because we are talking about some hypothetical population rather than an existing one, or because there are difficulties in locating many of the individuals in the population). Here Krumbein and Graybill (1965, 149–53) employ a useful distinction between a *target population* (which we require information about but whose members are not all available for sampling) and a *sampled population* (a subset of the target population which is available for sampling). Strictly speaking, probabilistic inferences can be made only from the sample to the sampled population, but these conclusions may be extended to the target population on the basis of substantive judgement *and only on that basis*. This kind of situation is very common in geography. We may use a sample of students (because they are available and responsive) to construct mental maps of areas, measures on actual town spacing (because they exist) as a sample of possible measures, meteorological stations (because they happen to have been set up), and so on. This often amounts to accepting an incomplete (and perhaps an inaccurate and inadequate) sampling frame as the basis for sampling.

In geographic sampling the map frequently functions as the basic sampling frame. Once we possess a set of co-ordinates it is possible to sample among them at random (using a table of random numbers for example) or to construct some more complex sampling design (Figure 19.2). The success of these methods ought not to blind us to the fact that using the map as a sampling frame amounts to accepting all the characteristics of the map within the sampling design. There are dangers in using a map constructed to represent physical distance as a sampling frame for analysing social interactions. Such a sampling frame is inappropriate and inadequate simply because it involves sampling non-Euclidean phenomena in a Euclidean frame. In a technical sense, therefore, the appropriateness of physical maps depends upon whether or not the auto-correlation function remains constant over the map surface or whether it changes systematically. When the form of the auto-correlation function is known it is possible to develop sampling designs on a Euclidean frame to counteract its effects. It might be appropriate to provide some map transformation from a Euclidean to a non-Euclidean surface before sampling. A random-sampling net cast over Harris's cartogram of retail sales in the U.S.A. (Figure 14.7) might yield more interesting information about retail trade than a simple random sample on a physical map of the U.S.A. Stratified random sampling (by area) with variable fraction (in proportion, say, to total population) on the physical map would, however, probably yield the same information just as efficiently.

(4) Sample data in geography

At the beginning of chapter 17 it was suggested that observation models could be designed to search reality for hypotheses or to structure reality to test hypotheses. The role of sampling in observation is simply to cut down the number of observations we need to make in either of these situations. It is generally agreed that a few well-selected observations can provide as much evidence for a hypothesis as a whole mass (above, pp. 136–7). The important thing is to possess a method of collecting the observations which ensures the maximum independence between the observation-collection procedure and the proposed hypothesis. Probability sampling, by anchoring the selection of observations to a purely chance set-up (above, pp. 236–9; 247) ensures this independence. More complex sampling designs are aimed at controlling the selection of observations in such a way that they can be used to test isolated hypotheses in essentially multiple-interaction situations. By using probability sampling it is also possible to map inferential problems into the calculus of probability and to use that powerful language to determine the level of confidence we can have in our results. Judgement sampling, on the other hand, cannot be used in such a fashion, but it is nevertheless of considerable utility since it allows the intensive examination of interactions and relationships.

Whatever approach we adopt to sampling, however, it is vital to realise that the conclusions drawn are entirely dependent upon the adequacy of the procedure used. Any sampling frame, such as the map, possesses certain characteristics, as does any sampling procedure. These characteristics provide us with a set of 'credentials', to use Stuart's (1962) term, and it is these credentials that provide us with the means for determining the validity of subsequent manipulations and conclusions. But the good employer of a technique will ask at the very beginning of an investigation whether its credentials are suitable for its employment in the pursuit of a particular objective.

III REPRESENTING DATA—THE MAP

Geographers possess a number of techniques for portraying, representing, storing, and generalising, information. Of these there is none quite so dear to the hearts and minds of geographers as the map. Hartshorne (1939, 249) even goes so far as to suggest the following 'rule of thumb' to test the geographic quality of any study:

if his problem cannot be studied fundamentally by maps—usually by a comparison of several maps—then it is questionable whether or not it is within the field of geography.

Wooldridge and East (1951, 64) quote with approval from H. R. Mill:

In geography we may take it as an axiom that what cannot be mapped cannot be described.

Sauer (*1963*, 391) writes:

Show me a geographer who does not need them constantly and want them about him, and I shall have my doubts as to whether he has made the right choice of life. . . . Maps break down our inhibitions, stimulate our glands, stir our imagination, loosen our tongues. The map speaks across the barriers of language; it is sometimes claimed as the language of geography.

Given the great esteem in which geographers hold the map as a means of description, analysis, and communication, it is rather surprising to find that many aspects of cartographic representation remain unanalysed. Most attention has been paid to solving the numerous technical problems facing the map-maker. There is a substantial literature dealing with such problems as map projection, map design, map symbols, and the like. Texts by Robinson (1952; *1960*) and Raisz (1948), and numerous articles (see the recent review by Board, 1967) deal adequately with these issues. This is not to say that all technical problems have been satisfactorily solved. Consider, for example, the problem of map projection which, in terms of designing projections from the spherical to the flat plane, appeared to have been solved by the late nineteenth century, but, as we have already seen (above, pp. 219–23) has been raised in a new and much more difficult form by the problem of map transformation. Nevertheless this substantial technical literature (much of which, it must be admitted, merely provides a manual of how to proceed), contrasts markedly with the almost total lack of consideration for the logical properties of the map as a form of communication. The map is a most complex device. Robinson (1965, 35) thus writes:

> The modern map, even a relatively simple, straightforward one, is an extraordinarily complex form of graphic expression because it is an artificial thing embodying, in addition to its visual complications, several transformations of reality in scale, shape, and symbolism that are quite beyond the normal experience of most people.

The map is, in short, a symbolic system. It is a complex language— 'the language of geography' perhaps—whose properties we know very little about. Thus Dacey (unpublished)[1] writes:

> Cartography does not provide a set of rules and principles forming . . . a mapping strategy. Neither does cartography provide a methodological statement relating the operations involved in map making, primarily those concerned with projection, design and construction techniques, to the conceptual and research processes in which the map has such a vital role. For example, it is easy to frame many pertinent questions about the selection of symbols, the assignment of information to symbols and the manner by which those symbols are related to the interpretation of map evidence; however, the only available answers invoke conventions, the sanction of prolonged use and appeals to the natural or obvious representation.

Dacey goes on to suggest that the map may be thought of as a language in the formal sense—it possesses *pragmatic* characteristics, *semantic* characteristics, and *syntactical* structure. This view of the map as a specialised language has some interesting implications. It is surprising to find, for example, how many philosophers of science, in seeking to explain the nature of scientific theories, have resorted to the map as an analogy. The map, they say, allows you to find your way around in reality, it allows you to say things about places you have never been to, and so on; likewise a theory helps you to find

[1] I am indebted to Dr Dacey for the loan of certain unpublished notes.

your way around and to say things about phenomena not yet observed (above, pp. 87–99; 169–72). The implication of such an analogy is that a map and a scientific theory are, in some respects, isomorphic. Both may be regarded as artificial languages for discussing the facts to be explained or described. Hence the pungency of a comment attributed to Ullman that 'the map was a theory which geographers had accepted' (Bunge, *1966*, 34).

Maps, like theories, are 'free creations of the human mind'—many a medieval map is quite uninhibitedly so. It is similarly open to any map-maker to create 'a system of apparent wisdom in the folly of hypothetical delusion' (above, pp. 87–8). Many a layman, and (dare it be said?) many a geographer, has been the victim of such hypothetical delusion. Yet geographers claim that their maps do bear some relationship to reality, that their maps do reflect and record reality faithfully in some respects, and that they can draw conclusions about reality by drawing conclusions from a map. Maps are therefore like scientific theories, controlled speculations. Yet we have no analytic philosophers to tell us what these controls are. The map is not automatically some objective representation of reality—though some would appear to think it is. John K. Wright (1966, 33) writes:

> Every map is thus a reflection partly of objective realities and partly of subjective elements, a circumstance that has been dealt with implicitly in all works on the art of cartography, but seldom considered explicitly as a subject in itself.

Map-makers—and, it should be added, map-readers—are human (to use the title of Wright's essay). The basic methodological problem which the map poses, therefore, is how to evaluate the respects in which a map reflects or represents reality and how to write out the rules which relate the reality represented to the symbolic form which we interpret. Here, of course, it may be argued that there are many situations in which the relationship is so obvious (and so controllable) that there is little or no use for such rules (most topographical maps fall into this category). But maps of socio-economic activity, flows, thematic maps, and the like do not fall into this category; for the relations they portray, Dacey (unpublished) writes,

> are artificial creations of the human imagination, the manner by which the physical reality is ultimately related to its symbolic construct is solely a function of a conceptual scheme. Under these conditions results that are obtained from a map are completely dependent upon the symbolic properties of a map and are independent of anything given in nature. Hence, results that are obtained do not necessarily have any reference or implications to that which is the reputed subject of the map and, thereby, the conceptual interpretation which is given to those results is ambiguous.

The map has validity (with respect to the real world) only if the conceptual schema which governs its construction itself has validity with respect to that same real world. The map is, therefore, simply a model of a theory about real-world structure. Constructing a map *without* explicit theory amounts to stating an *a priori* model; *with* explicit theory it amounts to stating an *a posteriori* model. The *a posteriori* model, however, can be related only to the domain of the theory which it represents. Using a particular *a posteriori* model to examine phenomena outside the domain of the theory which it represents amounts either to assuming the domain can be extended to phenomena not initially covered by the theory, or to using the *a posteriori* model as an *a priori* one. Consider the following example.

Maps represent, among other things, the relative location of objects in space. Most maps are constructed with reference to a Euclidean physical space and they are, therefore, *a posteriori* models of a physical theory about real-world structure—a theory, incidentally, that is generally accepted for the location of objects in physical space and forms for the geographer a quite self-evident theory about real-world structure. This 'obvious' theory and the *a posteriori* model derived from it have subsequently been used to map complex socio-economic relationships. This use amounts to postulating a Euclidean *a priori* model, or assuming that the relative location of objects in socio-economic space can be adequately described by a Euclidean theory of spatial structure. The nature of the spatial structure, however, is not given *a priori* and the determination of it is essentially an empirical problem. This problem has already been considered at length (above chapter 14), but the important point for our understanding of the map is simply this: the extent to which we can use the map to discuss the location of objects and events in space depends entirely upon the adequacy of the *text* linking map structure with real-world structure. If we possess no *text* then no inferences can be drawn about the real world. If we assume a *text* then we should be clear as to the nature of the assumptions and not infer with respect to the real world what we have already assumed in constructing a *text*. If we possess a *text* we should be prepared to state it explicitly so that we do not make inferences respecting domains not covered by that *text*.

What this amounts to, as Dacey (unpublished) points out, is a full discussion of the *semantics* of the map. Semantic issues are not simply related to the selection of a geometry but also relate to the symbols employed to represent phenomena. Here Dacey suggests that 'semiotics' or the theory of signs, has much to contribute to our understanding of the way in which the map can be used to convey different kinds of information. 'By subsuming the map symbol under the generic category of sign,' writes Dacey, 'the entire factual and relational content of the highly general theory of signs is made available as a basis for studying the map symbol.' We can thus apply formal semantics to understand the relationship between the map and the reality it is designed to represent. Carnap (1942, 24–5) suggests that the final aim of any semantic system is to establish *rules of truth*—that is, to establish a set of rules which allow us to determine whether a particular 'sentence' derived from a symbolic system, such as a map, is or is not true. Given such truth rules, it is possible to classify statements made from the map as true or false. Carnap points out that this procedure requires certain preliminary operations. First, a classification of signs is required. Second, rules of formation are required which state how and under what circumstances new signs and symbols may be formed from this initial classification (a simple example of such a formation rule might be the interpolation of a new contour line between two existing contour lines—the rules in this case state that interpolated contours should not cross etc.). Third, designation rules are required which relate the signs (map signs in this case) to some other sets of signs. These designation rules are of great interest for, as Carnap points out, there are no factual assertions in pure semantics, only conventions which relate one set of signs to another set. What this means in the mapping context is that the map signs do not represent the real world directly, but represent geographic concepts about the real world. The factual statement which a map makes is at two stages removed from the reality being described. Formal semantics simply discusses the relationship between geographic concepts and the symbolic representation in map form. From this it follows that if the geographic concepts are ambiguous and fuzzy, then the map statement, although it appears precise, will likewise be fuzzy or ambiguous. Thus a map of the distribution of agglomerations is only as good as our concept of an agglomeration. Here the process of definition becomes of considerable importance (see above, p. 304). The truth rules therefore allow us to identify necessary truths, given our prior conceptualisation of reality. Formal semantics provides us with a way of discussing the logical consistency of geographic statements and map statements. The degree to which the logically true statements are empirically true can be assessed only by evaluating the relationship between the geographic concepts and the reality these concepts are concerned to discuss.

It is also possible to discuss the syntax of maps. Here the concern is with the internal structure of the map statements and its basic form as an abstract calculus. There are some special features about map languages which are of some interest. In general, it appears that negative statements are not expressible in map languages (e.g. we can infer that something is not there only by noting the lack of any positive statement). Thus map statements have a number of peculiarities which in themselves are worthy of study.

The *pragmatics* of mapping amount, Dacey (unpublished) suggests, to the 'study of the relationships between the map symbol and the map-maker or the map symbol and the map-user'. This relationship is very important to our understanding of the map as a means of communication, for it is concerned with the way in which, for example, the map-user perceives the information which is fed to him. Again, it is a curious feature of the geographer's understanding of maps, that the general discussion of map design and map interpretation has not been accompanied by analytical and empirical investigation. In a rare exception, Ekman *et al.* (1963) have investigated certain aspects of map symbolisation by psychophysical methods. The basic concern of this investigation was to examine the relationship between the information input into the map, its symbolisation, and the interpretation of those symbols to yield information. The experiment specifically considered the relative efficiency of portraying information by way of volumes (e.g. spheres and cubes) or by way of areas (e.g. squares and circles). The study 'demonstrated that the estimates of the "volume" of these symbols merely reflect their perceived area', and therefore it can be assumed that portrayal of information by 'volume'-type symbols is likely to lead to a good deal of distortion in the interpretation of that information. Geographers have long been aware of this kind of problem, as the comments by Robinson and Wright, quoted above, show. Yet it is a problem that has been argued about rather than explicitly investigated. From the point of view of the map-user, the information which he derives from a map is a problem of perception of information contained in symbols. From the point of view of geographic form, this amounts to an ability to discriminate pattern, and to break down a multicomponent map into its constituent parts. The map is, in short, a stimulus which is designed to elicit a certain response. As map designers we have the ability to manipulate the stimulus, but that ability comes to nothing if we have no clear idea of the typology of the responses elicited. It goes without saying, of course, that it is difficult to design a coherent map without a notion of the responses required—in other words maps need to be designed for specific purposes. This is not to say that we should restrict the design of maps to those which have the very specific task of conveying unambiguously a restricted class of information. Maps have long fascinated people; and simply staring at them and living with them has led to the formulation of many an interesting geographic hypothesis. Perhaps, therefore, there is a case for designing maps to stimulate hypothesis formation—maybe even stimulate our glands, as Sauer might say. Here the work of psychologists on the reaction of people to pictures has a great deal to tell us about map construction. Simple psychophysical experiments on the perception of shape, line, orientation, and so on (Ekman, *et al.*, 1963; Beck, 1967) here provide us with some rudimentary ideas, but the perception of 'information' and 'structure' in multicomponent situations has also been intensively studied by psychologists (see the account by Garner, 1962). The evidence provided by such studies shows how ambiguity and uncertainty, and the consequent ability to discriminate structure, are dependent upon the stimuli and the cultural background of the subject from whom a response is being elicited. Segal *et al.* (1966) have demonstrated the considerable cross-cultural variation that exists in the perception of shape and form. Berlyne (1958) has also studied the preferences of subjects with respect to pictures of varying levels of complexity and found that in general people prefer to look longer at more complex pictures than they do at simple pictures. But the greater the complexity the greater the level of uncertainty and, hence, the greater ambiguity in interpretation. Too much complexity can saturate the 'channel capacity' of the human mind and merely succeeds, therefore, in eliciting a negative response. Ambiguity and uncertainty are not in themselves undesirable—a certain openness of meaning, as Kaplan calls it (above, pp. 304–6) is necessary if our understanding is to progress. It may even be possible to design maps deliberately that have the right amount of ambiguity to stimulate our glands and not too much to cause us to react entirely negatively. For most of us, Rembrandt and Cézanne would have made better cartographers than Jackson Pollock.

Maps have undoubtedly provided this kind of stimulus in the past. Most maps contain a certain element of pure poetry, and hence contain all of Empson's seven types of ambiguity, and more. But maps are also designed to convey specific information and to do so in such an unambiguous way that we are able to make specific decisions relating to real-world activity on the basis of the map evidence. The map functions as a communication system and it is relevant to ask how much noise that system can contain. The map makes a firm visual statement (colours change abruptly at boundaries, and so on) and this can often be deceptive. If, for example, we possess poor data, then how can we express notions about measurement error (above, chapter 17) in the map language? Choynowski's (1959) example of maps transformed from simple ratio measures to probabilistic measures of significance (Figure 15.5) is an excellent illustration of how different visual impressions can arise from various treatments of the input data to the map. A map can be no better than the data that is used as input to it and it is depressing to reflect that perhaps many a rational explanation has been found for map patterns that may be a substantial per cent noise. Ambiguity is often undesirable in the extreme. The map could scarcely function as an aid to navigation if it possessed a high level of ambiguity and noise. Within the limits of the map projection system and of data reliability, therefore, maps can be used to make firm and unambiguous statements about certain aspects of geographic reality. It is perhaps worrying to realise that many a planning decision is taken on the basis of map evidence. Town plans are thus drawn up on maps in which objects are located with respect to one another in a physical space rather than in a socio-economic space. Such a procedure is reasonable if the two spaces correspond, but they do not in many instances, and even if they do it is important to *show* that they do. Deciding the future distribution of socio-economic activity on the basis of a physical Euclidean spatial system does not seem a very realistic way to proceed when there is a considerable probability that socio-economic spatial interaction is best mapped into a non-Euclidean geometry. It would be tempting to relate many a planning disaster to this procedure. Certainly most town-planning appears geared to suit an inhabitant suspended some 10,000 feet up who is content to spend his life looking down—a position, incidentally, that the town-planner roughly finds himself in when looking down at a 6-inch to the mile map.

The map, it must be understood, is a model of spatial structure. Before we can accept that map as a theory about actual spatial structure (and therefore act upon the basis of that theory) we require to show that the model is *empirically* realistic with respect to the phenomena it is designed to represent. The choice of map, like the choice of geometry, is essentially an empirical problem. It cannot be determined *a priori*, nor can we afford to use a map relevant to one theoretical domain to discuss phenomena in a totally different domain without *showing empirically* that such a treatment is reasonable. The use of the map, like the use of any kind of model, poses a number of problems concerning inference and control. It is time, therefore, that these methodological issues were explicitly and comprehensively discussed.

IV REPRESENTING DATA—THE MATHEMATICAL REPRESENTATION OF PATTERN

It has been shown that the ability of the map-user to discriminate and evaluate the information contained in the map is not free from subjective elements and that the more the information contained in a map the more ambiguity and uncertainty there is likely to be as regards the interpretation to be put upon it. It is, however, possible to measure some aspects of map information and therefore it is possible to develop objective methods of map interpretation. These objective 'interpretations' of map information in fact provide us with higher-order information about the phenomena that have been mapped. It is therefore possible to think of geographic information as being provided at different levels in a hierarchy of generalisation. In this hierarchy, as Bunge (*1966*, 39) suggests, maps lie in an intermediate zone between pre-maps (raw information conveyed in a variety of ways) and mathematics (which provides a very general statement regarding the structure of the spatial information).

Maps have traditionally formed the main data-storage system which geographers possess. This use of the map as a locational inventory or record has now been challenged by the use of the computer tape which stores a much greater amount of information much more efficiently. Such a computer-storage system is still rather expensive to operate, but many planning organisations are introducing data banks to store this kind of information, and many national censuses are now developing schemes of this kind (Tobler, 1964; Hägerstrand, 1967). From this 'low-level' information it is possible to produce maps automatically by a computer-plotter machine. There are a number of methodological problems that arise, however, in the provision of automatic geographic information systems. These problems relate back to basic issues, already discussed in some detail, concerning geographic individuals and the development of space-time languages suitable for discussing geographical distributions. There are, associated with these problems, a number of technical difficulties of identification and map projection which have been discussed by Tobler (1964) and Kao (1963; 1967). Dacey and Marble (1965, 6) tackle the methodological problem, using the basic linguistic concepts of map form already discussed. They conclude

that maps, mathematical models, and other representations for spatial relations use a 'language', but that this language differs in quite important respects from everyday language, programming languages or the formal languages commonly developed by logicians. . . . An important difference is that neighborhood and juxtaposition in two dimensions is not a simple generalization of the concatenation of language. The similarities suggest that cartographic models require concepts involving higher dimensionality. It appears that the language of maps is a two-dimensional language and that the study of this language must take into account its two-dimensional structure.

Dacey (1965B) examines some aspects of such two-dimensional languages in further detail. The point here, however, is that *all* forms of geographic information (whether on computer tape, map, or in mathematical equation form) require to be analysed in terms of such two-dimensional languages. Several one-dimensional co-ordinate languages (see note at the end of this chapter and above, pp. 215–17) have been developed in geography, but there is a strong suspicion that many of the problems involved in communicating geographical information (and interpreting it) will not be solved until two-dimensional languages are developed. This is not to deny the great success of one-dimensional languages both as a vehicle for communication and as a means of generalising. To date, all our basic analysis of geographical patterns has been conducted using such one-dimensional languages. It has been shown that it is possible to generalise, either direct from the data or from a map, about spatial patterns in the context of ordinary co-ordinate languages, such as latitude and longitude. Such generalisations, however, necessarily rest on the employment of mathematical measures which require some rather strong assumptions if they are to be employed to discuss geographical patterns. On the other hand these mathematical methods allow us to make objective statements about pattern. It is thus possible to replace intuitive descriptions of, say, settlement pattern (using such vague terms as 'dispersed', 'nucleated', and the like) by an objective measure.

The mathematical method used to abstract generalisations about pattern depends upon the way in which we conceive the objects or attributes to be distributed in space. The employment of any mathematical procedure in geography requires the prior unambiguous conceptualisation of geographic reality in a manner that allows mathematical operations to be performed. It is possible to think of phenomena distributed in space in three basic geometrical forms:

(i) points;   (ii) lines;   (iii) areas;

to which we can add higher-dimensional forms such as:

(iv) surfaces;   (v) intensities;   (vi) flows;
(vii) associations.

By conceptualising the geographic pattern in one of these ways it is possible to translate information about that pattern into a mathematical language and use the properties of that language to generalise about the pattern. There are innumerable ways in which such generalisations can be abstracted. There is no point, therefore, in attempting to cover here every general measure of pattern, from the coefficient of localisation and other simple measures of concentration to the complex measures developed by way of spectral analysis (Haggett, 1965A, chapter 8, reviews some of these measures). It is, however, of interest to examine certain features of the mathematical generalisation of spatial pattern since this provides us with valuable insights into some primary methodological problems in geography.

Mathematical methods of describing spatial patterns may broadly be grouped under two headings (Harvey, 1968B):

(a) *Generalised mathematical methods* attempt to store as much information as possible about the pattern by way of some mathematical expression. An equation system can thus be devised that generalises about the data in some fashion and enables us to describe its general form. Perhaps the simplest way of generalising is to use some kind of filter to 'smooth' the spatial pattern and to derive a map statement that is a smoothed version of the original pattern. A number of ways of producing generalised descriptions have been formulated. Jenks (1963) has thus examined how smoothed generalised maps may be produced from an original data-set and Haggett (1965A, 153-4 and chapter 8) has also reviewed a number of such techniques. Tobler (1966B) has developed a more formal approach to such generalisation by applying a smoothing matrix to an initial-data matrix so as to produce maps that are successively more generalised (Figure 19.4). The advantage of this procedure, as Tobler points out, is that it is relatively easy to reverse the relationship and discover the initial map from the reverse application of the smoothing matrix. The relationship between the generalised surface and the initial surface is thus firmly established by way of the smoothing matrix.

Trend-surface analysis, in its various forms (see Chorley and Haggett, 1965B), also allows generalised statements to be developed about map pattern and the basic characteristics of that pattern to be preserved in the form of a mathematical equation whose parameters are empirically determined. It is thus possible to fit a polynomial expression of the form:

$$Z = a + bU + cV + dU^2 + eUV + fV^2 + gU^2 + \ldots$$

where Z is some variable which changes value over space and U and

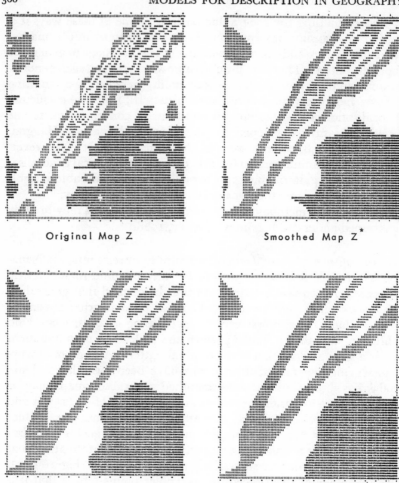

Original Map Z   Smoothed Map Z^*

Twice Smoothed Map Z^{**} Thrice Smoothed Map Z^{***}

Fig. 19.4 A series of map smoothings constructed by successive application of a smoothing matrix. The map becomes more generalised at each application (*from Tobler*, 1966B).

V are orthogonal locational co-ordinates. Fitting the model by least squares allows the actual surface to be decomposed into linear, quadratic, cubic, ..., components, with each successive component taking out a specified amount of the total pattern variation. This approach has been extensively employed to examine geological surfaces (Krumbein and Graybill, 1965, chapter 13) and has also been used in both physical and human geography to describe statistical surfaces (Chorley and Haggett, 1965B). The model is, however, a purely descriptive device. It assumes, for example, that all the parameters are linear (an assumption which is a very strong one) and the parameters cannot, without further evidence, be given any empirical interpretation.

An alternative way of expressing spatial surfaces is by way of two-dimensional Fourier analysis (Harbaugh and Preston, 1968; Casetti, 1966). Here the pattern is described by fitting mathematical expressions consisting of terms containing cosines and sines. The mathematical form is more complex, but in principle double Fourier-series analysis is no different from trend-surface analysis. It simply conceives of the height of a surface, z, as being a function of two variables, x and y, both of which are orthogonal to each other and both of which contain the sum of periodic sine and cosine functions (Harbaugh and Preston, 1968, 223). The surface is regarded, therefore, as being 'oscillatory in two mutually perpendicular directions'. The oscillations in double Fourier series are assumed to be regular and periodic. It is possible, however, to regard them as irregular and stochastic. In this case we are able to examine the nature of spatial pattern by way of two-dimensional spectral analysis (Bartlett, 1964; Bryson and Dutton, 1967). Spectral description of a spatial surface amounts to conceptualising it as a two-dimensional stationary stochastic series.

All of these techniques for describing spatial pattern by mathematical means have this in common: they simply amount to fitting an *a priori* mathematical model to a data set. This model has no necessary theoretical validity and cannot, therefore, be given any real-world interpretation without further evidence. It is possible, however, that these *a priori* models can be given a theoretical interpretation—this is an empirical problem—but even if they cannot be so interpreted they still perform a valuable function in yielding parsimonious and, in many cases, stimulating generalisations regarding spatial pattern.

(b) *Specific mathematical representations* relate the spatial pattern to a mathematical model based on some specific assumptions about process. Usually the process hypothesised is a random one. Dacey (1964A, 559) explains the meaning of this in some detail:

> To say that a distribution is random, in a non-technical sense, is to say that the pattern has no discernible order and that [its] cause is undeterminable. In the terminology of mathematical statistics, the term 'random' has a precise meaning which refers to the process generating a pattern, and the random pattern is the realization of a theoretical random process.

The theoretical random process thus provides a norm against which a particular pattern may be measured. This norm can be used to provide various objective measures of pattern. This kind of approach has been adopted to sequences of activities on a line (Getis, 1967A; 1967B) and to map patterns in which regions of different 'colour' (the colour representing a particular characteristic) can be studied by the use of contiguity measures (Dacey, 1968; Cliff, 1968). By far the most important methods, however, are those associated with the measure of point patterns by way of nearest-neighbour analysis and quadrat sampling.

Nearest-neighbour analysis combines geometry and probability theory (Kendall and Moran, 1963) in a way most relevant to the description of spatial point patterns. Geometrical probability refers to the probability of finding events in space. Consider a model in which every location in a given space has an equal chance of receiving a point. This model amounts to hypothesising a random process for assigning points to locations. If the area and the number of points to be assigned are given, then it is possible to derive various expected measures from the calculus of probabilities. Nearest-neighbour measures refer, as the name implies, to the distances between points and it is possible to calculate the distribution of distances to 1st, 2nd, 3rd . . ., nearest neighbour (over 1, 2, . . ., k, sectors if need be), given the hypothesised process. Actual-distance measures to nearest neighbours are then compared with random expectation. In the classic development of the nearest-neighbour technique by Clark and Evans (1954), a scale was constructed running from 0 (all points located at one point) through 1 (conforming to random expectation) to 2·1491 (a perfectly regular hexagonal distribution). Nearest neighbour provides a simple objective technique for measuring spatial pattern. Dacey (1962) has used it to show that a spatial pattern that Brush (1953), relying entirely on visual inspection, characterised as reasonably regular, is, in fact, rather close to random expectation. Since Brush had used the apparent regularity of settlement pattern as some positive evidence for the empirical utility of central-place theory, this objective test is of great importance. The nearest-neighbour method has thus been widely used in geography (Dacey, 1960; 1962; 1966A; Curry, 1964; Getis, 1964; provide a few examples) and has also been used to discuss problems in plant ecology (Greig-Smith, *1964*) and geology (Miller and Kahn, 1962).

Quadrat sampling, however, is concerned with the probability of finding 0, 1, 2, . . ., points in an area of a given size (usually called a quadrat). Again it is possible to calculate that probability under random expectation, provided that the mean density of points in the study region is known. It is then possible to measure deviation from random expectation and set up various measures of pattern. The details of these measures need not concern us, since full accounts are provided by Greig-Smith (*1964*) while Getis (1964), Dacey (1964A) and Harvey (1966B) provide examples of the use of quadrat sampling in geography.

Both quadrat sampling and nearest-neighbour measures provide us with a convenient way of describing objectively some of general characteristics of point patterns. These descriptions are constructed with reference to some hypothesised mathematical process. The question thus arises whether the hypothesised *mathematical* process can be interpreted in terms of some *geographical* process. Can the mathematical statement function as an *a priori* model about geographical process? It turns out that the particular mathematical law which describes random expectation in both nearest-neighbour-distance measures and quadrat counts is the Poisson law, and this, as we have already noted, is peculiarly appropriate for studying real-world processes which may reasonably be conceptualised in stochastic terms (above, pp. 267–9) Dacey (1964A 559) states:

> In terms of a map pattern, pure chance means that each map location has an equal probability of receiving a symbol. Since it is highly unlikely that geographic distributions, particularly locational patterns involving human decisions, are the result of equally probable events, it is expected that most map patterns reflect some system or order. It is for this reason that map patterns are examined for evidence of a spatial process. The search for a process may take many different paths. One procedure is to obtain a probability law that, on the one hand, accurately describes properties of the map pattern and, on the other hand, suggests properties of the underlying spatial process.

In this spirit Dacey (1964A; 1964B; 1966A; 1966B; etc.) has investigated the interrelationship between map pattern and hypothesised spatial processes in various empirical situations. Other examples are provided by Curry (1964) and Harvey (1966B).

This procedure of 'searching' map patterns with models constructed from some hypothesised stochastic process raises some thorny inferential problems. These problems are, for the most part, related to what we have already identified as the *scale problem* (above, (pp. 352–4). The various measures of pattern associated with nearest-neighbour analysis and quadrat sampling are not independent of the scale at which the point pattern is analysed. Different quadrat sizes yield different frequency distributions, and therefore provide different evidence for a hypothesised spatial process. Very small quadrats (small relative to the density of the point pattern) invariably yield a Poisson-like distribution and thus appear to suggest that the spatial process is random. Larger-size quadrats sampling the same point-pattern may produce a negative binomial distribution and suggest a different kind of spatial process (see Harvey 1968A). In general, therefore, inferences as to process derived from pattern analysis are not independent of the scale of analysis. Similar, although less serious, problems arise in nearest-neighbour analysis. Here the problem is the determination of the initial study region, the number and orientation of the k sectors, and the number of order-neighbours to be measured to. The inferential problem in this case is that a random distribution of distances to jth-order settlements over k sectors does not necessarily imply randomness in the distribution of distances to $j + 1$ order settlements over $k + 1$ sectors.

It is possible to consider these inferential problems in detail and in the process to express a series of basic methodological problems in technical terms. Consider quadrat sampling in which the inferential problems are at their most serious. A quadrat amounts to an arbitrarily imposed areal unit which is used to collect information about point pattern. Any size of areal unit can be chosen (pp. 351–4). A quadrat thus functions as an areal individual imposed upon a continuously distributed surface of point density. Now given the kind of probability model being fitted to that pattern, it is possible to lay down certain conditions which quadrat sampling should meet if inferences are to be made about the relationship between pattern and process. In a simple random model, for example, each quadrat must have an equal and independent chance of receiving a point. More complex models relax the notion of equality but independence remains an important criterion. This means that a reading (of any kind) made in a particular areal individual should be statistically independent of a reading in any other quadrat. It is possible to test for independence by examining the degree of auto-correlation in the spatially distributed population. In a technical sense the optimum size of areal unit is that in which the auto-correlation in the population lag 1, 2, 3, . . ., k steps is not significantly greater than zero. This condition pin-points the relationship between the nature of the attribute being measured and the size of areal unit identified. Since it also identifies populations whose characteristics are independent of each other, the spatial sampling problem is also dependent upon it (above, p. 363).

The appropriate size of areal unit depends on how the phenomenon being examined is spatially distributed and we can regard it, in theoretical terms, as that areal individual which minimises the degree of spatial auto-correlation in the data. This is not to imply that the areal individuals necessarily have to be of the same size in Euclidean terms, for on a Euclidean map it may be that varying areal-unit sizes will be more appropriate (and achieve a lower degree of spatial auto-correlation) because the phenomenon being examined is best analysed in non-Euclidean terms. This implies, incidentally, that the auto-correlation function on the Euclidean surface varies in its form over space.

It also turns out that there is a specific relationship between the auto-correlation function and the spectral-density function; the one is the Fourier transform of the other. It is thus possible to connect up one of the most general statements of map pattern with the specific measures of pattern which we are here considering. Underlying all our efforts to develop mathematical descriptions of spatial pattern is an attempt to identify the collections of spatial variances of different wave-length, amplitude, and frequency. Since it is the aim of spectral analysis to identify the contribution of a particular band of frequencies to the over-all variance in the spatial pattern, we may be able to identify, directly from the data, the most important wave-lengths (components) contributing to spatial variation. These significant wave-lengths identify the scale at which significant spatial auto-correlations may be found, and can be used, therefore, to identify the most appropriate size of areal unit to function as the basic geographical individual for the analysis of continuously distributed phenomena and grouped discrete phenomena.

Flat plane surfaces, regular geometric surfaces (both oscillatory and non-oscillatory), and the like, are relatively rare in geography. Usually the phenomena we are seeking to understand are distributed irregularly and confusedly over a kind of 'hummocky' space which is also frequently interrupted by major discontinuities such as topographic or political barriers. Our knowledge suggests that this surface cannot be conceptualised as purely random—'pure white noise', as the information theorists would call it. There are strong elements of organisation within geographic space and these can often be identified in a map pattern. The problem, however, is to identify the elements of regularity in what often seems a totally irregular spatial pattern. We have long sought to identify these regularities intuitively, by staring at the map and hoping that what we think we see is really there. We now possess objective ways of measuring pattern and these ways also relate operationally to some of the basic methodological problems with which geographical analysis is faced. The broad implications of this fact have, however, been discussed elsewhere (Harvey, 1968B).

These techniques of mathematical representation of map pattern thus integrate three important methodological problems—the scale problem, the nature of spatial pattern, and the relationship between spatial pattern and process. They provide some kind of framework for developing a deeper analysis of these three problems. Processes are relevant only at a certain scale of activity, and the relevant processes vary according to the scale of analysis chosen. Intra-city migration is probably explained by completely different variables to those relevant to inter-city migration. The relationship between process and spatial form is generally accepted as a fundamental concern of the geographer. Our descriptions of spatial form are entirely dependent on scale. And the relevant scale of analysis can be determined only in terms of the spatial variability and significance of a given process. There are therefore strong interdependencies between pattern and process and the only way we can avoid purely circular argument is to recognise very clearly the nature of these interdependencies.

(*Note:* Some confusion may arise from the use of the term two-dimensional in this context. Each simple language (space-time or substance) can have several dimensions, but in this case we are forming a complex language by bringing together two different languages into one information system.)

Basic Reading

Berry, B. J. L. and Marble, D. (1968). This excellent book of readings contains many articles of basic interest to all aspects of this chapter.
Blaut, J. M. (1959).
Board, C. (1967).
Haggett, P. (1965B).
Harvey, D. (1968B).

PART SIX

Models for Explanation in Geography

Chapter 20
Cause-and-Effect Models

The notion of *cause and effect* has been extremely important in the history of scientific investigation. There is, however, probably no other notion in the philosophy of science which has accumulated around it so much contentious argument. The confusions are legion. It is difficult, therefore, to cut through this mass of argument and dispute in order to show that cause and effect can provide us with a useful model for analysing geographic problems. There is no point in attempting a full discussion of the history of cause-and-effect argument or a full statement of the various nuances of meaning that are currently attached to the notion. Those interested are referred to the full discussion by M. Bunge (*1963*). But it is useful to clear away some of the semantic difficulties associated with cause and effect at the very beginning. M. Bunge (*1963*, 3-4) notes three principal meanings:

(i) *Causation*—the notion of a causal connection (sometimes called the causal nexus) which associates a particular event (or set of events) with a particular result (or set of results).
(ii) *The Causal Principle*—a law-like statement (i.e. a universal statement) cast in a cause-and-effect language.
(iii) *Causal Determinism* or *Causality*—a doctrine asserting the universal validity of the causal principle.

These three meanings are related. If we can establish the existence of 'causation' in certain cases, then causal laws might also be established. The formulation of numerous causal laws (all of which are demonstrably successful and in which we possess a good deal of confidence) may lead us to infer that causal determinism is the only principle upon which we can obtain realistic understanding of the world around us. We may also ask whether there are any other forms of explanation than cause and effect which are acceptable to us, and, further, raise the philosophical issue of whether the world of phenomena is *in fact* governed by cause-and-effect laws exclusively.

Now it would be impossible to trace all the arguments and counter-arguments which have raged around these questions. But given the views already set out regarding theories and models, as well as the general idea of explanation developed so far, it is possible to develop some ideas about cause-and-effect analysis that largely by-pass most of this argument. It is worth noting, for example, that much of the argument over cause and effect is metaphysical. In accordance with the general approach of this book we shall concentrate attention upon the logical issues involved.

If we accept that scientific explanation involves the construction of hypothetico-deductive systems, in which laws (or theorems) are deduced from postulates (or axioms), then it may be expected that cause and effect will provide us with an important rule of inference within such systems. At this level we may relegate cause and effect to the status of a useful rule of logical inference. Whether it is anything more than just this depends upon the nature of the correspondence rules which can be formed to link a theoretical structure, containing cause-and-effect rules of inference, to the real-world situations which we are seeking to explain. It seems reasonable, therefore, to restrict the discussion of cause and effect to two major issues: (i) the logical properties of cause-and-effect systems of analysis, and (ii) the development of correspondence rules that allow us to map real-world happenings into a cause-and-effect logical framework.

I THE LOGICAL STRUCTURE OF CAUSE-AND-EFFECT ANALYSIS

M. Bunge (*1963*, chapter 2) considers various forms of the causal principle and concludes that the most satisfactory way of stating it is:

If C happens, then (and only then) E is always produced by it.

which can be translated more simply as 'every event of a certain class C produces an event of a certain class E', or, more crudely, 'the same cause always produces the same effect.' Inherent in the notion of cause and effect is the idea of 'producing'. Not only is C always followed by E, but there is some *necessity* in the relationship. The difference between a causal relationship and one of constant conjunction is thus similar to the *prima facie* difference that Nagel (1961) asserts between *nomic* and *accidental* universality (above, p. 101) and it is similarly difficult to sustain in the face of close logical analysis. The causal principle, therefore, is nothing more than a simple but very important type of law statement. We may justify the use of cause-and-effect statements in the same way that we justify calling any statement a 'law'. This may amount to showing empirically that some mechanism connects the occurrence of event C with the subsequent occurrence of event E. In other cases we may feel justified in assuming some mechanism. From the logical point of view, however, the empirical status of the statement need not concern us. Once we write the logical relationship $A \rightarrow B$ (read 'A leads to, or causes B') with the condition that the relationship is non-reflexive, $B \nrightarrow A$, and asymmetric (B cannot precede A), and transitive (if $A \rightarrow B$ and $B \rightarrow C$, then $A \rightarrow C$), then we have established a set of rules of inference which can be used to describe interactions and relationships among any set of variables. There are three logical extensions which are of considerable interest. To demonstrate them it is useful to resort to set theory and generalise the causal model to sets of events. We may conceive of sets of events $(a_1, a_2, \ldots, a_n) = A$, $(b_1, b_2, \ldots b_n) = B$, and other sets, C, D, E, etc. Then a number of basic forms of the causal model may be identified:

(i) *Direct Causes* of the basic form $A \rightarrow B$.
(ii) *Causal Chains* of the form $A \rightarrow B \rightarrow C \rightarrow D \rightarrow. \ldots$
(iii) *Multiple Cause Structures* of the form

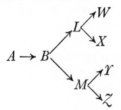

(iv) *Multiple Effect Structures* of the form:

$$A \rightarrow B \begin{array}{c} L \begin{array}{c} W \\ X \end{array} \\ M \begin{array}{c} Y \\ Z \end{array} \end{array}$$

Such logical structures are, to say the least, enticing. They certainly conform to many of our intuitive notions regarding what an explanatory model ought to look like. Apart from this, the very power of such structures suggests that they should play an important part in modelling explanations and analytic investigations. But the outsize importance of such structures in geographic thinking at the present time rests on the strong supposition that this logical form of analysis accurately reflects real-world mechanisms and processes. Whether or not this is true depends on the degree to which correspondence rules may be established between the abstract logical structure and real-world situations.

II THE APPLICATION OF CAUSE-AND-EFFECT MODELS

The problem of applying cause-and-effect logic in empirical research boils down to one of identifying the variables which are causally related and identifying the boundaries of the system within which we may apply cause-and-effect analysis. Bunge (*1963*, 50–1), following Russell (1914, 234–5) states that

the causal connection, being law-like, does not refer to isolated facts, but to facts belonging to certain *classes* or kinds—which takes care of the variations in the individual instances.

The problem is, then, to identify the set of events A (defined by some property or properties) which stands in such a relation to another set of events B that we may assert that A causes B. The identification of such sets is by no means easy.

It is implicit in the 'common-sense' version of the causal principle that the set of events A should be somehow different from set B. To state, for example, that farmers grow hops because they are hop growers scarcely conforms to our intuitive idea of the causal principle. On the other hand, we are faced with the problem that as soon as we connect two sets of events by way of the causal relation, we can, if we search, insert intermediate sets of events between them. For example, the sequence rainfall → wheat yields, may be converted into rainfall → soil moisture-content → wheat yields. In the first statement rainfall is regarded as the direct cause, but in the second it is treated as an indirect cause in a causal chain. Blalock (*1964*, 18) thus points out:

it will usually be possible to insert a very large number of additional variables between any supposedly directly related factors. We must stop somewhere and consider the theoretical system closed.

Thus a relationship which is direct in one system 'may be indirect in another, or it may even be taken as spurious'. As more and more variables are inserted, however, these variables will tend to be less and less different from each other, e.g. rainfall → soil moisture content → wheat growth → wheat yield. In many situations in reality we are dealing with a continuum of events which are not easily isolated from one another. The relationship between rainfall and soil moisture content may thus be regarded as part of the continuum that forms the hydrological cycle. The essential point here, however, is that the way in which a system is closed is largely an arbitrary decision, and in some cases it involves treating continuous systems as if they possessed discrete states which were recognisable in reality. In these circumstances, it becomes evident that the cause-and-effect model involves a good deal of *as-if* thinking in order for it to be applied to empirical situations. Where discrete states can be distinguished and shown to exhibit a causal connection, then the cause-and-effect argument may form the basis of some theory about reality. For the most part, however, it seems better to regard it as a convenient model for analysing real-world interactions.

The general application of cause-and-effect analysis also involves definition of appropriate sets of events. This is a classification problem and, as we have seen, classification either reflects a theory or acts as a stimulus to theory formation. In classifying according to properties, for example, the actual choice of properties depends upon some assumption regarding the importance of such properties in differentiating objects and events. Where elaborate theory exists these properties are usually shown to be important, but there are many situations where the classifications themselves are uncontrolled with respect to empirical data sets—these form the *idealisations* of much of social science and geography as well (above, pp. 91–6). Many of these idealisations are very difficult to justify except by introspection and intuitive appeal. Max Weber clearly recognised this problem. To Weber, the analysis of the evolution of social structures involved causal imputations, and the basic problem was 'to prove the existence of a causal relation between certain features of a given historical individual and certain empirical facts'. Such proof was possible, in his view, only if the problem could be reduced to one of the relationship between 'concepts' and 'ideal types' (Talcott Parsons, 1949, 610). The causal principle here becomes a schematic relationship between concepts, classes, and ideal types, and the adequacy of such a schematic representation depends very much upon the adequacy of the ideal types established. The applicability of cause-and-effect models depends both on the adequacy of the definition of the classes involved and on the choice of classes to be included.

Given this general conclusion, it is worth pointing out that no one causal structure has exclusive rule over a particular domain. If we wish to explain the behaviour of farmers, we may do so either by evolving structures that make reference to dispositions, motivations, and the like, or we may develop a structure that refers to environmental controls, market prices, and the like. Several different causal models may thus be used to explain the same set of events, and there is no need for these models to be mutually exclusive.

One of the conditions of the causal model is irreversibility, i.e. that if $A \rightarrow B$ then $B \nrightarrow A$. This condition involves problems of *identification*, by which is meant the problem of establishing what Simon (1953) calls the *causal ordering* among a set of variables. In many situations it is difficult to tell whether $A \rightarrow B$ or $B \rightarrow A$, or whether there is a two-way interaction between them. If it is stated, for example, that the development of hop growing in Kent was caused by the existence of a small-size, freehold farm structure, then it is possible to reverse the statement and have it make sense. In some cases a reverse statement does not make sense. It would be non-sensical to say, for example, that wheat yield $\rightarrow$ rainfall. If the reverse statement does make sense, however, we require some way of determining which variable causes which. In some cases it is possible to get round this problem by establishing a time-lag between the events. Causal explanation requires that B cannot precede A if $A \rightarrow B$. But the establishment of a time-lag is no more than a useful, if important, guide, the reason being that there may be complex sequences of events that allow B to precede A. Consider the case where we are trying to show a causal relationship between price level and the acreage of a particular crop. Suppose that quite regularly we find that a decline in acreage of the crop is followed by a drop in price. This instance does not make *economic* sense as it runs counter to the general expectation that price $\rightarrow$ acreage change. But suppose we interposed an intermediate term, such as expectation of future price levels? Then the system may make sense. The point here is that the time-lag between A and B assumes that A is a direct cause of B and it assumes that each variable has a discrete distribution in time. Again, continuously distributed variables are very difficult to discuss in the causal framework without making some strong assumptions, e.g. that price in the preceding six months causes output level in the next six months.

Simon (1953) and Blalock (*1964*) tackle the identification problem by way of an analysis of the structure of interlocking variables in a system. Simon's treatment is particularly interesting. He shows that if we possess a set of structural equations which link sets of variables, then if there is a unique causal ordering of the variables, the matrix of coefficients connecting the variables will be invariant under certain forms of transformation. If we can show empirically that the interrelations we are examining have this property, then the identification problem and the problem of causal ordering is effectively solved (Simon, 1953, 68–9). This treatment relies upon the assumed asymmetry of the relationship $A \rightarrow B$, but does not depend upon establishing a time-lag between the two sets of events.

To say that $A \rightarrow B$ implies that B could not exist without the occurrence of A unless there are other sets of events, Z, which are independent of A, but which could also give rise to B. Here again there is a complicated problem which can best be analysed by drawing attention to the difference between *necessary* and *sufficient* conditions for B to exist:

(i) A *necessary condition* is a state of affairs that would justify the prediction of the non-occurrence of an event. Lack of rainfall might be regarded as preventing the occurrence of hop growing in Kent (provided no other means of water supply is available). Necessary conditions are rather negative, they really define a set of constraints.

(ii) A *sufficient condition* is a state of affairs which would justify predicting the occurrence of an event. It is therefore more positive; if A is a sufficient condition then given that we have observed A we would automatically expect to observe B. If rainfall were a sufficient condition then we should expect to find hop cultivation wherever it rains.

Blalock (*1964*, 31) identifies four possible situations:

(i) A is both necessary and sufficient for B to occur (e.g. small-size freehold farms are essential for, and give rise to, hop growing).

(ii) A is a necessary but not a sufficient condition for B (e.g. small-size freehold farms are essential for, but do not give rise to, hop growing).

(iii) A is sufficient but not necessary for B (e.g. small-size freehold farms give rise to hop growing, but hop growing may be found where these types of farm are absent).

(iv) A is only partly necessary and/or sufficient for B (e.g. in most cases where we find hop growing we may attribute it to the effect of farm structure, or in most cases where we find a particular farm structure we shall find hop growing). This last form of statement amounts to a probabilistic version of the cause-and-effect model.

Blalock points out that situations (ii), (iii) and (iv) can be redefined to conform to (i) simply by insisting that A be *defined* as containing all necessary and sufficient conditions for B. Thus a situation found in (ii) amounts to one in which we are identifying *partial causes* in a multiple-cause structure in which some of the other partial causes are unknown. All that is required here is to identify those sets, say A_1 and A_2, whose joint occurrence $(A_1 \cup A_2)$ causes B. If we extend this idea of the intersection of many different sets of events (thus developing a highly complex multiple-cause structure) we come close to the basic assumptions of French geographical enquiry at the beginning of this century (Lukermann, 1965). In case (iii) the problem becomes one of defining A by the union of a number of sets such that A is necessary and sufficient for B. In case (iv) we may regard a probability statement of the form 'A causes B in 60% of the cases' as being a reflection of a situation in which the initial set, A, can be partitioned into two subsets A' and A'' such that $A' \to B$ and $A'' \longrightarrow\!\!\!\!/\, B$, and where A' occurs 60% of the time (Nowak, 1960). In all these cases it is possible to preserve the cause-and-effect framework in its most powerful form by redefining the system to conform to it. But it is clear that in doing so we are moulding operational definitions to fit the logical framework and conceptualising the situation in such a way that it fits the logical model. Such a procedure may be extremely useful, but it is clear that the cause-and-effect logic here functions as an *a priori* model for discussing reality rather than as a theory about reality.

III CAUSAL SYSTEMS

In spite of the difficulties which face the employment of cause-and-effect analysis in many empirical situations, the basic logic has proved extremely useful in analysing the structure of complex systems. A discussion of causal systems is partly necessitated, of course, by the nature of the basic model because

clearly, a causal relationship between two variables cannot be evaluated empirically unless we can make certain simplifying assumptions about other variables (e.g. no environmental forcings or postulated properties operating in unknown ways) (Blalock, *1964*, 13).

We can gain some insight into $A \to B$ type relationships only if we can assume that another variable C has no effect on B (its effect on A may be regarded as irrelevant), or if we can effectively control the impact of C on B. Isolating a single set of relationships (and eliminating interference from other variables) is one of the functions of experimental design, but in geographic investigation we can only occasionally experiment in the laboratory. We are thus faced with the difficult problem of collecting our data in such a manner that we maximise the amount of information we gain regarding the particular relationship under study. This involves minimising interference and this is usually done by randomisation procedures (see above, pp. 365–7).

This general problem of interfering variables, however, demonstrates that in most instances we are dealing with a complex causal system; and more recently the tendency has developed to deal with the whole system rather than with isolated parts of it. Of course the system has to be considered as closed at some point in an investigation. As M. Bunge (*1963*, 125–47) points out, almost any scientific investigation involves the isolation of sub-systems since 'the universe is not a heap of things but a system of interacting systems'. He then continues:

There are always connections among numerous sets of factors, never connections between single isolable events and qualities, as causality assumes . . . yet, the singling out effected by causal thought, while ontologically defective, is methodologically unavoidable; here, as everywhere else, the mistake consists not in making errors but in ignoring or neglecting them.

Consider the set of interlocking variables portrayed in Figure 20.1A, which is a hypothetical model of interactions in a farm economy closed around just six variables. Direct causes only are recorded. The same system may be represented by a diagram of 'flows' from causes to effects (Figure 20.1B). This is a characteristic kind of model which might emerge from a study of a particular situation. In this case we are merely hypothesising the relationships. In reality it is often very difficult indeed to establish the exact nature of the causal structure. Blalock (*1964*, chapter 3) discusses in some depth a method for evaluating such causal structures. The method begins by accepting a matrix of correlations between variables as basic data. These correlations cannot be regarded as evidence of direct causal connections because they are symmetric and because indirect causes may have a high correlation with effects. In Figure 20.1A, for example, there may well be a high correlation between transport cost to market and yield per acre, but this is because it acts through price and fertiliser input, rather than because it acts as a direct cause. But Blalock suggests that it is possible to evaluate the structure in this kind of situation by an examination of the partial correlation coefficients. The method will become clearer, however, if we examine it in the context of *recursive causal systems*.

Simon (1953) and a number of econometricians (such as Wold and Jureen, 1953; Wold, 1954; Strotz and Wold, 1960; Johnston, 1963, chapter 9), have attempted to analyse complex causal interactions by way of a set of recursive simultaneous equations. In general these may be written as:

$$X_1 = a_1 + b_{12}X_2 + b_{13}X_3 + \ldots + b_{1k}X_k + e_1$$
$$X_2 = a_2 + b_{21}X_1 + b_{23}X_3 + \ldots + b_{2k}X_k + e_2$$

.
.

$$X_k = a_k + b_{k1}X_1 + b_{k2}X_2 + \ldots + b_{k,k-1}X_{k-1} + e_k$$

The coefficients of this system, if they are invariant under certain forms of transformation, will indicate whether there is a unique causal ordering in it (Simon, 1953). But there are many situations in which

Causes ↓ / Effects →	1	2	3	4	5	6
1 Yield per acre				+*		
2 Fertiliser input	+*					
3 Labour	+*					−*
4 Gross farmgate receipts		+*	+*			+*
5 Transport cost to and from market		−*		+*		
6 Investment in machinery	+*		−*			

Fig. 20.1 (A)

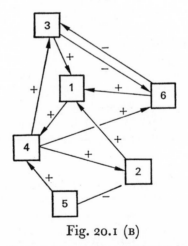

Fig. 20.1 (B)

Fig. 20.1 **A :** A tabulation of a set of hypothetical cause-and-effect relationships among a selected number of variables relevant to the study of farm yields; **B :** A graphical portrayal of the same relationships.

no unique causal ordering can be found. It is therefore simpler to consider a particular form of this structural equation approach. The following is condensed from Simon's (1953, 58) and Blalock's (*1964*) accounts.

Suppose we have the following causal system in which poor growing-weather $(X_1) \rightarrow$ small wheat crops $(X_2) \rightarrow$ increase in wheat price (X_3), and we assume the weather to depend only on a parameter, the wheat crop on the weather (plus some random shock), and the price on the wheat crop (plus some random shock). Assuming the relationships to be linear we have the following system of recursive structural equations:

$$X_1 = e_1$$
$$X_2 = b_{21}X_1 + e_2$$
$$X_3 = b_{32}X_2 + e_3.$$

This system requires, however, that X_1 has no effect on X_3 except through X_2, and it may therefore be regarded as a special case of the following system of equations in which b_{31} is set equal to zero:

$$X_1 = e_1$$
$$X_2 = b_{21}X_1 + e_2$$
$$X_3 = b_{31}X_1 + b_{32}X_2 + e_3.$$

If the parameters in this system are regarded as partial regression coefficients then:

$$X_3 = b_{31 \cdot 2}X_1 + b_{32 \cdot 1}X_2 + e_3.$$

Given the simple recursive model we are proposing, $b_{31 \cdot 2}$ should be equal to zero (apart from sampling variations). This means that there is zero correlation between X_1 and X_3 when X_2 is held constant. It should be possible, therefore, to evaluate such a causal model by showing that $b_{31 \cdot 2}$ is not in fact significantly different from zero. This forms the basis of Blalock's approach to the evaluation of causal models. There are, of course, complications and some difficulties with this approach and those concerned should read the rest of Blalock's analysis thoroughly.

Recursive causal systems are just one form of the structural equation approach. Another form involves the analysis of *reciprocal* causal systems. Here the model is designed to look at two-way interactions between variables but in order to do this it is necessary to examine a system over a number of time periods. We may thus build a special form of recursive system to deal with reciprocal interactions of the form

$$X_{t_0} \rightarrow Y_{t_1} \rightarrow X_{t_2} \rightarrow Y_{t_3} \rightarrow X_{t_4} \rightarrow \ldots$$

Such a model seems peculiarly appropriate for examining Myrdal's notions of circular and cumulative causation—a notion which Pred (1966; 1967) has followed up in some detail in the geographical context.

It is perhaps worth while closing this section on the analysis of causal systems, however, with a brief comment on the inclusion of probability statements within such systems. When dealing with aggregates of events, which can be treated statistically, it is possible to construct probabilistic causal models. In some cases this probabilistic element simply amounts to an error term or an interference which enters into the closed causal system from outside. Most econometric models are of this form. The terms can be partitioned therefore into a group of variables whose interactions are modelled deterministically and an error variable which is a catch-all term for measurement error, external interference, and the like. In other cases it is possible to construct causal systems in which the variables within the system are modelled probabilistically. Consider a causal-chain model in which the probability of event B following event A, $p(B \mid A)$, is known and the probability of event C following event B, $p(C \mid B)$ is also known, then by the multiplication theorem the probability of event C following event A is given by the product $p(B \mid A) . p(C \mid B)$, provided there is no direct connection between A and C. In this kind of model the deductive theorems of probability can be used in combination with the logical structure of cause-and-effect analysis to provide a convenient framework for analysis. Any more substantial element of uncertainty is difficult to combine with cause-and-effect logic, however, and non-deductive procedures certainly cannot be incorporated. But the general ability to extend cause-and-effect analysis to probabilistic situations has important philosophical implications since it does provide some kind of challenge to the metaphysical belief that cause and effect essentially implies determinism. The sense in which this is so needs to be explained.

IV   CAUSE AND EFFECT AND DETERMINISM IN GEOGRAPHY

Any causal system, if properly specified, is determinate in the mathematical sense. The logic employed in cause and effect is deductive and conclusions must, therefore, follow from the initial statements. This general rule applies as much to probabilistic statements as it does to deterministic ones, for given a set of initial statements which are probabilistic in form, then determinate deductions can be made as, for example, in the probabilistic causal-chain model mentioned above. That the *logic* of cause and effect implies *determinate* solutions has been used, however, as a support for the more substantive thesis of determinism itself. Such an extension is not supportable without further evidence, yet again and again in the history of science the existence of cause-and-effect logic and its importance in explanation has been used as direct evidence in support of determinism as a metaphysical position.

It would be misleading, however, to pretend that the exposition so far equips us to handle this complex issue. The idea of cause and effect has great psychological significance and it appears to be a basic primitive concept in our attempt to gain mastery over the field of experience. Piaget (*1930*) has, for example, shown how the idea of cause and effect develops quite early in children. The notion of cause and effect is also deeply embedded in language; it is thus very difficult to discuss anything without using terms such as 'determine', 'govern', 'affect', 'control', 'produce', 'prevent', 'give rise to', and so on, all of which tend to imply some kind of necessary causal connection. Indeed the whole idea of explanation itself is often regarded as being synonymous with 'establishing the causes of something'. The notion of cause is thus a very general one, but partly because it is so general it is plagued by ambiguity of meaning. It is important to recognise this for the argument that has surrounded the notion of cause in geography is a pale reflection of a vast and confused philosophical debate. Nagel (1961, 316) thus writes on the causal principle that

there is no generally accepted standard formulation of it, nor is there general agreement as to what it affirms. The principle is usually understood to have a wider scope than any special causal law. On the other hand, some writers take it to be a statement on a par with particular causal assertions, although affirming something about a trait pervasive throughout nature and not simply about features of a limited subject-matter. Others understand by it a principle of higher rank than are the specialized causal laws; and they maintain that the principle asserts something about laws and theories rather than about the subject-matters of laws and theories. Still others take it to be a regulative principle for enquiry rather than a formulation for connections between events and processes. Some regard it as an inductive generalization, some believe it to be *a priori* and necessary, and others maintain that it is a convenient maxim and the expression of a resolution.

Cause and effect thus means different things to different people and the use of a cause-and-effect framework evidently implies different things to different people. The geographic literature is replete with examples of all (and more than all) the attitudes described by Nagel. Most geographers assume that the diverse phenomena they study are somehow interrelated. Many assume that the interrelationships are causal. According to Hartshorne (1939, 67; 1959, 18), both Humboldt and Ritter 'presumed a causal interrelation of all the individual features in nature', while Hettner also emphasised the importance of understanding the causal connections that bind diverse phenomena together within regional complexes. The idea of cause and effect has in fact formed the basis for explanation in geography in the past, but each school of geography and each generation of geographers has tended to interpret the causal principle in a different way. To Vidal de la Blache explanation consisted of showing how a particular event stood at the point of intersection of complex causal chains (Lukermann, 1965), while the geographical determinists, such as Semple and Huntington, sought to show how human activity could be related back (often by a complex path) to the ultimate determinant, the environment. The geographical possibilists, on the other hand, seem not so much to have been at war with the causal principle as to have disagreed with the determinists over the identification of the right cause and the right effect. In retrospect it appears as if both camps were seeking to analyse in terms of one-way relationships what we now conceptualise as complex two-way interactions, their disagreement being over which was the cause and which the effect. The close association that existed, however, between determinism, environmentalism, causality, and cause-and-effect argument, often led to the abandonment in principle of causal arguments even if their use in practice continued unabated (with a mere change of notation, as it were, from explicit terms such as 'cause' to an implicit discussion of factors, relationships, and so on). Platt (1948) was so disturbed by the 'complex kind of determinism' which he could still detect in geographic work, that he was led to condemn the 'pseudo-scientific' use of cause-and-effect argument and to abandon the whole idea of explanation in geography.

Many of these general philosophical issues concerning cause and effect in geography were thrashed out in a series of articles in the 1950s. The contributions of Clark (1950), Martin (1951), Montefiore and Williams (1955), and Emrys Jones (1956) form a quite outstanding series which ought to have put the cause-and-effect argument in geography into perspective. Unfortunately, the really powerful contribution which these papers made to geographic methodology appears to have gone largely unappreciated. Sprout and Sprout (1965) have recently provided a summary of the argument in the context of political thought however. The key paper in the series is undoubtedly that by Montefiore and Williams which sets out to provide a criticism of Martin's views and to provide a consistent and fairly rigorous terminology for developing cause-and-effect explanation in geography. They begin by summarising Martin's argument under three headings:

(i) If we are to talk in terms of cause and effect at all, we must be Determinists, but (ii) we cannot help but talk in terms of cause and effect, therefore (iii) we cannot help but be Determinists.

This argument was used by Martin to show that determinism should be adopted as a basic hypothesis in human geography. Montefiore and Williams object to this conclusion on a number of grounds, the first of which concerns the way in which events may be categorised as causes and effects respectively. They take up Martin's comment that 'the same cause, if it really is identical and not merely similar, must always be followed by the same effect, without any room for doubt and choice', and point out that there is a choice involved, since it is always possible to define events so that they do conform to this particular constraint. They state:

The same cause *must* be followed by the same effect for the simple reason that one will refuse to call it 'same cause' if it is not. Is it possible then to provide criteria for deciding whether a 'cause' is the same independent of the question of the 'effect' by which it is followed? It is, of course, possible to do this in any given case, and thus to produce a genuine empirical hypothesis, but then we should no longer be able to guarantee its immunity against any counter-example. It is true that in the face of any counter-example that might actually crop up, we could always retreat into our first position of refusing to admit that the cause could have been the same; but in doing so we should once more be surrendering the empirical character of our assertion.

The general conclusion is that Determinism is not a hypothesis in the ordinary sense at all, for it cannot be subjected to either direct or indirect empirical test. It is essentially a working assumption (and an extremely important one, as Nagel (1961, 606) points out) which indicates 'a determination never to rest content with un-explained variations, but always to try to replace such variations within a context of wider generality.' Montefiore and Williams then go on to examine the nature of cause-and-effect arguments and to place these arguments in the context of explanation in general and of explanation in geography in particular. Their analysis builds on the notions of necessary and sufficient conditions and draws attention to cause and effect as a logical structure rather than as a universal empirical hypothesis. By discarding the general metaphysical questions of determinism and free will often associated with the use of cause and effect, they are 'able to discuss questions of causal influence in a language in which many of the traditional problems no longer naturally arise' (Montefiore and Williams, 1955, 11). Their general conclusion is that cause and effect is an important logical principle upon which empirical analysis may be conducted. They would not presumably differ too much with Blalock (*1964*, 6–7) when he writes:

One admits that causal thinking belongs completely on the theoretical level and that causal laws can never be demonstrated empirically. But this does not mean that it is not helpful to *think* causally and to develop causal models that have implications that are indirectly testable. In working with these models it will be necessary to make use of a whole series of untestable simplifying assumptions, so that even when a given model yields correct empirical predictions, this does not mean that its correctness can be demonstrated.

The trouble has been that geographers in *thinking* causally have then made these simplifying assumptions and, in many cases without realising it, have assumed further that a good performance of the model validates the assumptions and the mode of thinking. It does not. The argument over Possibilism and Determinism is an excellent example of what can happen when an *a priori* model is confused with a theory about reality without the necessary safeguards being enforced (above, pp. 144–54; 164–5). Here then is a classic case of a major methodological dispute within geography which simply arises out of the failure to distinguish between theory and model and a dispute that, for some fifty years, simply hinged upon the misapplication of the model concept in geographical thinking. It is not difficult, therefore, to agree with the conclusion of Montefiore and Williams (1955, 11), which is fully accepted by Hartshorne (1959, 156) that the issue of determinism and free will has not any practical relevance for working geographers. It is a matter of faith which has little to contribute to methodological understanding or to empirical study. Free from such metaphysical trappings, however, the cause-and-effect model has a vital role to play in the pursuit of explanation in geography.

V   CAUSE-AND-EFECT ANALYSIS IN GEOGRAPHIC RESEARCH

Partly because of the confusion of complex philosophical and logical issues, the cause-and-effect model itself has tended to be in disrepute in principle even if it has been used a great deal in practice. Geographers have not been alone in their furtive use of cause-and-effect analysis. Blalock (*1964*, 38) comments that

the statistical literature . . . is almost schizophrenic in its approach . . . there seems to be considerable confusion of terminology and almost a conspiracy of silence in dealing with the problem of causality. . . . . The literature on experimental designs, on the other hand, is filled with causal terminology. One takes out the effects of control variables, studies interaction effects, and assigns individuals to 'treatments'.

The disenchantment with explicit cause-and-effect models is, or at least was, perhaps understandable, but this scarcely justifies the continuing tendency to conceal causal forms of analysis under the thin veneer of terminological change or, in some cases, under the spurious cover of a functional or statistical analysis. We ought clearly to recognise that the causal model has been, and is likely to be, a basic model in geographic enquiry. It is perhaps true that geographers have often misused the model. This misuse stems largely from a failure to understand the logical structure inherent in the model and the empirical difficulties involved in mapping geographical problems into it. Too frequently, for example, one-way interactions are assumed among a set of variables when feedback effects are involved. Too little attention has also been paid to problems of identification and closure of the model.

But the model has provided an important basic framework for thinking about geographical problems and for thinking about possible solutions. Most geographers thus assume that geographical distributions have causes, and that it is the function of geographical research to identify these causes. This assumption has been productive. What it amounts to is an attempt to identify the particular intersection of sets of events which give rise to other sets of events, or, as with more traditional formulations, show how particular concatenations of individual events intersect to give rise to some new event. Consider the way in which we might set out to explain a particular industrial-location pattern. We might begin by considering a set of raw material deposits (A), a set of markets for the finished product (B), a set of locations at which labour supply is available (C), and then regard the intersection of these sets as giving rise to, say, an iron and steel works (X). In notation we have

$$(A \cap B) \cap C \to X.$$

This style of analysis is useful and underlies much of the theory we develop as well as much of our approach to empirical problems. Almost any elementary or advanced text will contain examples of it. Indeed the search for the 'factors' that 'govern' geographic distributions is nothing but an attempt to identify that intersection of sets of events which forms a unique set such that $A \to B$. This technique of investigation has more than mere academic interest. It is a basic framework for investigation and explanation.

More complex forms of the simple causal model are also employed. Causal chains, recursive systems, reciprocal recursive systems, and so on, have all been used, often implicitly, in a geographical context. Such models are particularly useful for examining the evolution of spatial systems over time. Thus much of Pred's (1966) work on spatial dynamics has been cast in terms of cumulative causal interactions. Similarly, process-response models may be regarded as a form of causal analysis which extends to reciprocal and recursive systems. In all of these cases, of course, certain assumptions have to be made, assumptions that concern causal ordering, identification of the variables, and closure of the system. In each case we are imposing something upon the empirical situation. There is nothing wrong with imposing such assumptions. The mistake comes in drawing inferences which neglect to take account of the *a priori* assumptions already made.

We may thus conclude that cause and effect provides a powerful model for the analysis of geographical problems. We may use the model to analyse individual occurrences, to indicate regular relationships, to examine dynamic systems, to construct theory, to state laws, and so on. But the model has its limitations. There are critical problems in mapping real-world problems into the model without doing too much violence to the empirical situation. The one major mistake to be avoided at all costs, is the false inference that because the model may be applied in a number of situations with a considerable degree of success, that it is the only model open to us to use in analysis and explanation and, further, that the real world is necessarily governed exclusively by the operation of cause-and-effect laws. To make such a false inference simply amounts to confusing an *a priori* model which helps us to understand reality with a theory about the nature of reality itself.

Basic Reading

Blalock, H. M. (*1964*).
Bunge, M. (*1963*).

Reading in Geography

Jones, E. (1956).
Martin, A. F. (1951).
Montefiore, A. and Williams, W. M. (1955).

Chapter 21
Temporal Modes of Explanation in Geography

The cause-and-effect explanations discussed in the last chapter possessed the property of temporal asymmetry. Such an explanatory form amounts to showing that one event necessarily follows another over time. It is a relatively short step from this simple form of cause-and-effect explanation to one that refers to a causal chain and, ultimately, to a chain that stretches back over the whole history of the universe, with each link showing a necessary and sufficient connection with the next. Such an extended form of causal explanation might sound ideal but it is certainly unfeasible. Indeed, for cause-and-effect explanations to function properly we require to specify some sort of closed system. Given our knowledge, it is unlikely that controlled cause-and-effect explanations stretching back over a long period of time will be devised in the near future. However, there are a number of what will be termed *temporal* modes of explanation that do seek to establish relationships back over a long period but deviate in some respects from the causal-chain ideal. These temporal modes of explanation are not, perhaps, as rigorous as we might like, but they seem to offer the only way of handling situations which involve reference to historical conditions.

There has been some controversy in geography as to the necessity of employing explanatory forms that refer to long periods of time or to origins. Any discipline, it may be argued, may specify the limits of the system within which it will seek to explain. In cause-and-effect analysis, for example, we presuppose that some reasonable judgement can be made regarding the closure of the system. Some geographers have thus argued that geography, being especially concerned with spatial relationships, should not seek to extend explanation over too great a time—in other words the system should be closed approximately around the present. Hartshorne (1939, 183–4), claiming Hettner as his major source of inspiration, thus writes:

> We may conclude therefore that while the interpretation of individual features in the geography of a region will often require the student to reach back into the

geography of past periods, it is not necessary that the geography of a region be studied in terms of historical development . . . Geography requires the genetic concept but it may not become history.

As the result, perhaps, of several major attacks upon Hartshorne's relative exclusion of temporal explanations, the *Perspective* shows some modification in this point of view, although still maintaining that the essential concern of geography is with present-day inter-relationships. It unfortunately seemed as if Hartshorne was proposing that geography should restrict itself to ecological or functional explanations (to be considered in the next chapter) to the exclusion of genetic forms of explanation. Within a year of the publication of the *Nature*, therefore, Sauer (*1963*, 352) was criticising Hartshorne because he did not

follow Hettner into his main methodologic position, namely, that geography, in any of its branches, must be a genetic science; that is, must account for origins and processes. . . . Hartshorne, however, directs his dialectics against historical geography, giving it tolerance only at the outer fringes of the subject. . . . Perhaps in future years the period from Barrows' 'Geography as Human Ecology' to Hartshorne's late résumé will be remembered as that of the Great Retreat.

This argument in favour of genetic forms of explanation received powerful support, particularly from geomorphology, which, under the Davisian influence, remained concerned with denudation chronology, and from historical and cultural geographers. Strong statements in favour of such an approach can thus be found in the writings of Darby (1953), Clark (1954), C. T. Smith (1965) and Harvey (1967A); although not all suggest that genetic explanation is the *only* appropriate form of explanation. This tends to be argued by the Davisian school and some historical geographers. Sauer (*1963*, 360) writes:

The geographer cannot study houses and towns, fields and factories, as to their where and why without asking himself about their origins. He cannot treat the localization of activities without knowing the functioning of the culture, the process of living together of the group; and he cannot do this except by historical reconstruction. If the object is to define and understand human associations as areal growths, we must find out how they . . . came to be what they are. . . . The quality of understanding sought is that of analysis of origins and processes. The all-inclusive object is spatial differentiation of culture. Dealing with man and being genetic in its analysis, the subject is of necessity concerned with sequences.

W. M. Davis (*1954*, 279) similarly writes:

The rational and modernized treatment of geographical problems demands that land forms, like organic forms, shall be studied in view of their evolution, while Wooldridge and East (1951, 82) come very close to committing what is known as the 'genetic fallacy' when they write:

> The best classifications, including those of land-forms are genetic, i.e., based on genesis. The best way, in general, of understanding anything is to understand how it has evolved or developed.

Others have sought to preserve some balance between the genetic and the functional mode of explanation. Thus Sorre (*1962*, 44) states:

> Within the group of natural and social sciences—to which human geography belongs—we use two types of explanation which are not opposed but complementary. Whatever the observed phenomenon may be, it is recorded in temporal series, it is the result of a long evolution, and it is explained by a series of anterior states. When we describe a phenomenon in these terms we give it a genetic explanation; let us say historical explanation for it is basically the same thing, history in a large sense being only the restoration of a succession. But the same phenomenon appears at the same time in a spatial context. It maintains multiple connections with its environment ranging from simple juxtaposition to causality. . . . Consequently, there is a place for an explanation drawn from the relations of a being to its environment—reciprocal relations because we are presented with a mass of complex actions, reactions, and interactions. This explanation, actualist by definition, is basically ecological.

C. T. Smith (1965) has similarly sought to reconcile these approaches. Enough has been said, however, to demonstrate the strength and importance of temporal modes of thinking in geography. The arguments by which such an approach may be justified require some clarification.

Proponents of the genetic approach have sometimes argued that genetic explanations are logically necessary in a discipline such as geography. Some have even suggested that it is the only acceptable form of explanation. The relevance and utility of genetic explanation can be assessed, however, only in terms of some given objectives. This point can be demonstrated with respect to the quotation from Sauer above (p. 408). The argument about studying houses and towns from a historical point of view is simply a matter of a preferred mode of approach. The only logical step in the argument presupposes that the objective is to 'understand human associations as areal growths', and naturally enough, if one is concerned with *growth*, some form of temporal explanation is unavoidable. Most logical arguments in favour of genetic explanations presuppose genetic objectives. These objectives are often implicit. Here the importance of terminology and classification to subsequent explanatory form can easily be demonstrated. It is surprising to find, for example, how many pleas in favour of the genetic approach in geomorphology seek to prove the appropriateness of the approach by referring to classes of events and objects in a terminology that is based on genesis (above, p. 348). The circularities in such argument should be obvious.

The justification for genetic explanations, therefore, must rest upon the identification of objectives which call for such an explanatory form. The argument regarding functional versus genetic explanation thus boils down to one of objectives. It is perhaps relevant, therefore, to discuss briefly how such objectives have developed in geography. Generally speaking they have been associated with views which place a very strong emphasis on the notion of development and change through time. This relates back, interestingly enough, to philosophical opinion regarding the nature of time itself.

I TIME

Time, like space, is a primitive form of human experience, although Meyerhoff (*1960*, 1) suggests that the experience of it 'is more general than space, because it applies to the inner world of impressions, emotions, and ideas for which no spatial order can be given'. There is a vast literature on the nature of time and upon the nature of development in time. It would be possible to treat time in even greater detail than was allotted to the consideration of space in chapter 14. Such a full discussion would be partly redundant since many of the conclusions drawn regarding space can be extended to time. In terms of space-time languages, for example, there appears to be no reason for regarding the x, y, and z co-ordinate dimensions as being radically different from the t dimension except in so far as time is irreversible (except in microphysics). Time, like space, is undoubtedly best conceived of as a relative rather than an absolute quantity. But time has emotive as well as scientific connotations. Given the emotion-charged debate over the significance of the genetic approach in geography, it is perhaps useful to begin by examining psychological and sociological aspects of concepts of time.

A Psychological and social time

Each individual experiences time and perceives time in some way, and attaches great importance to it. Meyerhoff (*1960*, 1) states:

> Succession, flux, change, . . . seem to belong to the most immediate primitive data of our experience; and they are aspects of time. . . . Time is particularly significant to man because it is inseparable from the concept of the self. We are conscious of our own organic and psychological growth in time. What we call the self, person, or individual is experienced and known only against the background

of the succession of temporal moments and changes constituting his biography. . . .
The question what is man, therefore invariably refers to the question what is time.

The importance of time to our conceptions of ourselves and of existence, coupled with the seeming helplessness imposed upon the human condition by the apparent inevitability of time flow, encompassing as it does our own death, makes the subject of time one of deep emotive significance. Temporal concepts emerge early in the young and are tightly bound up with the development of personality; the sense of time is buried deep in primitive myth and religious thought, and is deeply interwoven with the nature of social structures (Piaget, *1930*; Levi-Strauss, *1963*, 211–12). There are those who have sought to still time, to seek out some timeless world, to develop a metaphysic of some eternity outside time, religious or otherwise. 'To be conscious is not to be in time,' writes T. S. Eliot. Some have sought to show that time cannot flow. Zeno's 'paradoxes' of the arrow or of Achilles and the tortoise, sought to demonstrate that there was an inner contradiction in the concept of time flow; that if motion is movement from point to point (and how else could it be conceived?) then how can the arrow jump from one point of time to another without going through some timeless interval? Reichenbach (1956, 4) regards such philosophical and logical arguments as psychological projections, as defence mechanisms, 'intended to discredit physical laws that have aroused deeply rooted emotional antagonism'.

Others have sought to embrace time flow, have sought to turn the experience of time into a philosophy of being. Heraclitus thus believed that *Becoming* is the essence of living and being. 'All things are in a state of flux'; it is impossible to step into the same river twice. The philosophers of time flow, of *Being* and *Becoming*, of whom, probably, the most important to the twentieth century has been Bergson, have thus provided opposition to those who have sought to show time is illusion. There are many shades of opinion between. Kant characteristically tries to subsume both aspects of time in his transcendental idealist philosophy, a philosophy in which time, like space, is absolute but is also *synthetic a priori*.

The private and subjective experience of time has thus given rise to multiple interpretations of the meaning of time and to deeply held emotional attitudes—attitudes which have been explored in great detail in the twentieth century through the medium of literature and, more recently, through the medium of the cinema. Bergson thus suggested that the 'concept of physical time discarded what he believed to be the most essential qualities of time in experience, and their relationship to man' (Meyerhoff, *1960*, 138). These qualities of time have been the subject of literary investigation in the twentieth century—Proust, Joyce, Woolfe, Eliot, and the like, have all been deeply concerned with the meaning of time in private experience and the temporal aspects of *Being*.

The individual's perception of time is not independent of social and cultural concepts which, operating through language and social convention, allow the co-ordination of one individual's actions with those of others. These social aspects of time are built around what Hallowell (1955, 216) calls 'formalized reference points' to which past, present, and future may be related. These formalised reference points may be expressed in the calendar and the clock, the seasons, the life cycle, and so on. However they may be established,

it is impossible to picture any human society without them. In terms of individual experience, they are orientational. The individual's temporal concepts are built up in terms of them; he gets his temporal bearings by means of them, and his temporal perceptions function under their influence. It is impossible to assume that man is born with any innate 'temporal sense'. His temporal concepts are always culturally constituted.

The reference points established also vary from society to society and even within societies. Gurvitch (1964) attempts to classify some of these variations and to generalise about the kinds of social time which may be developed. He points out, for example, that the concept of time in peasant society, tied as such societies are to seasonal rhythms, is essentially different from the concept of time in an industrial society where technological change in itself promotes a far more progressive concept of time. Gurvitch thus emphasises the great multiplicity of temporal concepts and time scales. Levi-Strauss (*1963*, 301) similarly quotes the case of the Hopi kinship system which 'requires no less than three different models for the time dimension'. From the point of view of cultural anthropology and sociology, therefore, time is a measure geared to the process of living, and the time scales are as varied as the process of living itself.

These comments on the psychology and social aspects of our understanding of time are designed to illustrate the background to the substantive and methodological problems involved in treating time in geography.

(a) *The substantive problem* revolves around the elementary fact that the geographer is dealing with the results of human decisions which are partially dependent upon the perception of time by the individual making that decision. There is often a considerable difference between societies in which long-term investment is common (which requires a progressive view of time) and societies in which short-term attitudes are prevalent. Short-term thinking about agricultural land-use, for example, produces radically different agricultural patterns from long-term thinking. The social situation has a pervasive effect upon almost all aspects of the geographic reality we are dealing with, and notions of time are inherent in social structures. Consider, for example, the complex ramifications of different views regarding the 'right' time to get married—an important facet of social time scales. Consider the value placed upon time (however measured) in different societies and even within societies. To establish an adequate time-utility function is, thus, one of the basic problems facing cost-benefit analysis. Individual and social perception of time cannot be ignored in geographical analysis. This does not imply, however, that we are for ever doomed to some kind of formless relativism in our approach to social and individual activity in time. But we should be aware that activity can be understood only in terms of the social process and the social time scale, and that we cannot afford to ignore those scales in seeking out adequate explanations for particular geographical events. We can put time into our equations, but time is a parameter to be estimated in the context of the social situation rather than some quantity measured by a clock in Greenwich.

(b) The *methodological problem* is more basic, for it hinges upon finding an acceptable interpretation for time in geography. It has to be recognised that we, as individuals, are emotionally involved in choosing a concept of time, and that we are not free of certain predilections. How far emotion has affected geographical writing on time is difficult to determine. Certainly, many have sought to support the genetic viewpoint by referring to a particular concept of time-flow. Thus Darby (1953, 6) writes:

> Can we draw a line between geography and history? The answer is 'no', for the process of becoming is one process. All geography is historical geography, either actual or potential.

Sauer (*1963*, 360–1) similarly holds that:

> Retrospect and prospect are different ends of the same sequence. Today is therefore but a point on a line, the development of which may be reconstructed from its beginning and the projection of which may be undertaken into the future. . . . Knowledge of human processes is attainable only if the current situation is comprehended as a moving point, one moment in an action that has beginning and end.

Sauer here seems to be denying the philosophical possibility of timeless truths and to be embracing a philosophy of *Being* and *Becoming* worthy of Bergson and Heraclitus. Such arguments are, however, subjective and they are likely to produce emotional rather than rational reactions. This is not to deny the ultimate rule of value judgements over the selection of objectives. But it does suggest that we need to evaluate methodological argument about whether or not genetic explanation is important to geography against the psychological and philosophical background. Rational argument is possible only if we are prepared to recognise our own private predilections for what they are.

It is difficult, however, not to be subjective about time. Such subjectivity carries with it methodological dangers, for we are likely to impose a view of time developed as a response to our own private experience upon sociological and natural phenomena which may, or may not, conform to that experience. An *a priori* model of time must somehow or other be evaluated empirically. This has frequently proved difficult to do. The millions of years necessary for geological processes to complete their work, and the implication this carries for the age of the earth, the antiquity of man, and the descent of man, now so easily accepted, were matters of deep concern in the eighteenth and nineteenth centuries. It is surprising indeed to note how relatively recent what Toulmin and Goodfield (1965) have called *The Discovery of Time* has been. The Renaissance picture of man and nature was essentially static. Newton's mechanics made reference to time but the equations described an unchanging world that went on eternally revolving. Descartes provided a philosophy that appealed to timeless truths. The prevailing attitude to man's history was one in which development was not regarded as anything more than illusion. The prevailing paradigm in Western European thought was static. Only in the eighteenth century did this view begin to change substantially. By the nineteenth century genetic modes of explanation were definitely in vogue. The geographer's attitude to genetic explanation has not been unaffected by these changing modes of thought. It is therefore appropriate that we consider this discovery of time in science as a whole.

B The discovery of time in science

Leibniz had already suggested in the seventeenth century what later came to be known as the causal theory of time, which, in the context of physics, was to provide a criterion for establishing that time had *order* and therefore of necessity involved some kind of progression. But his views remained largely ignored. Vico similarly provided a turning-point in ideas regarding human history, countering the prevailing static view with a developmental one. But again his views did not at first gain credence. More influential, at the end of the eighteenth century, was Kant's cosmology:

The creation is never finished or complete. It has indeed once begun, but it will never cease. It is always busy producing new scenes of nature, new objects, and new Worlds. The work which it brings about has a relationship to the time which extends upon it. It needs nothing less than an Eternity to animate the whole boundless range of the infinite extension of Space, with Worlds, without number and without end (quoted in Toulmin and Goodfield, 1965, 133).

Such developmental views spread rapidly in the sciences, partly under the impact of empirical studies and partly through the growth of philosophies which resorted to appealing metaphysical concepts of growth and development through time. It was not long, therefore, before Hutton was proclaiming his celebrated conclusion to his empirical studies in geology:

The result, therefore, of our present enquiry, is that we find no vestige of a beginning—no prospect of an end (quoted in Chorley, et al., 1964, 46).

Lyell soon amassed sufficient empirical evidence to make Hutton's conclusions incontrovertible. The historical record which geology provided could not be ignored from the point of view of biology, which for some time before Hutton's statement had been toying with a developmental approach. The fossil evidence, together with the field evidence and medical evidence, could not be ignored, and in 1859 came the *Origin of Species*, probably the most influential book of the nineteenth century and one that was to have a tremendous impact upon social and political thinking and, naturally enough, on geography (Stoddart, 1966; 1967A).

Philosophy and history were not to be left out. Progressive political philosophies, and indeed political events, particularly the American War of Independence and the French Revolution, all seemed to be associated with a spirit that embraced development and change over time. The dialectical method in history—the notion of thesis, antithesis, and synthesis—dominated historiography in the work of Hegel and Marx; and although there was a reaction among the German historiographers, this was concerned to counter the notion of inevitability, historical determinism, and historicism, and did not challenge the notion of development through time itself. By 1900 sociology, economics, psychology, and indeed every academic discipline, had to some degree or other absorbed the idea of development, evolution, and change through time.

From the point of view of geography, these general developments were important in a number of respects. In the first place it was unlikely that geography would, or could, remain apart from such a major swing in the nature of thought. Ritter in 1833 pointed out the importance of the historical perspective in geography (Hartshorne, 1959, 83); and the historical perspective of geology had a considerable influence upon physical geography. By the end of the century Davis and Ratzel had both adopted evolutionary standpoints in their studies of landscape form and diffusion respectively—an evolutionary viewpoint that came to dominate geomorphology for half a century and had a profound impact on human geography through the work of Huntington, Griffith Taylor, Sauer, and others. In the second place it is perhaps to be expected that some of the excesses of the 'temporal change' school of thought should also be found in geography. These excesses go under a variety of names such as the *genetic fallacy*—the belief that the significance of something could be evaluated *only* by reference to its origins (a point of view the Victorians were particularly prone to) and *historicism*—the belief that the nature of something could be *entirely* comprehended in its development (Barraclough, 1955, 1). Such statements are not uncommon in geography, and even respected writers such as Sauer and Wooldridge and East (*cf.* the quotations on page 369) have been guilty in this respect. The fallacy comes in supposing that the genetic approach is the *only* one possible and that it produces *entire and complete* knowledge.

Perhaps the most important point about these 'background' features in developmental thinking, however, is that it is often difficult to distinguish between subjective (or simply assumed) properties of time and those properties that could be empirically distinguished. In general, the temporal systems of history (cyclical, linear, oscillatory, and so on) were assumed, and even the extensive empirical studies of Spengler and Toynbee assume rather than demonstrate the nature of historical time. In the natural sciences the situation was less clear. Temporal change and evolution could be established and in biology this appeared to take the form of an irreversible branching process—but there was considerable uncertainty, largely because of the difficulty of distinguishing the mechanism—a mechanism that was eventually provided by Mendelian genetics and subsequent developments in genetic theory. Out of this tangle of subjective and objective elements it was often difficult to choose rationally a coherent view of evolution and developmental change. Thus, as Stoddart (1966) is one of the rare articles on the interaction between modes of geographic thinking and general modes of thought, points out, evolutionary thinking in geography, heavily influenced as it was by Darwin's *Origin* took over the notion of a necessary evolution but somehow lost the idea of the chance mechanism (partly, it must be admitted, because of Darwin's own vacillations on chance mutation as the appropriate mechanism). It is perhaps important therefore, to try to set up some objective theory of time. Such a theory does *not* deny the importance and significance of time as a subjective experience (Bergson and James Joyce have settled that), but it does serve as some objective conceptual schema for dealing with the problem of time in the external world. The theory is based on physics, for, as Reichenbach (1956, 16) has pointed out,

> There is no other way to solve the problem of time than the way through physics. ... If time is objective the physicist must have discovered that fact, if there is Becoming the physicist must know it; but if time is merely subjective and Being is timeless, the physicist must have been able to ignore time in his construction of reality and describe the world without the help of time. Parmenides' claim that time is an illusion, Kant's claim that time is subjective, and Bergson's and Heraclitus' claim that flux is everything, are all insufficiently grounded theories. They do not take into account what physics has to say about time. ... If there is a solution to the philosophical problem of time, it is written down in the equations of mathematical physics.

What these equations tell us is not entirely incontrovertible nor is there full agreement on all aspects of interpretation. Thus Grünbaum (1963) finds plenty to disagree with Reichenbach (1956) about, as regards the nature of time. It is sufficient, however, for our purposes to note the two most important conclusions provided by the equations of macro-physics, namely, that time has *order* and *direction*. The first detailed empirical investigation of time came through relating temporal order to causal order to create the causal theory of time. Reichenbach (1956, 24) states:

> time order is *reducible* to causal order. Causal connection is a relation between physical events and can be formulated in objective terms. If we *define* time order in terms of causal connection, we have shown which specific features of physical reality are reflected in the structure of time, and we have given an explication to the vague concept of time order.

The causal theory of time, anticipated by Leibniz, became central to relativity theory and is generally accepted in modern physics as the macro-physical level. It relies of course on the temporal asymmetry of cause-and-effect relationships to give a firm empirical interpretation of time order. Time was provided with *direction* by the second law of thermodynamics which established certain irreversible processes and therefore provided a way of dealing empirically with our experience which shows that the past cannot be brought back. However, the equations that Boltzmann set up were statistical equations. Time could therefore be regarded as the movement from a condition where the outcome of a particular event is uncertain to a condition where the outcome is known. In terms of a closed system of the form specified in the second law, time therefore acted to maximise entropy (the state of disorder in the system). Reichenbach (1956, 55) states:

> The direction of physical processes, and with it the direction of time, is thus explained as a statistical trend: the act of becoming is the transition from improbable to probable configurations of molecules.

Physics therefore teaches us that time, in so far as it has order and direction, has them only in relation to process, and that there is, therefore, no absolute measure of time, only an infinite number of such measures, each associated with a particular set of processes. This poses an interesting problem regarding the actual measurement of time. Preferably some regular periodic process is used (the swing of a pendulum, the rate of decay of some radioactive substance, the rotation of the earth). An infinite number of such processes and time measures could be chosen (and indeed several different measures are referred to—e.g. sidereal time, astronomical time, etc.). Which should we choose? Nagel (1961, 180) suggests:

> We will seek as clocks periodic mechanisms which make it possible to compare and differentiate with respect to their respective periods an increasingly wider range of processes, and which enable us to establish with increasing precision general laws concerning the duration and development of these processes.

How far measures of time defined with respect to one process can be applied to the examination of other processes is therefore an empirical problem. There are conventionally established time scales for locating events (calendars, and so on) but it is worth while questioning whether these are appropriate for examining social processes which tend to vary greatly in terms of natural rhythm. It may even prove useful to invent our own time scales (resource-exhaustion time, amortisation time, erosion time, etc.) and to develop time transformations from one scale to another (Harvey, 1967A, 559). Most of our notions regarding time are akin to *a priori* models (they are intuitive and subjective) as compared with the *a posteriori* models which physics and astronomy provide. The use of such *a priori* models requires the usual care and 'eternal vigilance'.

II TEMPORAL MODES OF EXPLANATION

We can summarise the preceding discussion as follows:

(i) Temporal modes of explanation (usually called genetic **or** historical explanations if geography) are important in geography and provide a useful but not exclusive mode of approach, given objectives appropriate for such modes.

(ii) Temporal modes of explanation require an understanding of time.

(iii) The only way we can develop objective measures of time is in terms of some process.

(iv) From this it follows that rigorous modes of explanation cannot be developed without reference to process.

There are many modes of temporal explanation. Simple narrative is the weakest, evolutionary (with or without a mechanism), cyclical (with or without a mechanism), and the like, provide others of intermediate strength. The most rigorous is that defined by the scientific use of the term 'process' itself. It is therefore appropriate that we begin by examining process explanations in some detail.

A Process

The technical meaning of the term 'process' is rather different from its ordinary everyday interpretation. In the ordinary sense we frequently use 'process' to refer to any sequence of events over time. Such usage has little to recommend it in the context of explanation, for it fails to differentiate between *any* sequence of events (which cannot be regarded as explaining anything in particular) and a sequence which is connected by some established mechanism (and which can thus be regarded as explanatory). It is thus possible to differentiate between sequences that exhibit necessary and sufficient connections between the various stages in the sequence and those for which no such connections can be found. A process, in the scientific sense, can be given a more rigorous definition with the aid of some of the conclusions drawn from our analysis of the logical structure of cause-and-effect systems. A process law depends, therefore, on the specification of:

(i) the system (a closed system) within which the process law operates;

(ii) the relevant states of the system;

(iii) the relevant variables interacting within the system;

(iv) the parameters governing interaction among the variables and the direction of the interactions.

Process laws succeed in explaining only in so far as these features can be specified. Bergmann (1958, 93, 117) therefore puts the scientific process model in the following schematic form. Let $C = (c_1, c_2, \ldots, c_m)$ be a description of the system and $S^t = (x_1^t, x_2^t, \ldots, x_n^t)$ be the description of the state of the system at time t. Then given C and S^{t_1}, it is possible to predict S^{t_2} from the process law, but the reverse prediction can also be made (i.e. knowing S^{t_2} we can predict what the state of the system was at S^{t_1}). Thus 'if the system is known then any two of its states can by means of the process law be inferred from each other.' A process law thus defines what we shall later call the 'trajectory of a dynamic system' (below, p. 461).

This logical schema illustrates the meaning of process in its strict scientific sense. This meaning is typified by a markov process in which states of the system are identified, an initial state given, and a matrix of transition probabilities provides the necessary process law (see Harvey, 1967A for an elementary account of such structures in a geographic context). In empirical work, this view of process implies that process laws succeed in 'explaining' only in so far as the empirical situation can be regarded as isomorphic with, and hence mappable into, a logical schema that possesses these characteristics. This translation poses difficulties. It is thus not easy to identify the limits of the system being considered (i.e. to close it), identify the relevant states and variables, and to know everything there is to know about the interaction among the relevant variables. In controllable situations (e.g. under experimental conditions) these difficulties are not insuperable, but in most areas of geographical enquiry such a mapping requires some powerful assumptions. Bergmann (1958, 127) suggests that process laws, in the strict sense, cannot yet be identified in social science because the conditions necessary for such laws to operate have not been identified. Nevertheless, giving a strict specification of process has a number of advantages:

(i)   It provides a model framework for formulating process-type theory in geography. It also allows us to build process-type models for the purposes of prediction or experimentation. We may, for example, build a markov-chain model of migration or regional development and explore its characteristics analytically, or we may build simulation models of river patterns, urban growth, and so on (see Harvey, 1967A, for a review of such process models). In each case we are building a model which, although it may not be empirically realistic, has considerable analytic use. In some cases such models may yield valuable short-term forecasts and may also lead to the identification of theory.

(ii) The strict specification of process laws allows us to set up a norm by which we may evaluate the various attempts that have been made in geography to explain by stating some temporally ordered sequence of events. Such an evaluation is important, and we shall therefore attempt it in some detail.

B Temporal explanation

There are many examples in geography of descriptions and explanations that adopt a temporal approach. This approach is sometimes called *genetic*, but this title can be misleading. We have therefore used the phrase *temporal explanation* to denote all those forms of explanation that involve temporal relationships. There are many forms of such explanation and it is consequently difficult to classify them in any coherent way. We can classify, for example, in terms of

(i) Genetic explanations proper—i.e. explanation by reference to origin.
(ii) Evolutionary or developmental explanations, which seek to explain an event by way of the events that precede it.
(iii) Genetic and evolutionary explanations that refer to origin and subsequent development.

On the other hand it is possible to think of temporal modes of explanation in terms of whether they

(i) Assume no mechanism.
(ii) Hypothesise some mechanism (which may be probabilistic or deterministic).
(iii) State some mechanism for which there is detailed empirical evidence.

It would be equally possible to classify according to whether time is treated as

(i) Continuous (i.e. no natural breaks).
(ii) Discrete (i.e. discrete stages can be identified).

Actual explanations in geography often exhibit all of these features, and it is therefore difficult to pin them down. In discussions regarding the nature of historical geography a number of different approaches have been identified (Hartshorne, 1959, chapter 8; Darby, 1953; C. T. Smith, 1965). It is difficult to select coherently from these in order to portray some of the essential characteristics and shortcomings of temporal explanation in geography. In what follows, therefore, a number of types will be discussed, but in each case we are dealing with some 'ideal type' for it is impossible to consider all forms of temporal explanation that have been put forward in geography.

(1) *Narrative*

It is not usually thought that plain historical narratives are concerned with explanation. The aim is usually to describe the occurrence of events in terms of the time dimension. The historical narrative thus performs the same function for temporal relationships as the map does for spatial relationships. The aim of historical narrative may be purely descriptive but it is not free from explanatory elements. In the first place the selection of events is usually made according to some criteria of significance. Some events are regarded as significant and therefore included in the description and others are omitted as being irrelevant. Each generation tends to select according to different criteria. The medieval historian may simply have recorded tempests and calamities, the nineteenth-century historian the reigns of kings, the twentieth-century one the dates of important inventions. In the second place it is difficult not to avoid hints at relationships, associations, and even necessary and sufficient connections, in a historical narrative. Statements like 'The river silted up and subsequently the port of Rye declined' exhibit some causal connection. In other cases it appears as if appeal were being made to some general, although rather vague law; thus 'Nothing better illustrates the power of historical momentum than the growth of Bristol in the sixteenth century.' In other cases some more specific hypothesised law is referred to: 'Once Bristol had been established, it drew many functions to it by virtue of the economies of scale which resulted.' Or perhaps some law is implied: 'Lancashire possessed a damp climate and the cotton industry developed rapidly.'

It is very rare for historical narrative to be free from explanatory sentences. But most historical narratives are connected by occasional necessary and sufficient relationships, accompanied by jumps for which no such conditions can be found, and jumps that can scarcely possibly be connected by any explanatory argument at all. Narratives thus provide loose, weakly explanatory, non-rigorous modes of temporal explanation. In many instances they are the best (and most appropriate) mode of explanation, given the paucity of historical data (in many instances it is impossible to know what the necessary and sufficient connections were). The historian and the historical geographer are frequently forced to offer a relatively weak 'explanatory sketch' simply because of lack of information. There is no point in being rigorous with respect to explanatory form when the data do not warrant it. However, it has to be admitted that narrative falls far short of the ideal in terms of rigour, coherence, and consistency, and consequently suffers as a mode of description. Its explanatory role appears to be weak and incidental.

(2) *Explanation by reference to time or stage*

A narrative places events on some time scale. It is possible, however, to regard time itself as some kind of forcing variable. Consider W. M. Davis' (*1954*, 249) classic statement:

> All the varied forms of the lands are dependent upon—or, as the mathematician would say, are functions of—three variable quantities, which may be called structure, process, and time.

Here time is regarded as some independent variable which is, in itself, causally efficacious. This view of time is, as the preceding discussion has shown, misleading, since time is a parameter to be estimated and not an independent variable. This misconception of time is common in geography. Thus Stoddart (1966, 688) suggests that 'geographers interpreted the biological revolution in terms of change through time: what for Darwin was a process became for Davis and others a history'. There are many situations, however, in which it proves difficult to define an appropriate measure of time; and therefore some arbitrary time scale is set up which, presumably, reflects some unknown process. Davis thus sets up a cyclical time scale, that contains distinctive stages of youth, maturity, and old age (this, incidentally, is an interesting example of the personal experience of time being projected into a completely different domain of events). Explanation here amounts to locating a particular landscape on this time scale, for once it is so located we can *assume* the general nature of the preceding situations and, if we so wish, project the future. This is a very common mode of explanation. Explanations in terms of archaeological stages, economic-growth stages (such as those put forward by Rostow, 1960), population-growth stages (the so-called demographic transition is a good example here), are cases of such a mode from other disciplines. In geography the 'zones-and-strata' technique used by Griffith Taylor (1937), the 'sequent-occupance' concept put forward by Derwent Whittlesey (1929), and the 'cultural-stage' approach applied by Broek (1932) in his classic study of changing land use in the Santa Clara valley, are good examples. In many such studies the aim is not overtly explanatory, but they do explain implicitly. By setting up a sequence of stages it is implied that all areas must necessarily go through these stages, and in some cases it is also implied that the stages themselves are the result of some mechanism. In the Davisian system the process of erosion presumably provides the mechanism, in the zones-and-strata approach technological change (a learning process), might provide it. But, again, from the point of view of explanation, these mechanisms often remain assumed or inexplicit, and the main emphasis is on the stage itself as the explanatory variable.

This mode of explanation is relatively crude and frequently runs into difficulty. As long as phenomena can be unequivocally placed on the scale, the explanation at least appears efficient. The problem is that it is often difficult to locate phenomena in terms of their stage in some schematic evolutionary process. Rostow (1960) for example, finds that some countries are simultaneously in two stages of economic growth. An examination of historical development shows that some areas have missed out what is generally thought to be a vital stage (thus archaeological thinking has changed considerably regarding the transition stone age—bronze age—iron age). The most interesting case of this kind of problem is provided by denudation chronology which has been forced to admit interruptions in the cycle and, to deal with the complexities of actual landscapes, has sought to indentify cycles within cycles, sub-cycles, and so on, until the whole terminology of youth, maturity, and old age, might just as well be abandoned for one concerning the relative rates of uplift and erosion. Such stage-explanations, therefore, amount to presenting an idealised temporal sequence and they suffer, consequently, from not always being empirically identifiable.

Treating time or stage as an explanatory variable is not in itself unreasonable, provided that the process by which time is measured is known. In many cases it is not. In such conditions it is tempting to assume a mechanism and to show that there is some evidence for it. The mechanism is all-important and amounts to a full theory of growth, and it cannot therefore be treated in a cavalier manner. In practice, however, many stage-explanations simply amount to the application of an *a priori* model of time and an *a priori* model of a mechanism. In some cases the models are based on analogy, particularly with our psychological experience of time and with what we know about the life of organisms. It is therefore intuitively appealing (if rather naïvely anthropomorphic) to think of growth in time by means of terms such as youth, maturity, and old age. What seems a necessary sequence of events in our own lives is therefore assumed to be a necessary sequence in the development of landscapes or in the development of towns. Projecting such necessity into the domain of landscape evolution is a striking illustration of how some aspect of an *a priori* model may (unjustifiably) become a necessary characteristic of a theory of landscape evolution (above, p. 146–7). Stoddart (1966; 1967A) provides an excellent account of the role of organismic analogy in geographic thinking which shows how frequently we have made this mistake in seeking for genetic explanations.

Explanations in terms of temporal stage suffer from some methodological difficulty. The problem is not that it is wrong to refer to time or stage in this manner (indeed the methodological problem would be much easier if it were). It is, rather, that the validity of such a treatment depends entirely on:

(i)   the provision of the necessary mechanism, which also involves identifying fully the process in terms of which time itself may be measured;

(ii) defining the stages unambiguously so that we can locate any particular situation in terms of them.

These are empirical and operational problems, and until they are overcome, explanations in terms of stage or temporal sequence will lack validity. Unfortunately such models are intuitively attractive. It is all too easy to appeal to our sense of time passing, of *Becoming*, and of the inevitability of change in our own physical being. The statements already quoted from Darby and Sauer do precisely this. This is not to say that landscape is not *Becoming* or that the geographical world is not changing. From the point of view of explaining we need to show how it is changing and at what rate and the nature of the processes (conceived of in a scientific sense) which *cause* time to pass. To rest content with the idea that things of necessity change over time and that therefore the passing of time itself *causes* change, is to rest our analysis on a delusion. Explanations in terms of temporal sequence are therefore only as good as our ability to understand the nature of time itself with respect to the phenomena we are trying to explain.

(3) *Explanation by reference to hypothesised process*

Instead of assuming the properties of time itself (with only passing or partial reference to process) it is possible to assume a particular process and, given that process, generate an artificial time scale on which to locate events and hence explain them. It has been a feature of what are sometimes called 'historical' explanations, that they frequently assume some set of 'historical' laws that define some necessary and sufficient progression. The Hegelian and Marxist dialectic are classic examples of such an approach in history—a mode of explanation frequently regarded as deterministic and historicist in the extreme (Popper, 1957).

The classic case in geographic thought, however, is that of environmental determinism, particularly in its rather rigid early-twentieth-century form. Huntington and Griffith Taylor are probably the best representatives in the Anglo-American geographic tradition. The mechanism assumed in this case is simply the inevitability of man doing nature's will, and of adjusting the pattern of society so that it conforms to nature's 'design' (whatever that may be). There is an inevitable evolution towards that end. In this case, as indeed in the Marxist dialectic, the mechanism is subservient to some end-state of the evolution. Such explanatory frameworks are usually termed 'teleological', and will be dealt with in the next chapter. With or without some teleological assumption, however, it is possible to postulate some historical law that governs the transition of phenomena from one state to another. Such historical laws differ from scientific process laws in a number of important respects:

(i)   It is often impossible to assume (as is frequently done with physical processes) that the law itself is unchanging and that it can, therefore, be tested entirely from observation of present circumstances. Hutton was able to assume, with a fair degree of confidence, that the evolution of a land could be understood in terms of processes which can be observed at the present time. It is impossible to make the same assumption, at least in the present state of our understanding, about the evolution of a culture.

(ii) Such historical laws therefore appeal entirely to the evidence provided by history itself. Establishing a law of this kind therefore amounts to discerning some pattern in the historical record. Simply identifying pattern, of course, amounts to stating the kind of temporal sequence discussed in the previous section. The point here is that it may be possible to search history for a pattern that validates a particular mechanism that is set up *a priori*. Interpreting history frequently involves just this. The problem is that history has only one story to tell (out of an infinity of might-have-beens). It is therefore peculiarly difficult to validate historical laws or to validate the existence of an assumed mechanism.

It is not for us to consider the complexities of historical explanation which is, in any case, a subject of fierce debate (see above, chapter 5). However, it has to be recognised that explaining by way of some assumed historical mechanism is a chancy business. It may be that the evolution of spatial patterns in human geography has simply been the result of a growing adjustment of man to his natural environment, given particular levels of technology. But it would be extraordinarily difficult to prove such a postulated mechanism (and almost as difficult to disprove it). The kind of mechanism assumed, therefore, depends upon the predilections of the individual and upon the consensus of philosophical opinion among researchers in a given discipline. Without going into all the details, it is useful to demonstrate this by contrasting deterministic and probabilistic modes of envisioning mechanism and process.

The predominant mode of thinking about mechanism and process in the late nineteenth century was deterministic or, as it is sometimes called, mechanistic. There can be no doubt that this view of the world drew much from the inspiration of Newtonian mechanics, in which the behaviour of the planets could be predicted without any uncertainty from a fundamental set of differential equations. This mechanistic approach was particularly evident in the dialectical approach to history of Hegel and Marx, and in the sociology of Spencer. In geography the works of Ratzel, Huntington, Griffith Taylor, and so on, were all characterised by a basic assumption that the underlying mechanism governing the evolution of spatial patterns was a deterministic one. Such writers did not necessarily assume that they had identified the right mechanism or that their theses about development were empirically correct. They were seeking to find evidence for some assumed mechanistic process. Their work thus frequently appears to be special pleading for some assumed mechanistic process governing the evolution of spatial pattern.

It is possible to think of process and mechanism in terms of probability. It is curious to find the late nineteenth century in the social sciences governed by deterministic views drawn largely from Newton, when Boltzmann had already shown that time-order could best be conceived of as a transition from less probable to more probable configurations, a transition that was irreversible. Darwin had also suggested that the underlying mechanism in biological evolution was chance mutation, a point that all the social evolutionists, including those geographers who drew much from the organismic analogy, seem to have systematically ignored (Stoddart, 1966). Assuming a probabilistic mechanism is much more fashionable at the present time, and this assumption is far less demanding and far more flexible of application than its deterministic alternative. Thus the deterministic framework built up by Davis to explain landscape evolution (with an occasional accident interrupting its course) has been replaced by a probabilistic model which conceives of the physical landscape as the end-product of a series of irreversible steps governed by a mechanism that progresses towards maximising entropy; a process that has more in common with the second law of thermodynamics than it has with the Newtonian equation system (Chorley, 1962). Curry (1966A) has developed a similar view with respect to the evolution of the human landscape.

This probabilistic view poses a problem. It is often held that the geographer (and the historian) is concerned with particular events, a particular configuration of phenomena. Now it is a characteristic of probabilistic argument (see chapter 15 above) that the statement of a set of initial conditions in conjunction with a probabilistic law is not a necessary and sufficient condition for the result to be inferred.

The only situation in which such a result can be inferred is that in which we are dealing with large aggregates of events which can be conceptualised in terms of frequency probability. It is then possible to identify equilibrium states (as in markov-chain models of migration for example) and to determine the characteristics of those equilibrium states. Now the probabilistic concepts of process that entered into geography in the work of Vidal de la Blache were applied solely to individual events and not to aggregates. All that was possible under such circumstances was to state some of the necessary (but none of the sufficient) conditions for the occurrence of a particular event. It is possible, however, by adopting a subjective view of probability theory to discuss individual events in a more rational manner (Nagel, 1961, 563). Assuming a probabilistic mechanism thus poses some difficulties. Curry (1966A, 45) puts it this way:

Probabilistic reasoning has a new viewpoint to offer to the study of historical process in geography. If the development of landscape be a series of contingent events then in a certain sense the study of the historical record cannot reveal many of the generalities of process. Of course the development of particular places requires a particularist historical account without the need for generality of concept. The food, the shrine, the castle, the coalfield, the protestant ethic, the sea, appear in correct sequence to help explain Newcastle upon Tyne and another set of factors are used to explain another town. . . .
However, if we classify events sufficiently broadly, events which were previously regarded as different in kind now become only differences of degree. When we can specify the probabilities of occurrence of these degrees we can write a stochastic process of which the real history described above will be only one possible sample. . . . Clearly this is a valid procedure; whether it is a useful one depends on the question one asks of the landscape.

Assuming probabilistic mechanisms thus involves us in some conceptual difficulties. In general it adds nothing to the rigour of explanations of particular situations, and it can only add rigour if we are prepared to pay the price of mapping the varied phenomena we are dealing with into the calculus of probability. Assuming deterministic mechanisms does not involve this problem, but it has associated with it a number of other difficulties.

In both cases, however, it is important to remember that the mechanism is being *assumed*. The choice might be thought of as one between two different analogies—Newtonian mechanics on the one hand or the second law of thermodynamics on the other. In both cases, however, the real crux of the explanation comes in showing the isomorphism between the assumed mechanism (or the analogue model) and the real-world mechanisms which can be identified as generating particular situations through temporal change.

(4) Explanation by reference to actual processes

In this section we come close to considering those explanations which conform to the scientific meaning of process explanations. Let us suppose that we have strong empirical evidence for the existence of a given mechanism and that we are prepared, therefore, to accept it as operant in certain domains. Let us suppose also that we have a situation that falls firmly in the domain of the process identified. It is then possible to explain the situation in terms of that process. Evolutionary biology comes close to this at the present time, for a good deal of information has been built up regarding the actual mechanism governing inheritance, and it is reasonable to assume that this process has held constant through time. It is thus possible to explain present characteristics in terms of past characteristics plus this known mechanism (which is in fact probabilistic). More penetrating studies of process in geomorphology are similarly beginning to reveal a good deal of information which will presumably make assumptions about process more and more redundant. The present situation in human geography is less encouraging. We know very little about process and still assume mechanisms rather than investigate them. To this degree temporal modes of explanation are undoubtedly inferior in human geography, for there are few, if any, situations where we can point with reasonable confidence to a set of processes which we *know* are applicable and therefore can serve to explain a particular configuration of phenomena. This must in part be attributed to the general lack of a dynamic viewpoint in human geography in recent decades (Harvey, 1967A). The quite recent return to an examination of process in human geography is thus a welcome development. Work on perception, learning, searching, sequential decision making, environmental behaviour, and so on, is thus much related to this return to a closer look at actual processes. At least it indicates a welcome realisation that an understanding of process is the key to temporal modes of explanation. It is perhaps useful to take to heart the words of Leighly, uttered in 1940 as a criticism of the Davisian system, and to apply these strictures to human geography:

> Davis' great mistake was the assumption that we know the processes involved in the development of land-forms. We don't and until we do we shall be ignorant of the general course of their development. (Quoted in Wooldridge, 1951, 167.)

C The problem of time and explanation in geography

Cause-and-effect and process-type explanations have the property of temporal asymmetry. Indeed, it could be argued, and sometimes is, that *all* explanation is temporally asymmetric, since explanation ought to involve deducing some existing condition from a set of initial conditions and some process law. Such a view is not consistent with the broad interpretation given to 'explanation' in this book. However, a large class of explanations in geography involve temporal asymmetry, and it is generally agreed that such explanations are of great significance. It is possible to set up a rigorous model of such temporally asymmetric explanations and to show that a logically rigorous schema can be developed to link existing states with anterior states and, by regression, to genesis. There are, however, acute problems facing the application of such a logically rigorous model to an actual historical succession.

These problems are essentially the same as those which dog explanation in history itself, and they are present in all historical sciences such as geology. Historians and historical geographers certainly do not appear to attempt explanation in the rigorous sense and have developed their own special sense of 'explanation' which, it is sometimes claimed, is distinctive and completely different from explanation in the scientific sense. This is a controversial issue which we have already considered in some detail in chapter 5. It was there shown that the scientific model could not be rejected, but that there were considerable difficulties, conceptual and operational, in applying such explanatory forms in a historical context. The aim of this chapter has been to show that temporal modes of explanation conform, in some respects, to the scientific model, and that it is possible to regard all such explanations as approximations, necessitated by the contingencies of the situation, to the basic scientific model. The scientific model thus elaborates a logical schema which shows the necessary and sufficient conditions for a given state to be predicted from an anterior state. Most explanations in historical geography are concerned with establishing necessary conditions. It is fairly characteristic of such explanations to suggest that 'if it had not been for the fact that . . ., *x* would not have happened'. We are thus fond of pointing to Lord Nuffield's bicycle shop and house as a condition for the development of Lord Nuffield's factory at Oxford, to environmental conditions in the development of a given industry, to the existence of a harbour as a condition for the development of a port, and so on. It is relatively rare for sufficient conditions to be referred to.

This absence of the sufficient conditions in historical explanation has some serious implications. In the first place it relegates such explanations to being a rather inferior version of the scientific process model. This in itself is not perhaps so serious as the tendency to regard the necessary conditions as being in fact sufficient conditions. Much of the problem regarding environmental determinism in geography, as Montefiore and Williams (1955) point out, resolves itself into such a basic misunderstanding. It therefore seems unwise and somehow unsatisfactory to rest content with necessary conditions. It may be very difficult to identify the sufficient conditions, particularly in historical times, and particularly if we hold that explanation should be in terms of reasons, dispositions, and motives, of individual decision-makers. Identifying the sufficient conditions, however, amounts to identifying a process law. This conclusion is inescapable. Without such process laws, we are likely to confuse necessary conditions with sufficient conditions, erect pseudo-process-laws in the form of historical sequences or by way of assumed mechanisms, and generally misunderstand the true nature of the 'historical' explanations we are expounding. For this reason it is perhaps wise to regard most forms of temporal explanation as different from true scientific explanation. But they are, nevertheless, forms of scientific explanation, and may be evaluated in these terms.

Explanations that involve temporal asymmetry are of great importance in geography, for they are among the most persuasive forms of explanation that we possess. It is also important for geographers to come to grips with the concept of time and the meaning of temporal dependence. But in seeking for explanations that stretch back over time, we may choose among a variety of modes ranging from the logically rigorous process model to simple narrative. At the present time it seems that nothing useful can come from insisting that the only admissible form of explanation is that which is rigorously scientific and objective. We should, however, be prepared to admit the problems inherent in using less rigorous modes of temporal explanation. The problem is not, therefore, that we fail to be rigorously scientific and objective, but that we fail to acknowledge the respects in which we have been forced to compromise with this ideal, and hence fail to distinguish between permissible and non-permissible inferences, given the logic of the situation.

Basic Reading

Toulmin, S., and Goodfield, J. (1965).
Meyerhoff, H. (*1960*).
Nagel, E. (1961), chapter 16.
Reichenbach, H. (1956).

Reading in Geography

Hartshorne, R. (1959), chapter 8.
Harvey, D. (1967A).
Sauer, C. (*1963*), chapter 17.
Smith, C. T. (1965).
Stoddart, D. (1966).

Chapter 22
Functional Explanation

It is sometimes claimed that *functional explanation* (or its close relation *teleological explanation*) provides an alternative to cause and effect or the various temporal modes of explanation discussed in the previous chapters. It is also sometimes claimed that this form of explanation is uniquely suited to explaining the occurrence of some element or part-process within some complex functioning 'whole' such as a biological organism, a culture, an economy, or the like. It is consequently held that certain disciplines, such as anthropology, sociology, psychology, and biology, must necessarily resort to a different form of explanation and that they are methodologically distinct disciplines. Such a claim has not explicitly been made for geography, although there can be no doubt that functional explanation is extremely common in geographical analysis. London might thus be 'explained' by reference to its functions as an international monetary centre, as a capital city, as a manufacturing centre, as a port, and so on. Settlements may be 'explained' by their function in a central-place system. Towns, regions, communication systems, and the like, may all be 'explained' by reference to their function. It is also sometimes held in geography that the 'functional approach' is an alternative (and in some people's books a more important) approach to the study of geography to cause-and-effect and genetic approaches. The contrast between these approaches has already been commented upon in the preceding chapter. Consider, however, the statement by Philbrick (1957, 300–2):

is it not through functional connections established by people between people that the race as a whole has developed locational and internal organization with roots in specific places? . . . Is it not fundamental to human geography, therefore, to search out the inherent areal organization of society? Is not the functional organization of human occupance in area the basic subject-matter of human geography. . .?

Given the undoubted significance of the 'functional approach' in geography, the undoubted frequency of functional explanation of geographical phenomena, and the very strong claims made for functional explanation in certain other disciplines, it seems absolutely essential that we clarify the meaning of 'functional explanation' and, in the process, provide some evaluation of the role of functional thinking in geography. It is convenient to analyse this problem in three stages:

(i)   The logical form of functional explanation.
(ii) 'Functionalism' as a universal hypothesis.
(iii) 'Functionalism' as a working hypothesis.

A The logic of functional analysis

Given the considerable claims made for functional explanation it is hardly surprising to find that it has been subjected to searching logical analysis. It is thus considered in some detail by Braithwaite (*1960*, 322–41), Brown (1963, chapter 9), Hempel (1959), Lehman (1965), and Nagel (1961, chapter 12). The general conclusion to be drawn from these analyses is that the claims made by biologists, anthropologists, and sociologists, for functional explanation as a distinctive form of explanation cannot be substantiated in the face of logical analysis. These analyses do not deny that functional statements can provide explanations of a sort. Some, such as Braithwaite and Brown, suggest that there are many circumstances in which functional explanations may be appropriate and provide reasonable answers to questions. Others, such as Hempel and Lehman, consider functional analysis as providing rather weak forms of explanation. All, however, deny that functional explanation is different in kind from other forms. It is, apparently, better viewed as some kind of *approximation* to more efficient forms of explanation—an approximation often necessitated by the complexity of the phenomena being investigated.

Hempel (1959, 278) characterises the nature of functional analysis as follows:

> The kind of phenomenon that functional analysis is invoked to explain is typically some recurrent activity or some behavior pattern in an individual or a group; it may be a physiological mechanism, a neurotic trait, a culture pattern, or a social institution, for example. And the principal objective of the analysis is to exhibit the contribution which the behavior pattern makes to the preservation or the development of the individual or the group in which it occurs. Thus functional analysis seeks to understand a behavior pattern or a sociocultural institution in terms of the role it plays in keeping a given system in proper working order and thus maintaining it as a going concern.

It is useful to replace Hempel's psychological, anthropological, and sociological examples, by a geographical one. Consider, therefore, a central place or a market. This has a function within an 'economy' (which is the system), for markets and central places fulfil a need, namely the distribution of goods and services in an efficient manner. They function to keep an economy in 'proper working order'. It may be possible to develop a more penetrating analysis of the functions of markets and central places. Merton (*1957*) thus distinguishes between *manifest* and *latent* functions—it may be that central places *appear* to exist to facilitate exchange of goods and services but that they *really* satisfy some basic psychological or sociological urge in man concerned with gregariousness. If this could be shown to be so, then the exchange function would be a *manifest* function (rather like rain-making ceremonies) whereas the social function would be the *latent* function (like reinforcing group identity by participation in rain-making ceremonies). Now there can be no doubt that questions of this kind lead to interesting lines of enquiry. The problem we are here concerned with, however, is whether such questions also lead to distinctive modes of explanation. Suppose therefore that we attempt to explain a given trait i (e.g. a market) in a system s (an economy) at a certain time t. Hempel (1959, 283) represents such an explanation as follows:

(i) At t, s functions adequately in a setting of kind c (where c defines a set of environmental and internal conditions).
(ii) s functions adequately in c only if a certain necessary condition, n, is satisfied.
(iii) If trait i were present in s then, as an effect, the condition n would be satisfied.
(iv) *Hence*, at t trait i must be present in s.

Hempel goes on to consider in detail the logical difficulties inherent in this form of analysis. He points out, for example, that (iv) can be obtained only if (iii) states that a given need n can be satisfied *only* by a given trait i. In other words, if goods and services could be distributed only by way of markets. Such an assertion is dubious, and most functional analyses readily admit functional alternatives, functional substitutes, or functional equivalents. In general there are a range of traits, $i_1, i_2, \ldots, i_n$, all of which are capable of fulfilling a given need. The premises of the argument outlined above provide us with no grounds for expecting any particular i out of this range of alternatives. Hempel thus concludes that functional analysis 'affords neither deductively nor inductively adequate grounds for expecting i rather than one of its alternatives'. Functional analysis can provide, therefore, only a weak kind of explanation by showing that at least one out of the range of traits must exist to fulfil a particular need. This is a rather unilluminating conclusion, for it amounts to stating that goods and services must be distributed in an economy *somehow*, for otherwise the economy will not function. Hempel suggests that the appeal of functional analysis is 'partly due to the benefit of hindsight', in that we already know that markets do exist. Functional analysis provides a convenient method for identifying some of the necessary conditions for the existence of central places, but it cannot, apparently, provide necessary and sufficient conditions for explaining a given trait, such as the occurrence of central places.

This analysis of the logical deficiencies of functional explanation does not in any way vitiate the use of functional statements in explanation. Taken in conjunction with certain other specific conditions, function statements may indeed be used in the course of offering an explanation. These conditions have already been hinted at in the use of phrases like 'recurrent activity', 'proper working order', and so on. Brown (1963, 110) suggests that functional relations are a sub-class of causal ones:

> The difference between the two sorts of relations is that functional relations hold only between traits within a specified system of a certain type—a self-persisting one—while the class of causal relations is much larger. It includes these and others as well. . . . Functional relations, then, are certain causal ones which operate within self-persisting systems.

Brown thus indicates that 'functional' explanations can be reduced to some more effective 'non-functional' explanation in the context of self-persisting and self-regulating systems (a view that Nagel, 1961, 402–6 also appears to accept). This automatically removes the special nature of functional explanation from view, but at the same time it plunges us into a consideration of the nature of self-regulating systems—indeed the whole onus of developing an appropriate mode of functional explanation is thrown on to the specification of self-regulating systems. Nagel (1961, 406–21) provides an extended general analysis of such self-regulating, or, as he terms them, goal-directed, systems. The simplest example is provided by a thermostat arrangement for heating water, in which the function of the thermostat is to register any variation in temperature of the water and to take appropriate action to keep that temperature at the same level. The general characteristic of such systems is the phenomena of negative feedback and there are, of course, plenty of examples in geography where such phenomena may be observed (for example, any form of adaptive behaviour which functions to maintain the *status quo*). This loose description is not sufficient, however, for us to justify the use of function statements in explanation. Before we can use such statements in fact, we require to specify the precise nature of the self-regulatory system and to show, by way of empirical test, that this specification is a reasonable one. Thus Hempel (1959, 290) states that

if a precise hypothesis of self-regulation for systems of a specified kind is set forth, then it becomes possible to explain, and to predict categorically, the satisfaction of certain functional requirements simply on the basis of information concerning antecedent needs; and the hypothesis can then be objectively tested by an empirical check of its predictions.

Further, as Lehman (1965) insists, we require a precise meaning to be given to the notion of a system that is 'in proper working order' or is 'functioning properly'. There are, for example, a number of ways in which we might regard an economy as being in good working order; and if we have some criterion of efficiency (e.g. a solution that minimises effort), it may be possible to show that market-places must occur. A function can be discussed, therefore, only in relation to self-organising systems whose properties are known (this ought to include some information on the tolerance limits within the system). But given this information there is no need to resort to functional explanations, for such explanations simply constitute a variety of causal-law explanations operating in a systems context. Lehman (1965, 16) thus suggests that there is a tendency for function state-ments to be eliminated from biology and for them to be characteristic of disciplines in the early stages of development.

These views regarding the logic of functional analysis should, it must be emphasised, only be related to the version of functional analysis defined by Hempel. There are a number of variants of this form of analysis. Indeed, part of the difference of opinion on the logic of functional analysis must be attributed to different interpretations given to the term 'function' itself. The term is highly ambiguous, and Nagel (1961, 522–6), for example, discusses at least six meanings that can be given to it. Hempel's usage corresponds to the sixth usage given by Nagel—the contribution that some item makes toward the maintenance of some given system. But it is also possible to speak of function as a relationship (e.g. a mathematical relation or an observed regularity connecting two variables), as an indicator of the utility of something, and so on. This need not lead to confusion regarding functional explanation, were it not for the fact that the interpretation put upon function can yield various forms of func-tional explanation. Hempel thus regards functional explanation as a form of *teleological* explanation, while both Nagel and Braithwaite dis-cuss teleological explanations separately from functional explana-tions, while noting that they are very similar.

Teleological explanation is itself an ambiguous phrase. Braithwaite (*1960*, 324) is content to define it as a form of causal explanation in which the cause lies in the future or in some end that may be future or present. Functional explanation as we have considered it so far is a particular form of this. It is possible to argue in favour of teleological explanations in a number of ways. We can maintain that:

(i)   There are 'final causes' which are God-given or nature-given. This is a metaphysical assertion which is incapable of empirical test and which has no relevance to empirical understanding.

(ii) There is some degree of purposive behaviour on the part of individuals or objects. My staying in Bristol may thus be associated with an intention on my part to finish writing this book. Such purposive behaviour may seem a reasonable assertion about individuals, but many would maintain that it is needlessly anthropomorphic to assert that cultures have intentions, economies have intentions, and so on.

(iii) Teleological explanations are in principle (as both Nagel and Braithwaite show) reducible to ordinary causal explanations (a conclusion which Hempel's analysis also indicates).

Out of these we must necessarily reject (i) as unscientific (which is not to say it is untrue), while (ii) provides us with a rather weak form of explanation (see above, p. 57). The problem with (iii) is that the circumstances for such a reduction are not in general known. Braithwaite (*1960*, 334–5) thus concludes that

in general irreducible teleological explanations are no less worthy of credence than ordinary causal explanations. . . . It seems ridiculous to deny the title of explanation to a statement which performs both of the functions characteristic of scientific explanations—of enabling us to appreciate connexions and to predict the future.

Braithwaite's conclusion is not at variance with the conclusions drawn by other analysts, except that it shows a greater preparedness to accept teleological explanations as first-stage approximations to some more rigorous scientific explanation form. The general consensus being that teleological and functional explanations cannot be regarded as different in kind from other varieties of explanation, it seems that the only important decision we need to make regarding such explanatory forms is how long and under what circumstances we are willing to put up with first-stage approximations of this sort. It may well be that rough functional and teleological explanations are the best we can achieve in the given state of our understanding. But in principle there seems no reason why we should not resort to more complete explanatory forms. Even with these, functional and teleological explanations may still have a role to play since they can, as Lehman (1965) concludes, often provide a quick and efficient way of answering a particular kind of question.

B The philosophy of functionalism

Logical analysis suggests that functional and teleological explanations cannot be regarded as being different in kind from other forms of scientific explanation. The claim that they are, however, has been quite common in the past. These claims amount to stating *a priori* some functional or teleological philosophy. Teleological philosophies, which are stated in terms of some envisioned end, have been extremely important in the history of scientific thought, and they are not unknown in geography. Ritter, according to Hartshorne (1939, 59–62), regarded the earth as being divinely planned and man's actions were consequently directed to realising that plan. Griffith Taylor's determinism with its emphasis upon nature's plan has a similar teleological ring to it. The point about such philosophies is that they are empirically untestable and, hence, can add nothing to our understanding of the world and may, indeed, detract from it by assuming the efficacy of explanations which cannot be shown to be true. Since science is concerned with empirically testable statements it is clear that teleological philosophies, which may or may not be justified (according to one's taste) have nothing whatever to do with explanation in geography.

This dismissal of teleological philosophies in general does not entail, however, the dismissal of a particular form of teleological philosophy which we will term *functionalism*. Functionalism does not provide us with a concise coherent philosophy regarding the nature of the real world. It provides us, rather, with a convenient umbrella term with which to characterise rather varied viewpoints that have something in common. In some disciplines a particular version of functionalism has been erected into a philosophical orthodoxy, but in each case it has assumed a different guise. In the twentieth century functionalism has tended to mark a reaction against crude cause-and-effect determinism (causalism) and nineteenth-century positivism. It has sought to replace the language of cause and effect by a language stressing interrelationships, and to provide an alternative mode of explanation to those mechanistic forms characteristic of physics. This tendency was most marked in biology where, it was pointed out, complex organisms are analysable only in terms of 'indivisible wholes' and that something is lost in taking them apart and examining their constituent parts separately. Such a claim has more substance than 'mere' teleology, in that there are many situations in which it is difficult to show empirically that something is not destroyed by taking it apart. There are many phenomena, thus, that appear to be best described with reference to 'indivisible wholes'. Functionalism in this sense appeared, therefore, to provide a reasonable philosophical assumption in a discipline such as biology.

This philosophy proved attractive for other disciplines and by way of the organismic analogy functional modes of thought penetrated many of the behavioural sciences. Malinowski, in anthropology, thus embraced functionalism and expounded its philosophy. He regarded it as a way to 'break out of the strait-jacket of nineteenth-century historicist theory without getting hopelessly bogged in empirical detail' (Leach, 1957, 136). His fellow anthropologist, Radcliffe-Brown, similarly embraced functionalism, although he gave it a rather different interpretation. Radcliffe-Brown's (1952, 178–80) point of view is worth considering as an example of such functional modes of thought in the social sciences. He begins by recognising that 'the concept of function applied to human societies is based on an analogy between social life and organic life.' After a detailed discussion of this analogy he concludes:

> The concept of function as here defined thus involves the notion of a *structure* consisting of a *set of relations* amongst *unit entities*, the *continuity* of the structure being maintained by a *life-process* made up of the *activities* of the constituent units.
>
> If, with these concepts in mind, we set out on a systematic investigation of the nature of human society and of social life, we find presented to us three sets of problems. First, the problems of social morphology—what kinds of social structures are there, what are their similarities and differences, how are they to be classified? Second, the problems of social physiology—how do social structures function? Third, the problems of development—how do new types of social structure come into existence?

Defining questions in this manner, of course, partly predetermines the explanatory form to be used in providing answers. If we ask how social structures function, then we may expect answers that contain law-like statements governing interrelationships between the elements as they function within that social structure. We may even be prepared to call such statements functional laws. This structural-functional approach has been widespread in anthropology and has, on occasion, been raised to the status of orthodoxy. Radcliffe-Brown, however, saw functionalism as a working hypothesis and did not fall into what might be called the functional fallacy—the view that *everything* in a society has a function and that it could therefore be understood only in terms of that function. Nevertheless it is impossible to avoid making statements about the social structure or system being dealt with. In short, by posing questions in the language of functionalism it is impossible to avoid making some assertion as regards the nature of the system under investigation. The essential difference between methodological and philosophical functionalism is that the latter makes such assertions on *a priori* metaphysical ground, whereas the former (as Hempel indicates in the passage quoted above, p. 437) relies upon assertions that can in some measure be empirically and objectively evaluated.

It is not our concern to trace the movement of functional thinking, to assess all the nuances that became attached to it. It is significant, however, that functionalism became an explicit philosophy in sociology, particularly under the influence of Merton (1957) and Talcott Parsons (1951) (who are as different in their views as were Malinowski and Radcliffe-Brown), and that functionalist approaches were prevalent in psychology and economics (see the reviews contained in Martindale, 1965). In short, functionalism has been a very significant element in the intellectual climate of the twentieth century. It is also significant that the situation at the present day appears, in general, to be rather different in all disciplines from that of some three decades ago. The reason for such a general change is not far to seek. Functionalism was originally construed as an alternative, and more palatable, philosophical position to mechanistic determinism. Functional analysis seemed to hold out the prospect of conducting detailed investigation and constructing theory without resorting to cause-and-effect language and the causalism that was then associated with it. More recently there has been a reassessment of functionalism as a philosophy, as a theory of how societies or organisms actually work. This has come in part as a result of logical analysis of the sort we have already considered, but it also reflects a changing climate of opinion as regards functionalism in the various disciplines. Functionalism is often now regarded as a convenient method (which produces fruitful results) rather than as a necessary philosophy. At the same time it has become clear that cause and effect can be separated from causalism as a philosophy, that process models can be freely used without implying historicism, and therefore there is less pressure to regard functionalism as an alternative philosophy to causalism or historicism.

Geography has never claimed for itself a philosophy of functionalism in the way that anthropology and sociology have. Yet much of the empiricial work in geography could well be construed as functionalist in form, while there are numerous methodological statements that appear to embrace functionalism as a philosophy. Vidal de la Blache (1926, 7–9) thus suggests that 'the phenomena of human geography are related to terrestrial unity by means of which alone they can be explained.' He continues by expounding an ecological approach to the study of interaction between man and environment. Brunhes (1920, 14–15) similarly emphasises interrelationships within some complex system:

One cannot be content with the observation of a fact by itself or of an isolated series of facts. After this initial observation, it is important to place the series back in its natural setting, in the complex ensemble of facts in the midst of which it was produced and developed.

The general similarity in philosophical outlook between the French geographers and the functional anthropologists, such as Malinowski, has been noted by Wrigley (1965, 15). In part this similarity stems from the same basic reaction against determinism and positivism and it is hardly surprising, therefore, to find that 'possibilism' and 'functionalism' have much in common. Consider Wrigley's (1965, 8) portrayal of the method of Vidal de la Blache with its emphasis on the two-way interaction between man and environment:

Man and nature become moulded to one another over the years rather like a snail and its shell. Yet the connexion is more intimate even than that, so that it is not possible to disentangle influences in one direction, of man on nature, from those in the other, of nature on man. The two form a complicated amalgam. . . . The area within which an intimate connexion between man and land had grown up in this way over the centuries formed a unit, a region. . . .

This is similar to Malinowski's functionalism, which

aims at the explanation of anthropological facts at all levels of development by their function, by the part which they play within the integral system of culture, by the manner in which they are related to each other within the system, and by the manner in which this system is related to the physical surroundings. (Quoted in Nagel, 1961, 521.)

To Malinowski the culture formed an indivisible whole with respect to which events could be explained; to Vidal de la Blache the region formed a similar 'whole' with respect to which the parts could be explained. In both forms of analysis there is an emphasis upon equilibrium states and adaptive behaviour. This notion of symbiotic equilibrium is highly developed in French regional geography, in the work of those who adopted the 'ecological' approach and in the writings of those, such as Forde (1934), whose vision spanned geography and anthropology and brought explicitly to the former much of the latter's functional method and philosophical outlook. Perhaps the most interesting application of this point of view is in the conceptualisation of the region as a 'functional whole'. The term 'region' has taken on various connotations during its long and turbulent use in geography, but there have undoubtedly been occasions when the region has been regarded as a 'thing' that was something more than the mere sum of its parts. This 'reification' of the region as a functional unit is no different in kind from the reification of cultures, societies, political units, and the like. Once regions, cultures, and political states, are regarded as things, it is possible to attribute behaviour to them and, particularly by way of the organismic analogy, to talk of them as if they were living organisms. Hartshorne (1959, 136) provides a good illustration of this approach.

> In determining that an area is a functional region, the student is reconstructing an existing areal synthesis . . . in the degree to which an area is a functional unit, it constitutes a *whole*; for its unity has the structure of totality, or is more than the sum of its parts. . . . In the respects in which it is a functional unit, but only in those respects, it represents an areal feature in reality, to be discovered and analyzed by the geographer.

Not all discussions of functional regions take this particular approach, of course, for it involves making an assertion that is incapable of empirical test. It is possible to regard a region as comprising complex interactions among phenomena without attributing 'holistic' qualities to it. The retreat from the view of the region as more than the sum of its parts, and a distinctive thing, thus corresponds to the retreat from *gestalt* metaphysics in psychology, from 'holistic' metaphysics in sociology and anthropology. One of the effects of ecological and, more latterly, systems concepts, has been to facilitate this retreat.

The functionalist philosophy has tended to remain implicit rather than overt in geographical thinking. Geographers use the term 'function' frequently, and often resort to functional modes of analysis. Methodological functionalism may be distinct from philosophical functionalism, but it is frequently difficult to tell whether or not that distinction has been made in geographical enquiry. Indeed, the historians of geography may well come to regard that period from 1920 to 1940 that Sauer characterised as the Great Retreat, as the era of methodological functionalism if not of philosophical functionalism. Certainly Hartshorne's (1939) discussion has a strong functionalist emphasis. Later commentators, such as C. T. Smith (1965, 130) have thus seen the clash of opinion between Sauer and Hartshorne in the 1940s as a clash between functionalist and genetic philosophies, between those who would require that we show how the snail became moulded to its shell over time, and those who were concerned to describe areal interconnections as expressed in a system of regions. It will remain a challenge to the historian of geography to sort out these cross-currents of opinion and nuances of interpretation.

C 'Wholes'

Hartshorne, in the passage on functional regions quoted above, suggests that a functional unit constitutes a whole which is in some way more than the sum of its parts. In doing so he refers to a general

philosophical problem that has been a source of major controversy
in organic biology and in psychology (where *gestalt* psychology provides a whole school of thought stemming from the belief that there
are certain indivisible units). Functionalism is thus sometimes associated with a doctrine of 'wholeness'. In geographic thinking such a
doctrine has not been a central one (Hartshorne, 1939, 260–80),
although it has had some impact by way of the cruder applications
of the organismic analogy to, for example, regional and political
wholes. It has also been used in connection with *gestalt* concepts
to describe the reaction or relation between man and environment
(Kirk, 1951; Hartshorne, 1939, 276). More recently the concept of
organic or functional wholes has been raised in a new and rather
interesting form through the application of systems thinking to geography. It is perhaps useful to take a closer look at such concepts.

Nagel (1961, 380–97) shows that the exact meaning of the statement that 'the whole is more than the sum of its parts' cannot be
determined, simply because the terms 'whole', 'part', and 'sum', are
variously interpreted and hence ambiguous (he lists eight possible
meanings given to the term 'whole' for example). Given this ambiguity, it is impossible to prove or deny that objects or systems
actually exist which are indeed more than the sum of their parts.
Such an assertion is in part a matter of *a priori* belief and thus belongs
more to the realm of metaphysics than to that of logic. Nevertheless
there are numerous situations in which it is common to talk of such
functional wholes. The distinctive feature of such situations is that
the behaviour of some system is not determined by the individual
elements within it, but rather that the behaviour of the individual
elements is determined by the intrinsic nature of the system itself
(i.e. the whole). The individual elements may thus show a high
degree of mutual interdependence, but to demonstrate this empirically is not to demonstrate that the whole determines the part. Some
writers have suggested however, that any system that exhibits a
high degree of internal organisation should be deemed as a functional
whole. Nagel (1961, 393) thus concludes that

although the occurrence of systems possessing distinctive structures of interdependent parts is undeniable, no general criterion has yet been proposed which
makes it possible to identify in an absolute way systems that are 'genuinely
functional' as distinct from systems that are 'merely summative'.

Now systems that can be regarded as the sum of their parts are
analysable in terms of those parts. This Nagel terms an 'additive'
form of analysis. If functional wholes exist, some other 'non-additive'
form of analysis would be necessary. Nagel thus goes on to examine
the suitability of additive forms of analysis and contrasts the particle

physics of classical mechanics with the field approach of electro-
dynamics. This example is of considerable interest, for thinking
in terms of fields of influence, hinterlands, and so on, is very common
in geography, while more recently the field concepts of psychology
(which are closely related to *gestalt* psychology in spirit) have been
explicitly introduced into geographic theory (Wolpert, 1965; Berry,
1966; Harvey, 1967B). The question whether such field phenomena
should be regarded as functional wholes incapable of additive analysis is thus of some significance. Nagel does not provide a firm answer
to this problem. He points out that it may be convenient to treat
field phenomena as functional wholes and to develop non-additive
methods of analysis to deal with them. This does not mean that field
phenomena must necessarily be so regarded; and indeed in certain
ways they may be treated additively without any loss of information.

The question of functional wholes, indivisible units, and the like,
cannot therefore be solved by logical analysis. Yet that analysis
yields sufficient understanding for Nagel (1961, 397) to assert with
some confidence that the issue 'cannot be solved, as so much of the
extant literature on it assumes, in a wholesale and *a priori* fashion'.
The position we adopt with respect to such functional wholes may be
incapable of full empirical support, but it cannot, either, ignore the
results of empirical and logical analysis. These suggest that it may
be safest to construe the differences between units conceived of
as functional whole and units conceived of as summative units, as
differences of degree rather than of kind.

D Methodological functionalism

Whatever may be said about the logic of functional explanation or
functionalism as an *a priori* assumption, there can be no doubt of the
very substantial achievements and insights that have been gained
through adopting functionalism as a working hypothesis. To attack
functionalism as a philosophy is not, therefore, to attack it as a
methodology. Indeed, it has been claimed that the heuristic rules
that form the basis of functionalist methodology are extremely useful
and need not be associated either with functionalist philosophy or
functional forms of explanation (Jarvie, 1965). It is perhaps in this
light that we should accept Philbrick's (1957) plea for a greater use
of the functional approach in geography.

The methodological strength of functionalism really lies in its
emphasis upon interrelatedness, interaction, feedback, and so on, in
complex organisational structures or systems. This kind of approach
to problems has been extremely rewarding where it has been freed

from metaphysical connotations. It thus seems very valuable to enquire how central places function in an economy, simply because it opens up a whole range of questions which, if we can provide reasonable answers, will yield us deeper understanding and greater control over the phenomena we are examining. The world is manifestly a complicated place, and the heuristic strength of functionalism is that it directs our attention to this complexity. As a philosophy it assumes that this complexity is essentially unanalysable into constituent parts—a point of view that seems unscientific since it provides licence for the employment of intuitionism in (who can tell?) revealing or specious form. As a methodology it provides us with a series of excellent working assumptions about interactions within complex systems. But as the analyses in the previous two sections have shown, functionalism cannot suffice either as philosophy or as a form of logical analysis. In a discipline still heavily dependent upon 'first-stage approximations' a strong case might be made for a fuller employment of methodological functionalism as a heuristic device, and even for explanations in terms of function. But first-stage approximations must presumably give way at some stage to full-blown theory. The danger we must here avoid at all costs is that mortal inferential sin of erecting an *a priori* functional model into full theory without knowing it and without the necessary confirmatory evidence.

Basic Reading

Hempel, C. G. (1959). This is an essential reference.
Brown, R. (1963), chapter 9.
Nagel, E. (1961), pp. 380–446 and 520–35.

Reading in Geography

Philbrick, A. K. (1957).

Chapter 23
Systems

In the last chapter it was suggested that the valid use of function statements was entirely dependent upon the objective specification of some *system* (above, p. 435). This conclusion is of considerable significance, for it indicates that the rendering of an explanation of *any* type may be contingent upon the specification of a *system*, and that therefore our understanding of the process of explanation itself may be contingent upon our understanding of the concept of *system*. Gouldner (1959, 241) writing in a sociological context puts it this way:

> The intellectual fundament of functional theory in sociology is the concept of a 'system'. Functionalism is nothing if it is not the analysis of social patterns as parts of larger systems of behavior and belief. Ultimately, therefore, an understanding of functionalism in sociology requires an understanding of the resources of the concept of 'system'. Here, as in other embryo disciplines, the fundamental concepts are rich in ambiguity.

Any explanation involves the isolation of certain events and conditions and the application of certain law (or law-like) statements to show that the events to be explained must necessarily have occurred, given that certain other events or conditions occurred. Isolating events and conditions in this way amounts to defining a *closed system*. Hempel's deductive model of an explanation, for example, corresponds to what Ashby (*1966*, 25–6) calls a 'state-determined' system in which 'a particular surrounding condition . . . and a particular state determine *uniquely* what transition will occur.' This relationship between the closure of some system and the operation of explaining is an important one. Hagen (1961, 145) thus writes:

> For use in analysis, a system must be 'closed'. A system which is interacting with its environment is an 'open' system: all systems of 'real life' are therefore open systems. For analysis, however, it is necessary in the intellectual construct to assume that contact with the environment is cut off so that the operation of the system is affected only by given conditions previously established by the environment and not changing at the time of analysis, plus the relationships among the elements of the system.

447

Systems analysis cannot proceed, therefore, without abstraction and without closure. As Ashby (*1966*, 16) points out, any real system will be characterised by 'an infinity of variables from which different observers (with different aims) may reasonably make an infinity of different selections'. In reality any system is infinitely complex and we can only analyse some system *after* we have abstracted from the real system. Systems analysis relates to the abstraction rather than to the reality. It is therefore advisable to think of systems not as real things at all, but as convenient abstractions which possess a form which facilitates a certain style of analysis. Now this necessity for abstraction and closure has been noted with respect to both cause-and-effect analysis (above, p. 392) and process explanation (above, p. 419). Thus the rigorous application of cause-and-effect analysis requires the definition of some closed specified system. When we write $A \longrightarrow B$, for example, we assume that A and B are distinctive elements that stand in a particular relationship to each other and that there are no other elements interfering with the relation (i.e. the system is closed around A and B). One of the main difficulties of using cause-and-effect analysis in real-world situations is the isolation of the cause and the effect from a complicated environment which is bound in some way or other to interfere with the simple relationship we are trying to examine. It is thus the function of experimental-design procedures to try and create systems that are reasonably isolated from their environment so that the necessary closed-system analysis can proceed. In non-experimental situations, however, we seek to impose a closed-system pattern of analysis upon data generated from situations in which we have no operational control over the interfering variables. Here statistical versions of cause and effect become particularly useful, for they allow us to represent interference from the environment as random noise. In the regression version of cause-and-effect analysis, for example, this interference is contained in the error term in the equation $Y = a + bX + e$. It is useful to think of e as representing the environment of some system closed around Y and X. The same kinds of statement could be made with respect to process models and their various time-oriented approximations.

The general relationship between the concept of an explanation (whether conceived of in cause and effect, process, or functional terms) and the concept of a system provides a strong justification for examining the nature of systems. Ashby (*1966*, 28) has thus remarked that the 'property of being state-determined must inevitably be of fundamental interest to every organism that, like the human scientist, wants to achieve mastery over its surroundings.'

A further justification for regarding the system as the key to explanation arises from major applications of the concept in all areas of empirical investigation during the twentieth century. From both a methodological and an empirical point of view, therefore, the concept of a *system* appears absolutely central for our understanding of explanation in geography.

In this chapter we shall consider the meaning of the term *system*. Analytic frameworks based on systems concepts have grown very rapidly in science as a whole, and there seems little doubt that systems analysis will prove one of the major methodologies of the last half of the twentieth century. Yet in spite of the burgeoning growth of systems thinking in all disciplines, systems concepts remain rather ambiguous. In spite of their ambiguity, however, it seems almost impossible to proceed without them. At a time when geography appears to be moving towards the adoption of a new 'system-based paradigm' it seems doubly important to attempt some evaluation of systems concepts, if only because of this ambiguity. Further, the theory of systems, particularly the interpretation given in *General Systems Theory*, has been acclaimed by some as a basic framework for unifying all scientific thinking. At times it has seemed as if *General Systems Theory* was about to proclaim for itself some logically necessary view of the universe—rather as causalism, historicism and functionalism did before it. Thus, a metaphysic based on *General Systems Theory* has its attractions for some. But most of the work in systems analysis has come from those concerned with it as a methodology pure and simple. Nevertheless it is useful to identify some points where misunderstandings can arise and to attempt some evaluation of the claims of *General Systems Theory* as a unifying framework for all scientific thought and, in particular, for geographic thought. We shall, however, begin by discussing systems analysis and pointing to its applications in geography before going on to examine these broader philosophical issues.

I SYSTEMS ANALYSIS

The concept of a system is not in any way new. Newton wrote on the solar system, economists have written on economic systems, biologists on living systems, plant and human ecologists have used system concepts, and geographers have certainly made considerable use of the notion of a system ever since the discipline originated. Although systems concepts are very old they have tended to remain on the fringe of scientific interest—acting as contraints almost—rather than being the subject of intensive investigation. The modern emphasis on systems as an explicit (and indeed central) item for analysis may be seen as part of a general change in emphasis from the study of very simple situations in which the interactions are few, to situations in which there are interactions between very large numbers of variables. It has been pointed out (von Bertalanffy, 1962, 2; Ashby, *1963*, 165–6) that classical science was concerned almost entirely with simple linear causal chains or with unorganised complexity (such as that characterised in the second law of thermodynamics). Indeed, Ashby (*1963*, 165) suggests that the success of science for the last two hundred years or so must be attributed to its exploitation of 'the many interesting systems in which interaction is small'. We have already noted how, in the behavioural and biological sciences, the complexity of interaction poses special problems for the application of classical methods—indeed, the functionalists were adamant that the classical concepts of science were inappropriate for the study of complex systems. The interest in these complex systems has grown rapidly in the twentieth century, but the problems they posed seemed scarcely tractable without some conceptual and technical breakthroughs. The necessary technical apparatus to handle systems has been steadily built up out of the mathematical development of communications engineering, cybernetics, information theory, operations research, and the like. Thus:

> Since 1940 . . . a serious attempt has been made, aided by the new techniques, to grapple with the problems of the dynamic system that is both large and richly connected internally, so that the effects of interaction are no longer to be ignored, but are, in fact, often the focus of interest. . . . So has arisen systems theory—the attempt to develop scientific principles to aid us in our struggles with dynamic systems with highly interacting parts (Ashby, *1963*, 166).

The intimate connection between systems analysis and the analysis of complex structures makes this approach very attractive to those disciplines dealing with phenomena that are highly interconnected. Given the multivariate nature of most geographical problems, it is hardly surprising that systems analysis provides an appealing framework for discussing these problems.

A The definition of a system

The nature of systems thinking can most easily be characterised by examining the definition of the term 'system' itself. We can attempt definition in a number of ways, but since it is best to think of a system as an abstraction, it seems logical to attempt a syntactical or mathematical definition of its first, and then go on to discuss the various problems that arise in finding operational interpretations for the abstract terms used in the mathematical definition. Klir and Valach (*1967*, 27–54) provide a mathematical definition using the tools of set theory. The set of objects (identified by a set of attributes of objects) contained within some system S may thus be represented as a set of elements $A = \{a_1, a_2, \ldots a_n\}$. To this we may add an extra element a_0 to represent the environment. We can then introduce a set $B = \{a_0, a_1, \ldots a_n\}$ which includes all the elements within the system plus an element that represents the environment. The interactions and relationships between these elements can then be examined. If we let r_{ij} represent the relationship between any element a_i and a_j (the case where $r_{ij} = 0$ would be the case where a_i has no effect upon a_j, for example), then we can denote the set of all r_{ij} ($i, j = 0$, $1, \ldots n$) by R. The definition of a system is then contained in the statement that every set $S = \{A, R\}$ is a system.

This definition therefore provides us with a firm definition of a system which we may generally interpret as constituting:

(i) A set of elements identified with some variable attribute of objects.
(ii) A set of relationships between the attributes of objects.
(iii) A set of relationships between those attributes of objects and the environment.

This abstract construal of a system has a number of important advantages. It allows, for example, the development of an abstract theory of systems which is not tied down to any one particular system or set of systems. This theory provides us with a good deal of information about the possible structures, behaviours, states, and so on, that might conceivably occur, and provides us with the necessary technical apparatus for dealing with interactions within complex structures. Systems theory is thus associated with an abstract mathematical language which, rather like geometry and probability theory, can be used to discuss empirical problems. But in order for empirical problems to be treated in this language, we require a number of assumptions and conceptual adjustments to our view of the substantive problem in order to complete a satisfactory translation into the abstract language that systems theory provides. In this, therefore, systems theory poses the same kinds of problem as does the use of any *a priori* model language for discussing substantive problems (above, pp. 151–4; 158–61).

(1) *The structure of a system*

Given the definition of a system it is possible to elaborate somewhat regarding its *structure*. This is composed essentially of *elements* and *links* between elements.

(a) *The element* is the basic unit of the system. From the mathematical point of view an element is a primitive term that has no definition (rather like the concept of point in geometry). Mathematical systems analysis can thus proceed without further consideration of the nature of elements. But the use of the mathematical theory of systems to tackle substantive problems depends entirely upon our ability to conceptualise phenomena in such a way that we may treat them as elements in a mathematical system. Put the other way around, it depends upon our ability to find a substantive interpretation for the mathematical element. It is not always easy to find such interpretations which we can agree upon as reasonable and unambiguous. There are two basic problems here. First the scale problem has to be faced. The substantive interpretation of an element is not independent of the scale at which we envisage the system's operating. The international monetary system, for example, may be conceptualised as containing countries as elements, an economy may be thought of as being made up of firms and organisations, organisations themselves may be thought of as systems made up of departments, a department may be viewed as a system made of individual people, each person may be regarded as a biological system, and so on. The definition of an element thus depends on the scale at which we conceive of the system, or as Klir and Valach (*1967*, 35) call it, the resolution level:

> Every element is characterised by forming, from the point of view of the corresponding resolution level (at which the system S is defined), an indivisible unit whose structure we either cannot or do not want to resolve. However, if we increase the resolution level in a suitable manner . . . the structure of the element can be distinguished. In consequence, the original element loses its meaning and becomes the source of new elements of a relatively different system, i.e. of a system defined at a higher resolution level.

In substantive terms, therefore, we face the problem that systems may be embedded in systems, and that what we choose to regard as an element at one level of analysis may itself constitute a system at a lower level of analysis. A car may be an element in the traffic system, but it may also be regarded as constituting a system. This feature of systems analysis carries with it some problems. Blalock and Blalock (1959) thus point out that there are two ways in which we can conceive of an element at some higher level in the hierarchy of systems. It can be regarded as an indivisible unit that acts as a unit (e.g. we may think of a firm deciding or responding) or it may be regarded as some loose configuration of lower order elements (e.g. individuals within an organisation interact with individuals in other organisations). These two interpretations are portrayed in Figure

23.1. Blalock and Blalock (1959) point out the considerable confusion that may result from a failure to distinguish between these
two points of view and suggest that major controversy in sociology
regarding the nature of 'the social system' can be attributed almost
entirely to semantic difficulties surrounding these two different

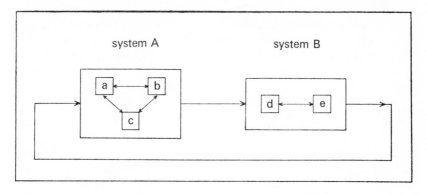

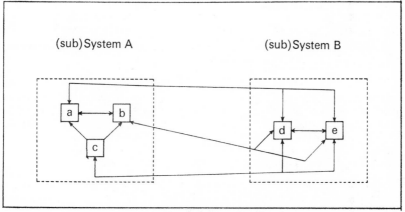

Fig. 23.1 Two different views of inter-system interaction. The upper
shows system A and system B interacting as units, with smaller system
interactions going on within each system. The lower diagram shows
systems A and B interacting at lower levels (*after Blalock and Blalock*,
1959).

approaches. In spite of these difficulties the notion of systems embedded within systems within systems *ad infinitum* is an attractive one.
It poses no mathematical difficulty for we can simply group elements
into a hierarchy of 'classes' with each higher-order class forming an
element in a higher-order system (there is good use for the set-
theoretic procedures outlined above, pp. 327–30, in this respect).

It also provides an appropriate methodology for dealing with the kind of complexity that is imposed on geographical method by the scale problem itself. We have long been aware of this scale problem, and the systems approach which is flexible with respect to

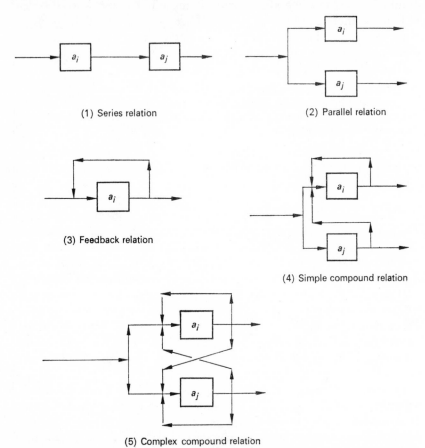

(1) Series relation

(2) Parallel relation

(3) Feedback relation

(4) Simple compound relation

(5) Complex compound relation

Fig. 23.2   Diagrams to show the kinds of relation that may exist between the elements within system (*after Klir and Valach, 1967*).

scale therefore provides an appropriate framework for analysing processes which we know are not independent of scale in their contribution to spatial variation (above, pp. 352–4; 383–6).

The second problem of some significance in our search to give an interpretation to the mathematical concept of an element is simply the identification problem (also a problem we have met before, above, pp. 215–17; 351–4). Having defined a system at some scale how can we recognise elements within it? From the geographical

point of view this amounts to defining individuals at a given scale—
a problem which is in itself quite difficult to solve in certain cases
(e.g. continuously distributed phenomena) and self-evidently soluble
in others (e.g. farms and other discrete phenomena). But from the
point of view of mathematical systems theory an element is a variable.
It follows, therefore, that in seeking for a translation of the mathematical element in geographical context we must construe the element as an attribute of some defined individual rather than as the
individual itself. Kuhn (*1966*, 50) thus states that

> the elements of systems are states or conditions of things, not the things themselves.
> In systems involving persons, it is not the *person* who is an element, but his state
> of hunger, his desire for companionship, his state of information, or some other
> trait or quality relevant to the system.

In defining and identifying elements, therefore, we not only require
to define individuals, but to have some sound procedure for measuring their attributes (chapter 17).

(b) *The relationships or links* between the elements provide the other
component in the structure of a system. Three basic forms of
relationship can be defined (Figure 23.2). A *series relation* is the
simplest and is characteristic of elements connected by an irreversible
link. Thus $a_i \rightarrow a_j$ forms a series relation and it may be observed
that this is the characteristic cause-and-effect relation with which
traditional science has dealt. A *parallel relation* is similar to multiple-
effect structures in that both a_i and a_j are effected by some other
element a_k. A *feedback relation* is the kind of link that has been newly
introduced into analytic structures (mainly through the development of cybernetic analysis by Wiener, *1961*, and others). It describes
a situation in which one element influences itself. Thus the value
assigned to an attribute of an object is affected by that value itself.
It is possible to combine these relationships in a number of ways (see
Figure 23.2), so that two elements may be connected in various
different ways simultaneously. The links thus form a kind of 'wiring
system' connecting the elements in various ways. This emphasis on
connectivity between elements can be discussed in terms of topological relations and it is not surprising to find, therefore, that graph
theory is an important descriptive device in the analysis of the structure of systems.

(2) *The behaviour of a system*
When we speak of the *behaviour* of some system we are simply referring
to what goes on within the 'wiring system' that makes up its structure. Behaviour has to do therefore, with flows, stimuli and responses,

inputs and outputs, and the like. We can examine both the internal behaviour of some system or its transactions with the environment. A study of the former amounts to a study of the functional 'laws' that connect behaviour in various parts of the system as discussed in the last chapter. Most analyses of behaviour tend to concentrate on the latter aspect, however, and it will be examined in this context here. Consider a system that has one or more of its elements related to some aspect of the environment. Suppose the environment undergoes change. Then at least one element in the system is affected and effects are transmitted throughout the system until all connected elements in the system are affected. This constitutes a simple stimulus-response, or input-output system without feedback to the environment:

$$\text{stimulus} \atop (\text{input}) \;\to\; \boxed{\text{system}} \;\to\; {\text{response} \atop (\text{output})}$$

The mathematical meaning of behaviour can easily be defined. We can define a vector of inputs $X = (x_1, x_2, \ldots, x_n)$ which function as partial stimuli to the system, and a vector of outputs $Y = (y_1, y_2, \ldots, y_n)$ which refer to the response of some system. The behaviour of a system can then be expressed in general as the transformation T of the vector X into the vector Y:

$$Y = T(X)$$

In other words the behaviour is described by the equations (deterministic or probabilistic) that connect the input vector with the output vector (Klir and Valach, *1967*, 31–2). The simplest example of this is provided by input-output analyses of economies, in which a vector of final demands (e.g. derived from exports, home consumption, or however) is related to a vector of final outputs in various sectors in the economy. In this case the system is composed of all inter-industry links within the economy and the effects are traced throughout the economy to yield final outputs. The system is represented in this case by the matrix of technological coefficients.

The mathematical aspects of the behaviour of systems will not be considered further here, but it is useful to mention a practical constraint imposed by the feasibility of mathematical manipulation. In general it is easiest to handle linear systems. Much of the literature on systems behaviour, therefore, is concerned with linear systems, and in practice it proves necessary to conceptualise (or measure) real-world relationships *as if* they were linear even if they are not. This constraint is not one imposed in principle, it is simply related to the relative difficulty of handling large systems of non-linear equations. In practice we may be able to observe the transformation by recording inputs and outputs in tabular form, yet fail to identify any mathematical function which fits that transformation. Ashby (*1966*, 21) suggests that simple mathematical functions are rarely found in practice. It is useful to differentiate, therefore, between situations in which we can handle the behaviour analytically and situations in which we possess observations of input and output only.

(3) *The boundaries of a system*

It is possible to investigate the structure and behaviour of some system only if the *boundaries* of that system are first identified. Mathematically this is no problem, since the boundaries are given by defining certain elements as being in the system and other (relevant) elements as belonging to the environment. In order to use the mathematical properties of systems analysis, however, we require some operational method of defining boundaries. This is no easy problem. In some cases the boundaries are fairly self-evident (particularly when we are dealing with systems that are discrete and have well-defined connections with their environment, e.g. a firm within an economy). In other cases we are forced to *impose* boundaries in some fashion, and in doing so employ our own judgement as to where the system begins and ends. Forrester (1961, 117–18) points out that such judgement decisions do not mean that the choice lacks 'a foundation of fact or contact with reality'. We may evaluate the choice of system boundaries by referring that choice to the objectives of an investigation together with our experience of such systems. The choice of boundaries, however, may have a considerable impact upon the results obtained from systems analysis. A good example is provided by input-output analysis of an economy, in which the final demand (the stimulus) can be conceptualised in a variety of ways. If it is treated solely as exports, for example, then the household sector and labour inputs become a part of the network of interactions within the economic system. In some studies, however, household demand is treated as a component of final demand and therefore lies outside the system (Isard, 1960, chapter 8). The predictions of final output will vary according to which of these choices is made. Both approaches are reasonable; and which is chosen depends mainly on what is feasible and what we wish to investigate.

(4) *The environment of a system*

In general terms the *environment* of some system may be thought of as everything there is. But it is useful to develop a much more restricted

definition of an environment as a higher-order system of which the
system being examined is a part and changes in whose elements will
bring about direct changes in the values of the elements contained
in the system under examination. Again, there seems no objective
way of defining an environment in such a way that everyone will
agree that a particular definition of the environment is correct.
Defining the environment in an operational sense poses as much
difficulty as defining the system contained within it. This is not to
say that the definition is entirely arbitrary. Perhaps the best way to
approach this problem is to ask what are the *relevant* elements in the
environment for the operation of some system? We may then close
off some 'meta-system' which is made up of relevant elements of the
environment in interaction with the system under consideration. The
irrelevant elements in the environment (and they are presumably
infinite in number) are then discarded. Again we must abstract and
close our model before any analysis can proceed. In most cases it is
surprising how easily agreed definitions of the environment can be
arrived at, e.g. that the economy constitutes the environment of a
firm, etc. In other cases the environment is made up of two rather
different systems intersecting with respect to the system under investigation. A farm system, for example, may have for its environment
the biosphere and the economy. This flexible approach to the concept of environment in systems analysis is particularly useful to geography, which has made considerable use of the notion of environment but has tended to use it in rather a rigid fashion. It is useful
at this juncture to clarify the usual meaning of the terms 'open' and
'closed' system. The former refers to models which incorporate interaction with the environment, while the latter refers to models built
without any interaction. The terminology is a little unfortunate since
analysis requires closure of the meta-system in any case.

(5) *The state of a system*

In general the *state* of a system may be thought of as the values which
the variables take on within the system at any particular point in
time. Now it is possible for the variables to take on a large number
of values, so that the term 'state' is often used in a more restricted
sense to refer to 'any well-defined condition or property that can be
recognised if it occurs again' (Ashby, *1963*, 25). It is therefore useful
to differentiate between *transient* states and the various types of
equilibrium states which have distinctive properties. These will be
examined later. One further point about the concept of state, is that
the values taken on by the variables correspond to the outputs and
that 'the concept of state and of output thus tends to merge into one,

and it is not incorrect to use these terms interchangeably, at least for heuristic purposes' (Rosen, 1967, 167).

(6) *The parameters of a system*

Certain variables may be unaffected by the operation of a system. The values taken on by these variables may be determined by the environment or by the output of some other system and they therefore serve as basic input to the system, and they remain unaffected by the interactions which go on within it. These inputs are usually termed the *parameters* of the system, and in operational systems models they require estimation either from theoretical considerations or from an empirical investigation of the interaction between some system and its environment. In a gravity model, which may be regarded as a very simple system, the distance parameter may be set at 2 on *a priori* grounds or it may be estimated from actual behaviour (see above, pp. 110–11).

B The study of systems

In the previous section a number of concepts were developed to enable us to say something about systems in general. Systems analysis has yielded insights into the structural characteristics and the behaviour of complex interacting phenomena, and systems concepts therefore provide an appropriate conceptual framework for handling substantive geographical problems. It will be useful to consider some of the general notions that have arisen from the study of systems under three headings: types of system, organisation and information in systems, and optimality in systems. A fourth important topic, that of systems modelling, will be left until a later section.

(1) *Types of system*

There are various ways in which we could classify systems. We could differentiate between closed and open systems, between man-made and natural systems, and so on. Rather than attempt an exhaustive classification of systems, however, we will concentrate attention upon those types of system which have something new to tell us regarding the analysis of complex interactions. Those systems, such as *simple-action systems* (Kuhn, 1966, 39–40), which express traditional modes of analysis, such as cause and effect, in a systems framework will therefore be ignored. Most that is new in systems analysis has to do with systems which are homœostatic, self-regulatory, adaptive, and particularly with systems that incorporate some form of feedback.

(a) *A homœostatic system* is one that maintains 'a constant operating environment in the face of random external fluctuations' (Rosen, 1967, 106). Such systems resist any alteration in environmental conditions and exhibit a gradual return to equilibrium or steady-state behaviour after such an alteration. The displacement of a spring, for example, will be followed by a series of oscillations until eventually the spring returns to a stationary state. In this case the phenomenon is known as 'dampening'. Variable pressure from the environment may mean that a spring will never be stationary, but that large-scale pressures will be dampened over time until the spring is in a steady state, oscillating very slightly in response to normal pressures from its environment. This latter interpretation of a homœostatic system refers, of course, to open-system analysis, and is associated with this important concept of a steady state—a concept which has great significance in the study of fluvial and other geomorphological processes (Chorley, 1962; Leopold and Langbein, 1962).

Some homœostatic systems contain a negative feedback mechanism to enable them to return to equilibrium or steady state. In such systems, the larger the original disturbance, the larger is the force exerted in returning the system to its steady state. Rosen (1967, 113) suggests that two kinds of mechanism can be studied here. The first, that of a 'servo-mechanism', provides a kind of tracking system that monitors changes in the environment and imparts information to the system, which acts accordingly. The second, a 'regulator', operates to keep the behaviour in the system constant in the presence of environmental variation. Servo-mechanisms and regulators are similar in many respects and they will be treated as the same here. Such negative feedbacks are extremely important and are 'fundamental to all life and human endeavour' (Forrester, 1961, 15). A classic example is provided by the process of competition in space which leads to a progressive reduction in excess profits until the spatial system is in equilibrium. Any disturbance (such as an increase in population or market demand) provokes a reaction until the system is once more in equilibrium. In the real world the competitive process does not operate this perfectly, but most analytic approaches to the study of spatial systems incorporate such a negative feedback effect and postulate equilibrium in some form or other.

(b) *Adaptive systems* are similar to homœostatic systems in many respects, but possess some special characteristics. Rosen (1967, 167) provides the following definition:

> An adaptive system is one for which there exists for each possible input a set of one or more preferred states, or preferred outputs. The adaptive character of

the system means that, if the system is not initially in a preferred state, the system will so act as to alter its state until one of the preferred ones is achieved.

The study of such systems provides a mode of approach to systems that are usually thought of as 'goal-seeking' or 'teleological' in the sense discussed in the previous chapter. Such systems clearly rely upon feedback mechanisms of some kind in order to achieve the preferred state. This feedback may operate in a number of ways. Most analytic studies have conceptualised the problem by postulating that feedback effects the condition of the environment and thus alters the inputs until the desired response (or preferred output) is achieved. Another possibility is that feedback affects the parameters of the system itself. There seem to be many situations in which the latter occurs, but such feedbacks are difficult to handle (Rosen, 1967, 168). An example of feedback via parameters in geography might well be the variable distance function in interaction models. In this case the preferred state of, say, journey-to-work systems is one in which demand for labour at employment opportunities is met by supply of labour from residences. If a rehousing project takes residences further away from employment opportunity then the system adapts by altering the parameter of the distance function. Adaptive systems are thus of considerable interest in geography, but in studying them we need to identify the meaning of 'preferred state' and all that it entails.

(c) *Dynamic systems* may be regarded as a separate class of system. There is some confusion surrounding the term 'dynamic', particularly in the social sciences (Hagen, 1961). Both homœostatic and adaptive systems show a change of state over time as they move towards steady or preferred state. In a truly dynamic system, however, feedback operates to keep the state of the system changing through a sequence of unrepeated states usually termed the *trajectory* or *line of behaviour* of the system (Ashby, *1963*, 25). Feedback may, for example, cause new preferred states to be identified (this is a characteristic of the learning process itself). Economic-growth models, such as the circular and cumulative causation model, may be regarded as dynamic systems. We may classify the various kinds of dynamism which a system's behaviour may exhibit. In some cases we may identify oscillatory behaviour, but in others the line of behaviour may be totally unstable and in such cases the system may become explosive. Richardson (1939), for example, examined the escalation of arms races and succeeded in identifying situations in which the system became so unstable that it exploded into war.

(d) *Controlled systems* are those in which the operator has some level of control over the inputs. Such controlled systems are, of course, of great interest in systems engineering and are the major concern of cybernetics (Wiener, *1961*; Ashby, *1963*). Systems-control theory provides a good deal of insight into the behaviour of systems, and is not irrelevant to the application of geography to substantive problems. Particularly in the field of planning, government, at both national and local levels, controls some of the inputs into the economic system and manipulates these in order to try and achieve some desired level of output. Monetary and budgetary policy are thus used to stimulate or dampen home demand, while at the local level the investment in roads, utilities, public housing, and so on, which is controlled by local government, provides an important means for varying the inputs in order to achieve certain goals (outputs). Most urban-model building has been concerned with exploring the possibility of controlling certain inputs and thereby directing the urban system's growth in a manner that achieves a desired set of goals. In most situations we have control over certain inputs while others are impossible or too expensive to manipulate. In seeking to maximise agricultural output, for example, we may be able to control the water input by irrigation, but we must do so in a situation where other aspects of the biosphere remain uncontrolled. Partial control-systems are thus of great interest. Perhaps the applied geography of the future will be in terms of developing models of controlled systems to demonstrate how the spatial system can be organised by manipulating a few key regulators.

(2) *Organisation and information in systems*

The twin concepts of *organisation* and *information* are exceedingly important in systems analysis. They provide the necessary concepts for discussing certain aspects of the behaviour of systems in a general and objective way. The concept of organisation can best be examined by way of an example. Consider a system containing n elements that behaves in such a way that if we know the value of one element in the system we can predict the values of all the others. Such a system is highly organised. Consider a similar system in which even though we know the values of $n - 1$ elements, we still cannot predict the value of the nth element. Such a system is disorganised. Information 'may be regarded as the measure of the amount of organisation (as opposed to randomness), in the system' (Klir and Valach, *1967*, 58). Information theory was initially developed in relation to communications engineering, but it has since been extended to deal with problems in psychology (Attneave, 1959), biology (Quastler, 1965), economics (Theil, 1967), and numerous other natural and social sciences. Information theory in fact makes use of a basic mathematical notion taken from the second law of thermodynamics, by regarding information as a quantity analogous to *entropy*. An increase in entropy amounts to a change from a more highly organised state in some physical system to a less organised state. When entropy is maximised the system is least organised. This concept of entropy is generalised in information theory.

As used in information theory, entropy has no relation to its established use in physics, and an abstract, statistical definition is introduced for it. If out of n events, each can occur with the probabilities $p_1, p_2, \ldots, p_n$, where $\sum\limits_{i=1}^{n} p_i = 1$ (i.e. some of the given events are bound to occur), then the expression $H = -\sum\limits_{i=1}^{n} p_i \log_a p_i$ is called entropy. (Klir and Valach, *1967*, 61.)

When all the p_i are equal H is at a maximum, which simply means that when all events are equally likely maximum entropy occurs. This leads to the fundamental mathematical operation of maximising the entropy function. In a specified system we can find the maximum value of H (this will obviously depend upon the size of n in the above equation). This value can be used in a number of ways. Suppose we specify a system in which the output variable can assume certain discrete equi-probable states no matter what the initial input. We have now specified a theoretical system in which H is maximised. Any deviation between the behaviour of this theoretical system and that of an observed system is then used to indicate regularity of behaviour in the observed system which then, in turn, may be used to infer some degree of organisation in the observed system. In this case we are using the entropy function as a random expectation model with which observed behaviour can be compared. But we can seek for other interpretations. If all events are equally likely, then we have maximum uncertainty regarding the behaviour of some system. When H = O we have absolute certainty. The entropy concept may be introduced here to measure our degree of uncertainty. Since science is often regarded as being about the reduction of uncertainty this measure has some use. It has thus been used by Theil (1967) to assess the efficacy of economic forecasting techniques and it can be used to answer such questions as 'how much information do we gain from a forecast?' or 'how much do forecasts reduce our uncertainty?' An interesting property of the entropy function is that it seeks to make all outcomes equally probable when maximised. Suppose that we build a system which possesses a certain degree of organisation. If we seek to maximise entropy in the behaviour of that system we can do so only while observing the various constraints we have built into the model. Maximising the entropy function in this case simply amounts to a procedure for obtaining unbiassed estimates regarding the most probable behaviour of the system. Leopold and Langbein (1962) use this kind of approach when they use the concept to identify the most probable profile for a river, while Wilson (1967) has recently used the same procedure in the context of an urban system and shown that the gravity model can be regarded as the most probable outcome in a regional system which operates with certain basic constraints.

From this discussion we may conclude that the meaning of terms such as entropy, organisation and information can vary according to the kind of model we specify. In maximising the entropy function we thus find several different interpretations for the same mathematical operation. In this the entropy concept does not differ from any other mathematical concept.

(3) *Optimality in systems—the allometric law*

A third aspect of systems analysis which is of great importance is the notion of optimality in systems. We noted above that functional explanations depended upon the definition of systems that are in some sense in good working order (above, p. 434). It is possible to define good working order by developing notions regarding optimality in systems. Optimality has been an important assumption of science. Rosen (1967, 1) suggests that 'the idea that nature pursues economy in all her workings is one of the oldest principles of theoretical science.' Optimality principles are not simply convenient tenets of faith. Physical systems have been shown to operate according to such principles, while the extensive study of the 'least-effort' principle in human behaviour by Zipf (1949) suggests that such principles are not entirely inappropriate in the study of social systems. It is also possible to provide a deeper rationale for the search for optimality principles. In biology the mechanism of natural selection exerts a pressure upon the structure, function, and performance of organisms which tends to force them to conform to optimality principles. Rosen (1967, 7) writes:

> On the basis of natural selection, then, it may be expected that biological organisms, placed for a sufficiently long time within a specific set of environmental circumstances, will tend to assume characteristics which are optimal with respect to those circumstances. This means that organisms tend to assume those characteristics which ensure that, within the specified environment, there shall be no selective disadvantage vis-a-vis the other organisms with which they compete.

A similar argument can be formulated to justify the employment of optimality principles in those aspects of human activity that involve competition, adaptation, and survival. In capitalist societies firms compete. They survive, adapt, or go out of business. In terms of spatial organisation it is competition in space (among locations) that provides a justification for seeking for optimality principles.

It is important to clarify the meaning of optimality in such situations. It is possible to attach meaning to optimality only if there is a class of competing solutions and costs can be attached to each solution. From the mathematical point of view there are numerous techniques (varying from the theory of maxima and minima of mathematical functions including the entropy function, to game theory and linear programming) which can be used to solve such problems once the necessary information is known. There are clearly many operational difficulties in identifying all reasonable competing solutions and in identifying the costs associated with them. Nevertheless it is often useful to establish the principle of optimality in some system, even if it is operationally difficult to identify the actual optimal solution, for without some notions with respect to optimality we cannot make any statements whether or not a particular system is in good working order. Theoretically we may construe optimality as a cost minimisation (or in economics a profit maximisation) problem. Such a view of optimality is extremely useful to our understanding of homœostatic and adaptive systems. The phenomena of negative feedback in homœostatic systems, for example, involves the minimisation 'of one or more specific quantities which measure the difference between the desired behaviour of the system and its actual behaviour' (Rosen, 1967, 114). Adaptive systems may similarly be regarded as seeking to minimise the difference between some preferred state and an actual state. One interpretation of the mathematical operation of maximising the entropy function is that it models a progression to minimum organisation in the state of some system. The optimality principle underlies, therefore, our understanding of the behaviour of most types of system.

There is not space to consider in any detail the implications of the optimality principle for the analysis of the structure and state of systems. It will be useful, however, to consider one example, partly to demonstrate that optimality does have implications for form, and partly because the example itself is inherently interesting. If we take measurements on some parts of a system and compare them with measures taken on other parts of the system (or over the whole system) we frequently find a basic mathematical relationship holds good, namely the *allometric law*, which has the form $y = ax^b$. This law is an example of what Rosen (1967) calls a *form functional*, and it is a good example too of what is meant by a functional law. It has been shown in many areas of biological and physical research that the form functional described by the allometric law very frequently holds. It holds, for example, for the relationship between trunk width and length of different species of animal, and for the growth of organisms over time. The latter case, that of the allometric growth law, specifies how measures made on one part of a system must change proportionately to the growth of the system as a whole. This allometric law has recently been introduced into geography and shown by Nordbeck (1965), Woldenberg and Berry (1967), and several others, to be characteristic of both human and physical geographical systems. The allometric law promises to become an integral part of geographic theory, since it can be used to account for such phenomena as population-density gradients, the rank-size rule, and so on. Nordbeck thus shows that the area of a city is related to its total population by the allometric law, and this suggests that a process of growth over time can be connected mathematically to size in space. If the size of a city is restricted then so is its total population. Here the functional relationship between a temporal process and a spatial form (see above, p. 127) is very clearly demonstrated. This then poses the basic problem of accounting for the allometric law itself. Rosen (1967, 80–6) suggests that this, and other functional laws, can be expressed in terms of optimality, and that there is a strong relationship between optimisation and the law of allometry. It may be, therefore, that the rank-size rule is an expression of some kind of optimisation principle in social organisation. The exact nature of this principle is still obscure, but these general results generated by Rosen are interesting enough to follow up.

The principle of optimality is well-documented and well-founded, and has a great deal to offer us in our attempts to grapple with complex systems. This is not to say that all systems are in an optimal state with respect to their environment. They obviously are not. But there can be no doubt of a pervasive underlying tendency for many systems to move towards optimum efficiency. In any case optimisation techniques seem to hold out to us the only criteria for judging whether or not a system is working well, and they are therefore vital to us in our search for rigorous explanations. The optimality principle has much to offer, and we should be foolish not to study it in considerable detail in seeking to apply systems analysis to geography.

C Systems analysis in geography

The application of formal systems concepts has been a relatively recent feature of research in geography. Given the complexity and interaction with which most geographers deal, it is difficult to see how we can avoid using techniques and terminology specifically developed to deal with such complexity and interaction. The notion of a system is not, however, in any way new to geographic thought. It probably has as long a history as the concept of system in science as a whole. This long history of systems thinking in geography is encouraging in that it suggests that many substantive geographical problems lend themselves naturally to being formulated in systems terms.

The history of systems thinking in geography is very much bound up with the functional approach, with the organismic analogy, with the concept of regions as complex interrelated wholes, and with the ecological approach to geography. It is possible to identify elements of systems thinking in the work of geographers such as Ritter, Vidal de la Blache, Brunhes, Sauer, and so on. But, as in the rest of science, the concept of a system has tended to remain on the periphery of geographic thought rather than at its very centre. During the last few decades the focus has changed to make the concept of the system of much greater significance. In the same way that considerations of logic in functional analysis (and cause-and-effect analysis) lead to the concept of the system, so it seems, the various pathways of geographic thought lead inevitably to systems thinking. Perhaps the best defined of these paths runs from those who, at the beginning of the century, sought to give a new focus to geographic thinking by interpreting geography as a form of human ecology, to those who now seek to interpret geography in *ecosystem* terms. This path is by no means straight. The implicit ecological approach of Vidal de la Blache (above, p. 441) differs somewhat from Harlan Barrows' (1923, 3) claim that geography is the science of human ecology. Barrows' view is scarcely the same as the ecological viewpoint expounded by Eyre and Jones (1966) or Brookfield (1964), while Stoddart (1966; 1967A) is critical of all of these interpretations. The terms 'ecology' and 'system' are clearly as rich in ambiguity in geography as they are in other areas of science. In many cases the ambiguity in geography is partly the result of drawing inspiration from different areas of science. Human geographers frequently look to notions developed in human ecology (Schnore, 1961); and even within human ecology there has been considerable change and variation in ideas (Theodorsen, 1961). Other geographers have looked to the biologists and plant ecologists. Stoddart (1967A, 524) thus draws considerable inspiration from Tansley in developing the concept of the ecosystem as a fundamental organising concept in geography, which, he claims, appeals because

First, it is *monistic*: it brings together environment, man and the plant and animal worlds within a single framework, within which the interaction between the components can be analysed. Hettner's methodology, of course, emphasises this ideal of unity, and some synthesis was achieved in the regional monographs of the French school, but the unity here was aesthetic rather than functional. . . . Secondly, ecosystems are *structured* in a more or less orderly, rational, and comprehensible way. The essential fact here, for geography, is that once structures are recognised they may be investigated and studied, in sharp contrast to the transcendental properties of the earth and its regions as organisms or organic wholes. . . . Third, ecosystems *function* . . . they involve continuous through-put of matter and energy. To take a geographical example, the system involves not only the framework of the communication net, but also the goods and people flowing through it. Once the framework has been defined, it may be possible to quantify the interactions and interchanges between the component parts, and at least in simple ecosystems the whole complex may be quantitatively defined. . . . Fourthly, the ecosystem is a type of general system, and possesses the attributes of general systems. In general systems terms, the ecosystem is an open system tending towards a steady state under the laws of open-system thermodynamics.

If geographers can reasonably conceptualise in systems terms, the problems with which they deal, then, it is clear, the whole power of systems analysis can be brought to bear on such problems. For the most part systems analysis in geography has not gone very much beyond the stage where we are exhorted to think in terms of systems. Statements by Blaut (1962), Chorley (1962) Ackerman (1963), Berry (1964A), and Stoddart (1967A), have thus directed our attention to the need for a reformulation of geographic objectives in systems terms—a viewpoint that has almost received official support in the United States through the report of the National Academy of Sciences on geography (N.A.S., 1965). In some cases systems concepts have been used to develop new theoretical formulations in geography. Chorley's attempt to reformulate thinking in geomorphology in terms of open-system thermodynamics, Leopold and Langbein's (1962) use of entropy and steady state in the study of fluvial systems, and Berry's (1964B) attempt to provide a basis for the study of 'cities as systems within systems of cities' by the use of the twin concepts of organisation and information in spatial form, are examples of this process. More recently Woldenberg and Berry (1967) have used systems concepts to analyse central-place and river patterns, while Curry (1967) has also attempted to analyse settlement-locations patterns in a systems framework. In general the search for predominant modes of spatial organisation in geography has much to learn from systems analysis and in particular from information theory, which has the power to distinguish the degree of organisation (the degree of departure from randomness) and thus to provide objective measures of pattern and a technique of pattern recognition. Those geographers who focus attention upon spatial organisation invariably invoke systems analysis as Haggett's (1965A) account of locational analysis in human geography demonstrates.

Yet it seems that the employment of systems concepts and systems analysis has not yet achieved powerful operational status in geography. In part this must be attributed to the complexity of systems analysis itself, which, if it is to be fully employed, involves mathematical techniques beyond the reach of most geographers. The solution to this difficulty is, of course, for the geographer to learn more mathematics, but this is perhaps easier said than done. But there are other difficulties. Operationalising systems analysis involves many evaluative judgements regarding the closure of the system, the definition of the elements, the identification of the relationships, and so on. The greater our experience of some problem and the more information we possess, the easier it is to make such evaluative judgements with some degree of confidence. Our general lack of experience with systems analysis, together with the relatively weak development of theory, does not allow us to make such evaluations with any degree of confidence, except in those cases where we can easily make assumptions regarding the structure and behaviour of some system. We are, in short, very much in the stage of *a priori* model use in our attempts to apply systems concepts to geography. But experience in cognate disciplines such as economics (e.g. Orcutt, *et al.*, 1961), psychology (Miller, 1965), political science (Deutsch, *1966*), urban economics and planning (Meier, 1962), business economics (Forrester, 1961), and so on, suggests that the attempt to use systems concepts will be worth while, if only because it provides the necessary framework for asking the kinds of question that seem particularly relevant to the study of the 'organised complexity' with which geographers deal. But we should be foolish to think that we require merely to wave the magic wand of systems analysis over the subject-matter of geography for all to be revealed. As Hagen (1961, 151) has pointed out, each discipline has to grope its way through its own substantive problems before it can see how and why systems analysis is relevant and applicable:

This is why each discipline slowly and stumblingly rediscovers concepts concerning method already discovered long ago in other disciplines—why, for example, economics clumsily and painfully groped its way to the concept of marginal productivity and only subsequently realised that it was merely applying elementary calculus to its problems. . . .

Systems analysis thus provides us with a convenient calculus for examining geographical problems. But to use that calculus we require geographical concepts that allow us to find an interpretation for that calculus in geographical context.

II SYSTEMS MODELLING AND GENERAL SYSTEMS THEORY

If we compare different systems in nature, we often find pairs of systems which resemble each other from a certain point of view. In such cases we may sometimes consider one of these systems as the model of the other system and duly utilise it in this function. . . . The similarities between different systems have been utilised intuitively by man since ancient times, in science as well as in the fine arts. At the present, however, science is getting more and more interested in this field, both in a theoretical sense . . . and in the practical application of models. (Klir and Valach, 1967, 92.)

We know that many natural systems are similar to each other, that we can often use the same analytic frameworks (say calculus or probability theory) for dealing with very disparate phenomena, and that the history of science is crammed with examples of disciplines that have 'groped their way' to formulations long ago discovered by other, more advanced, disciplines. The question has thus arisen whether the progress of science might be more efficient if some basic theory could be devised which covered isomorphisms between systems, and if criteria could be developed to facilitate the transference of ideas and formulations across disciplinary boundaries. The call for such a theory has been answered by the development of a theory of general systems. This theory has been expanded by some into what is usually termed *General Systems Theory*, which is an attempt to unify scientific knowledge and procedure through the use of general-systems concepts. This latter use of general systems amounts to providing a basic philosophy (as opposed to a methodology) for science as a whole and is concerned to lay the foundation for a new paradigm in scientific thought. It is important, however, to evaluate general-systems concepts from various points of view. We shall therefore begin by considering the problem of systems modelling, proceed via general systems as a methodology, and close by examining *General-Systems Theory* as a philosophy.

A Systems modelling

Consider two systems, S_1 and S_2, whose structure and behaviour are known. By this we mean that we can identify the elements in each system, the links, the outputs, and so on. Given this information, it is possible to establish in what respects S_1 and S_2 are similar and under what conditions S_2 can be used as a model of S_1 and vice versa. Now there may be several respects in which S_1 and S_2 are similar and several in which they are dissimilar. They may exhibit similarity of structure, of behaviour, of composition of the elements, of mathematical form of the relationships, and so on. It is therefore important to distinguish the respect in which we are modelling the system, whether, for example, we are modelling its structure, its behaviour, or the system as a whole. We will first examine the most general case and then examine the more specific aspects of systems modelling.

(1) *The model of a system*

Klir and Valach (1967, 108) point out that the principle of modelling S_1 and S_2 is based on the concepts of *isomorphy* and *homomorphy*. Two systems are isomorphic if the elements in S_1 can be uniquely assigned to the elements in S_2 and vice versa, and if for every relationship (r_{ij}) in S_1 there exists an exactly similar relationship in S_2 and vice versa. The isomorphic relation between two systems is symmetrical, reflexive, and transitive. A system through which water flows may thus be made isomorphic with a system through which electric current flows. Two systems are *homomorphic* when the elements in S_1 can be assigned uniquely to elements in S_2 but not vice versa, and the relationships in S_1 can also be assigned uniquely to relationships in S_2, but not vice versa. A good example of homomorphic relationships is that between a map and the countryside, in that every element on the map can be assigned to an element in the countryside (but the countryside contains many elements not recorded on the map) and every geometric relationship portrayed on the map also holds in the countryside (with respect to physical distance that is) but there are many geometrical relationships which actually exist in the countryside which are not portrayed on the map. In such systems the relationships between them are not symmetrical, reflexive, or transitive. We may treat the map as a model of the countryside, but we cannot treat the countryside as a model of the map.

Isomorphic models of total systems are particularly useful where we can construct them, since it frequently happens that one kind of system (an electric circuit, say) is much easier to construct and use than another (a water supply system, say). But most relationships tend to be homomorphic, and these involve greater problems of control, in that the mapping of the original system (the countryside) into another system (the map) imposes some difficulties, since we must needs be certain that the output of the modelling system is also characteristic of the output of the original system. This brings us to the question of partial models.

(2) *Partial models of systems*

Consider the situation in which the structure and behaviour of S_1 cannot be related to the structure and behaviour of S_2 (through ignorance or inability). It may then be possible to take certain aspects of a system, its output for example, and seek to model S_1 by S_2 simply with respect to this aspect of the system. The commonest case of such systems modelling is that which relies upon a comparison of the behaviour of systems. We shall therefore restrict attention to this aspect of systems modelling here. Consider two systems S_1 and S_2, into which we put an exactly similar stimulus and from which we get exactly similar responses, no matter how we vary the stimulus. S_1 and S_2 exhibit similar behaviour, and even though we know nothing of their structure we may use S_1 as a model of S_2 and vice versa, provided we are concerned simply with behaviour. Now it is rare for us to be able to identify systems into which we can put an exactly similar stimulus. In most cases it is necessary for us to translate one input stimulus (an electric charge, say) into another kind of input stimulus (pouring in water, say), and to translate the output response in a similar manner. This involves what is sometimes termed 'input and output mapping', by which is meant the transformation of inputs of a certain type into S_1 into inputs of a different but equivalent type into S_2 and a similar transformation of the outputs. To use such partial models to solve problems, predict, and explain, requires some degree of control (either by way of theory or by way of calibration and experiment) for us to be certain that the result obtained by way of the model is characteristic of the system being modelled. In general it requires some definition of equal behaviour (Klir and Valach, *1967*, 105-7). Given the necessary controls, there can be no doubt that partial models are extremely valuable.

(3) *Black boxes and white boxes in systems modelling*

Systems analysis frequently makes use of the concept of a black box, which is a system whose internal characteristics (structure and functioning) are unknown, but whose stimulus-response characteristics can be studied in detail[1]. If we know the input–output relationship characteristic of a given black box, it may be possible to seek out some model of that system, whose structure is known, which achieves precisely the same behaviour. The inference is often then made that the black box in fact has the same structure as the model (usually called the white box) or at least that it is reasonable to proceed as if it did. The replacement of black boxes by white boxes in this way (and the attempt to infer structure from behaviour) is an extremely important part of systems analysis which we avoided in the general discussion of systems analytic procedures since it involves the problem of systems modelling. It is usual, however, to make use of some formal mathematical theory of organisation of systems in searching for white box substitutes for what are initially black boxes. Thus Quastler (1965) discusses cybernetic systems, game theory, decision theory, and communications theory, as white boxes which can perhaps replace many of the black boxes which the biologist is faced with when he attempts to grapple with complex aspects of biological behaviour and organisation. In this case systems modelling is concerned to set up some artificial mathematical structure which succeeds in behaving like the original structure. This is, of course, the concern of cybernetics, with its emphasis upon machine intelligence, fundamental theories of organisation, and the like. In any complex situation where we know very little or where it proves impossible to discern the structure of the organism without destroying it, the black-box approach to systems modelling will have important applications. If it is to fulfil this function effectively, however, we require some measure of control regarding the similarity of the two systems.

B General-systems theory as methodology

The process of modelling one system by another, whether totally or partially, has gone on in science (and in geography) since time immemorial. The extent of such systems modelling has tended to pass unnoticed until relatively recently, and certainly it has not been the focus of detailed methodological study. Von Bertalanffy's (1951; 1962) work has drawn attention to the extent of isomorphism between systems designed to deal with very different domains. His approach is largely empirico-intuitive. He examines various types of real system identified in many different disciplines and shows that many of these systems have quite surprising similarities: in short, our knowledge of diversified domains of events can be organised parsimoniously by way of a few systems concepts. Further, it is possible to use the information on the properties of one real system to elucidate the properties of another (presumably little known) real system, to model one by another, and to draw conclusions about one real system by way of another. This leads to the concept of a general system which is one that subsumes the characteristics of the various real systems and therefore allows us to discuss systems problems in a unified analytic framework. A general system is a kind of higher-order generalisation about the variety of real systems which individual disciplines have identified. This notion of a general system is important in itself, but von Bertalanffy uses it as a tool in an attempt to unify science. This last aspect of von Bertalanffy's argument involves us in philosophical problems which we will consider in the next section. We shall here restrict attention to the methodological aspects of general systems.

Mesarovic (1964, 4) characterises general systems theory as follows:

> Since it can be argued that both science and engineering are concerned with the study of real systems and their behavior, it follows that a general theory should be concerned with a study of general systems. . . . It suffices for the present discussion to consider a general system as an abstract analogue or model of a class of real systems. General systems theory is then a theory of general models.

Mesarovic then goes on to consider such a theory in linguistic terms, on the grounds that general-systems theory should provide some kind of higher-order language (a meta-language) for discussing the various languages that constitute the theories of science (above, pp. 179-81). Now it should be apparent from Mesarovic's definition that general-systems theory, in the sense in which he uses the term, is concerned not merely with isomorphism and analogy in systems analysis, but with setting up some general theory from which the characteristics of the various systems can be deduced. It is concerned with the deductive unification of systems analytic concepts. It thus seeks to do for systems analysis what Klein's analysis did for geometry (above, pp. 202-5)—namely, provide a framework for relating individual systems and types of system within a unified hierarchical structure. Such a structure is useful in that it allows us to understand better the relationships that exist between the various types of system, to state categorically the conditions under which one system approximates another, and to identify types of system that may be useful to us even though we have not yet identified real systems to match them. General-systems theory under Mesarovic's interpretation (and this is the predominant interpretation in cybernetics, mathematical systems analysis, and so on) should provide us with well-developed syntactical structures ready-made for the discussion of substantive (geographical) problems, provided, that is, that we can map the substantive problem into the appropriate systems language. It also should provide us with the necessary criteria for undertaking systems modelling in controlled fashion. This analytic-deductive approach to the study of general systems contrasts very much with von Bertalanffy's, although in certain respects the aim is the same. At the present time, however, general-systems studies do not provide us with a unified deductive framework for understanding systems. They have provided us, however, with a deeper understanding of certain kinds of system, such as homœostatic, adaptive, and dynamic systems, together with a general understanding of the nature of recurring system states, such as steady-state or dynamic behaviour. Both the empirico-intuitive approach of von Bertalanffy and the analytic-deductive approach of Mesarovic therefore provide us with methodological insights of considerable importance for the application of systems concepts to substantive geographical analyses.

C General-systems theory as philosophy

It is perhaps useful to begin an examination of general-systems theory as a philosophy by recapitulating some general points that have been made with respect to systems analysis:

(i) All explanation is contingent upon the definition of some closed system within which analysis can proceed.

(ii) It is possible to pursue our understanding of phenomena by way of systems analysis since it is possible to operationalise systems concepts with, in many cases, a reasonable degree of certainty.

(iii) Systems analysis is peculiarly appropriate for the study of complex interacting phenomena, and it encompasses the formulations of traditional science while extending an analytic framework to disciplines (such as geography) where interactions are so complex that they have not been analysable by way of traditional scientific techniques.

(iv) Numerous isomorphisms exist between systems designed to deal with very different domains of phenomena.

(v) The theory of general systems provides an inductive and analytic approach to the unification of much of our thinking about systems and for an understanding of isomorphisms between systems.

(vi) The breadth and sophistication of systems analysis are tempered by the large element of ambiguity that surrounds systems concepts, particularly as regards their application in the social sciences.

These statements regarding systems analysis seem unexceptionable. It is tempting, therefore, to suggest that all explanation must be rendered in systems terms, that we can understand phenomena only in so far as we can conceptualise phenomena in systems terms, and that general systems theory provides the ultimate key to a unified science. Such suggestions contain false logic in the same way that the argument for the 'genetic' approach or the 'functional' approach contained fallacious arguments. Future generations may well come to identify a 'systems fallacy' in much the same way that we now identify a genetic fallacy or a functional fallacy (above, pp. 416; 446). General systems as a philosophy therefore requires some careful evaluation.

Von Bertalanffy proposes to use a general-systems approach to put back together again those pieces of reality that science has dismembered for the purpose of analysis. He argues that science has tended to involve increased specialisation and that this intense specialisation has led to disciplines' developing in isolation. Yet each discipline tends to develop similar explanatory structures, to identify similar system structures, and to 'grope its way' to similar system formulations. Struck by the numerous isomorphisms that can be demonstrated to exist between systems identified in disparate fields, von Bertalanffy goes on to propose *General-Systems Theory* as the key theory to explain such isomorphisms (and to identify others), and hence to provide a unifying framework for our understanding of nature and the reality around us. *General-Systems Theory* is, according to von Bertalanffy, a theory about reality—a calculus with an interpretation—which has the properties of what Ackoff (1964, 58) calls a metatheory—'a theory to explain disciplinary theories'. Now it should be clear that this version—which we will term *General-Systems Theory* (GST)—is rather different from the syntactical approach to a theory of general systems as developed by Mesarovic and others. The first proposes a theory about reality, the second, a unified framework for discussing the various syntactical structures used for modelling real systems. Von Bertalanffy's presentations are not always free from ambiguity, and on occasion he appears to take both views. Nevertheless it seems that it is GST rather than the theory of general systems that has come in for the heaviest criticism. Buck (1956) calls it 'naïve and speculative philosophy', while Chisholm (1967) has characterised it as an 'irrelevant distraction' to geography. In both of these discussions the tendency is to reject the theory of general systems as irrelevant by presenting arguments against GST—a procedure that is clearly fallacious. Chisholm, for example, appears to regard the presentations of Mesarovic (1964) and von Bertalanffy as 'isomorphic' which they are not. Given these criticisms, it is important to identify the real nature of GST and to assess the real grounds for accepting or rejecting it. Ackoff (1964, 54) seeks to reject GST but in doing so clarifies the nature of GST as developed by von Bertalanffy:

> Bertalanffy accepted the current disciplinary structure of science as a starting point. By seeking structural isomorphisms among the laws established by the various scientific disciplines, he hoped to find a more general theory than could be produced by any one discipline. Hence Bertalanffy implicitly assumed that the structure of nature is isomorphic with the structure of science. Nothing could be further from the truth. Nature is not disciplinary. The phenomena and the problems which nature presents to us are not divisible into disciplinary classes. We impose scientific disciplines on nature; it does not impose them on us. Some of the *questions* that we ask of nature—in contrast to the problems it presents to us—can be classified as physical, chemical, biological, and so on; but not the phenomena themselves.

Von Bertalanffy thus attempts to unify the body of facts, laws, and theories, produced by science, under one integrating theory. One problem with such a theory is that

> Its validity would depend on the deducibility of disciplinary theories from it and, hence, in turn, on the validity of the disciplinary theories. It is thus twice removed from the experimental and applied aspects of science.

A second problem is that the theory must explain in some way or other the isomorphisms that exist between system structures developed in diverse domains. The syntactical approach of Mesarovic explains this in terms of the analytic constructs which we can most conveniently bring to bear on reality, rather than in terms of the structure of reality itself. But von Bertalanffy seeks to explain such isomorphism with reference to reality itself. This is a very important difference. Mesarovic seeks to construct a theory of general models— i.e. a theory to explain scientific constructs or a theory to explain our *as-if* thinking about reality. Von Bertalanffy seeks to construct a general theory about reality itself. Now clearly billiard balls are not the same as atoms, nor are people the same as molecules. Von Bertalanffy therefore seeks for some deeper theory of similarity between real systems, and postulates GST as a kind of *a priori* model of what such a theory might look like. This approach raises some important issues regarding isomorphism and analogy in science, about the meaning to be attached to analogy, and about the necessity for such analogies. Some have been very critical of von Bertalanffy's approach on this score. Hempel (1951, 315), for example, writes:

> It does not seem to me . . . that the recognition of isomorphisms between laws adds to, or deepens, our theoretical understanding of the phenomenon in the two fields concerned; for such understanding is accomplished by subsuming the phenomena under general laws or theories; and the applicability of a certain set of theoretical principles to a given class of phenomena can be ascertained only by empirical research, not through pure system theory.

Buck's (1956) and Chisholm's (1967) criticisms of GST amount to a 'so-what?' reaction to the discovery of isomorphism as well as invoking the well-known dangers of argument by analogy.

Von Bertalanffy (1962, 8–9) replies to Buck's criticisms by pointing out that isomorphism and analogy are extremely important in science and that without analogy science would be almost impossible. At this point it is clear that the argument moves towards an argument about the philosophy of explanation itself. Hempel, for example, belonging to the deductive-predictive school (above, pp. 13–14), argues against a mode of explanation by way of system models, but others, such as Workman (1964) would probably support, either totally or partially, the rendering of answers to questions by way of isomorphism and analogy (above, p. 154). This argument is essentially irresolvable except by reference to our own beliefs. It is a philosophical rather than a methodological problem. In general, therefore, the particular view we take on GST depends upon our own beliefs, upon our own philosophy, upon our own 'world image', as Kuhn (1962) would call it. This is not to say that people cannot be persuaded of the virtues of GST. Probably the most persuasive arguments are those developed by Boulding (1964), who seeks to demonstrate the virtues of adopting GST as a point of view—as a basic working hypothesis—on the grounds that it opens the channels of interdisciplinary communication (which most would regard as a good thing), that it provides scientific endeavour with a unifying vision, and that it provides an imaginative conceptual framework for formulating the questions we ask of the complex world around us. Like any philosophy, however, GST has its dangers. The dangers of false analogy are very persistent—the remedy here, Boulding (1964, 36) suggests, is 'not *no* analogy but *true* analogy'. Indeed GST provides us with a very far-reaching *a priori* model regarding real-world structure. There are many who resist such an *a priori* model, particularly if postulated in such a way as to exclude other possible models. Even if we accept GST as an *a priori* model, however, we need to observe the methodological rules that govern the application of all such *a priori* models to real-world situations. Certainly much of the work that has been done on systems analysis—and particularly on the problems of isomorphism and homomorphism—helps to clarify the application of systems models to real situations. In addition, the methodological analysis of systems structures provides us with criteria for evaluating how far one system may be used to model another.

Perhaps the most persuasive argument of all in favour of GST would be a positive indication of its utility by way of research results, new understanding, and the like. Here it seems that von Bertalanffy's ideas have not been very productive. GST is not widely accepted in science (although it is clearly fashionable among certain groups of geographers), nor has it yielded very much in the way of results. Yet movements very similar to GST have undoubtedly yielded a great deal. The strong current of interdisciplinary research—particularly the trend to greater unification in the social sciences (typified in Kuhn's (*1966*) approach)—has much in common with GST. Similarly, Ackoff (1964) notes the tendency for the techniques of operations research—such as inventory-control theory and queueing theory—to provide basic theories which are systems-oriented and interdisciplinary in character. The black-box-to-white-box transformations envisaged in biology by Quastler (1965) have been extremely useful and have shown the utility of systems modelling as a mode of approach to explanation, while cybernetics has taken advantage of such transformations to develop elaborate control systems for automata. Systems analysis itself clearly provides us with the most sophisticated mode of approach to explanation and to understanding, in the sense that it has applicability to a far wider range of phenomena than have more traditional scientific approaches.

Systems analysis, the theory of general systems, and *General-Systems Theory*, are relatively new departures in scientific investigation. It would be surprising if there was no ambiguity, no confusion, no misunderstanding, no break-down in the communication process. In arguing about systems analysis in geography, or in any other context, it is important to be clear as to the precise nature of the concepts we are criticising. *General-Systems Theory* may be an 'irrelevant distraction', as Chisholm (1967) suggests, but it would be foolish to infer from this that the theory of general systems is irrelevant or that it is impossible or misleading to apply the concepts of open-system thermodynamics, as Chorley (1962) does, to geomorphological processes. These are all separate questions and must be treated as such. Whatever our philosophical views may be, it has been shown that methodologically the concept of the system is absolutely vital to the development of a satisfactory explanation. If we abandon the concept of the system we abandon one of the most powerful devices yet invented for deriving satisfactory answers to questions that we pose regarding the complex world that surrounds us. The question is not, therefore, whether or not we should use systems analysis or systems concepts in geography, but rather one of examining how we can use such concepts and such modes of analysis to our maximum advantage. In this there remains a great deal to do, both in methodological analysis and in empirical research and investigation. This chapter has sought to provide the basis for the methodological analysis of system concepts, and it is to be hoped that such an analysis will serve to guard against false argument and false inference in the application of systems analytic techniques to the conduct of empirical enquiry.

Basic Reading

Ashby, W. R. (*1966*).
Blalock, H. M., and Blalock, A. (1959).
Hagen, E. (1961).
Klir, J., and Valach, M. (1967).
Mesarovic, M. D. (ed.) (1964).
Rosen, R. (1967).

Reading in Geography

Berry, B. J. L. (1964B).
Chorley, R. J. (1962).
Nordbeck, S. (1965).
Stoddart, D. (1967A).
Wilson, A. (1967).
Woldenberg, M. J., and Berry, B. J. L. (1967).

Chapter 24
Explanation in Geography
A Concluding Comment

There is a tendency to look for some conclusion to a book in much the same way that one waits upon the punch-line of a good joke or story. The detailed methodological conclusions are contained in the body of the text and there seems little point in attempting to summarise them here. But there are certain recurrent themes in this work and in this concluding comment I want to discuss how these themes relate to one another and how we can use methodology to promote a more concise philosophy of geography.

What I have sought to do in this book is to provide a series of spring-boards for future methodological analysis and, I hope, some rough and ready guidelines for the conduct of empirical research in geography. Some of the spring-boards may appear to have very little spring in them, others seem to form an intellectual trampoline which, without adequate control, may send one spiralling off into heady abstraction only to land with an uncomfortable thud on the *terra firma* of geographical reality. Yet the world of ideas, images, abstractions, concepts, and the like, are as much a part of the *terrae incognitae* of the geographer as is the world of direct experience. We cannot explore these worlds without taking risks, without taking a leap from some spring-board of thought even though we do not know what lies beyond. It is not the role of the methodologist, therefore, to curb speculation, to decry metaphysics, to chain the imagination, although many would attack him because they fear that he might try to do all these things and worse. But at some stage we have to pin down our speculations, separate fact from fancy, science from science fiction. It is the task of the methodologist to point to the tools which can be used to accomplish this, to assess their efficiency and worth. In doing so the methodologist must be critical. The danger is, however, that what the methodologist says may become orthodoxy, that methodology (rather than the methodologist) may curb speculation, dampen intuition, dull the geographical imagination. I have tried to avoid this danger by separating speculative philosophy in geography from methodology as far as possible. My own objective has been to give the geographer free reign in his choice of objectives, in the belief that geography always has been and always will be what those who call themselves geographers choose to do. Unfortunately it is not always possible to separate philosophy and methodology since at many points the two become so interwoven that they effectively merge into one. I cannot claim, therefore, to have written a book on methodology which is untainted by any philosophy. Nevertheless, there are many methods which can be evaluated independent of their philosophical connotations. This is particularly the case when we examine the syntax of some explanatory form for we can show that a particular logical form—say cause and effect or systems analysis— has certain qualities which allow it to be used to accomplish certain tasks independent of the philosophy of the investigator. This kind of methodological freedom is, in my view, very important. It gives us tremendous flexibility in our approach to the study of geographic phenomena, and also provides a necessary vocabulary for reasonably objective discussion of geographical problems.

But geographical problems cannot be solved by the mere selection of some logically consistent methodology. Something more is needed. This 'something more' amounts to an adequate philosophy of geography. We cannot escape this since in any study we need to make assumptions. A very important value judgement is involved in deciding what questions to assume away and what questions to investigate. Further, it is often impossible to make a good methodological decision without making prior philosophical decisions regarding the goals and objectives of a particular investigation. This proposition applies to theory construction, model use, the selection of an appropriate language, classification, measurement, sampling, and so on. It should be clear, therefore, that an adequate methodology provides a *necessary* condition for the solution of geographical problems; philosophy provides the *sufficient* condition. Philosophy provides the steering mechanism, methodology provides the power to move us closer to our destination. Without methodology we will lie becalmed, without philosophy we may circle aimlessly without direction. I have mainly been concerned with the nature of the power devices which are available to us. But I would like to close by returning to the interface between methodology and philosophy.

I began by setting up a simple structure for explanation in which initial conditions and covering laws are brought together to allow the deduction of the event to be explained. The main difficulty in operating such a structure is that we need adequate laws (we may prefer to avoid the term law and talk of generalisations, principles, law-like statements, and so on, but for the sake of exposition we will call these statements laws from now on). The question then arises as to how we can discover and control the use of law statements. Laws, it is generally agreed, should be *reasonable* with respect to experience and *consistent* with respect to each other. Some also prefer them to be *powerful* (very general) on the grounds that explanation will be a much more efficient process if we possess statements of great generality. It is generally agreed that we approach this problem by constructing and validating a theory. But theories are speculative constructs and speculation is a metaphysical and philosophical enterprise whether we like it or not. The geographer cannot proceed, therefore, without some clear notion as to what it is he is supposed to speculate about. Even if we choose to view a theory as an abstract syntactical structure (a calculus or a specially constructed language), we still have to face the problem of interpreting that theory by way of a *text* which, among other things, relates that theory to a particular *domain* of events. The philosophical implication is that the geographer *needs to identify the particular domain (or set of domains) with which he is specifically concerned*. Methodological considerations lead me to conclude, however, that it is easiest to identify such a domain (or set of domains) when we possess a well-articulated and well-validated theory. Hence the significance of the comment (p. 129) that the nature of a discipline can be discerned through an examination of the theories which it develops. The point of view of geography is thus embodied in geographic theory, and the subject-matter is identified through the texts of these theories which relate them to given domains. I speculated a little regarding the nature of geographic theory and suggested that we possess indigenous theory regarding spatial form and derivative theory concerned with temporal process and that general theory in geography amounts to a theory examining the interactions between temporal process and spatial form. This suggestion is undoubtedly controversial and will be disputed by many. But I am prepared to put it forward as one basic tenet of geographical thought. Since academic activity frequently involves division of labour, individual geographers may specialise in various aspects of this general theory and some may not be aware of the general structure to which they are contributing. This view of general theory in geography helps us to pin down the point of view of the geographer but it does not help a great deal in establishing the domain of geographical thought. I find it difficult to state with any certainty the domain of geography, but a number of issues emerged in subsequent chapters which at least provide certain clues as to how we might set about solving this problem. At various points, for example, three interrelated problems in methodological analysis emerged which could only be solved by making some prior philosophical decision. These problems concerned:

(i) The nature of *geographical individuals*.
(ii) The nature of *geographical populations*.
(iii) The problem of *scale*.

The first two problems cannot be resolved without a firm definition of the domain of geographical investigation. The third has, I suspect, even greater significance. In the chapter on systems we hit upon a very telling idea. Systems, it was there suggested, are made up of individuals (or elements), but if we choose to alter what was termed the resolution level these individuals themselves could be treated as systems containing lower-order individuals. The philosophical implication of this is that the definition of an individual depends upon the particular resolution level or scale at which we choose to work. I think it useful to speculate a little regarding the typical resolution level of the geographer. He is not concerned with the spatial patterning of crystals in a snow flake (the resolution level is too high) nor is he concerned with the spatial patterning of stars in the universe (the resolution level is too low), although both can be of interest to him from the abstract mathematical point of view. Geographers tend to pick a resolution level that lies somewhere between these two. Characteristically geographers tend to work with human and physical differentiation at the 'regional' level although it is difficult to pin this down with any precision. In technical language what I am trying to say is that the geographer tends to filter out small-scale variation and large-scale variation and to concentrate his attention upon systems of individuals which have meaning at a regional scale of resolution. It may be that the domain of the geographer can best be approached by an analysis of the particular resolution level at which he works rather than by an examination of the kind of subject-matter he discusses. At the present time, for example, I suspect that human geography is rather uncomfortably sandwiched between microeconomic theory and macroeconomic theory in terms of resolution level (although fortunately the structure of geographic theory appears to be quite different). It would be surprising if the typical resolution level remained constant over time, however, since new disciplines emerge and old ones change their particular resolution level. I suspect, thus, that geography is moving into a phase in which it ranges less over a whole spectrum of resolution levels mainly because other disciplines have encroached. International trade and inter-cultural differences are firmly in the hands of economists and anthropologists, social interaction on housing estates firmly in the province of the sociologist. It is perhaps easier now to identify the typical resolution level of the geographer than it was some thirty years ago. But here I am speculating. Nevertheless I am prepared to suggest that another basic tenet of geographic thought is that its domain is defined in terms of a regional resolution level. Any phenomena that exhibits significant variation at that resolution level is likely to be the subject of investigation by the geographer.

The problem of scale can thus be used to get at certain philosophical problems regarding the nature of geographical individuals and geographical populations. But, as I have already remarked, the closer definition of the domain of geography must wait upon the statement of well-articulated and well-validated geographic theory. Since theory is only weakly developed in geography, the strategic question then arises as to how best we can pursue the creation of an adequate corpus of geographic theory. Here the role of the model, and particularly the *a priori* use of mathematical models is of major interest. Mathematics provides us with a vast number of already constructed calculi which we may use as *a priori* models in the search for geographical theory. I have tried to emphasise the dangers of *a priori* model use, but if these dangers are avoided, I believe that one of the most effective strategies we can currently adopt is to seek out possible interpretations for what we believe to be interesting and appropriate mathematical calculi. Geometry and probability theory are two obvious candidates here, but there are plenty of others, which we can perhaps most easily identify by looking at the kind of mathematics usually employed in the study of various kinds of system. But the complex relationship between geographical conceptualisations on the one hand and syntactical mathematical systems on the other, provides us with a whole series of challenging questions. I attempted to indicate the kind of thing involved in the chapters on geometry and probability. What, for example, is the appropriate geometry for discussing a particular process operating on a particular surface? What interpretation can we give to the sample space of probability theory in geography? How can we apply techniques of statistical inference in geography and, further, what is the population from which we are presumed to be sampling? These, and other questions, are not simply a matter for methodological analysis although clearly formal semantics has a great deal to contribute to our understanding of the relationship between geographical concepts and mathematical structures. Philosophical decisions are also involved. Suppose, for example, that we can envisage vast application for a particular mathematical calculus (say, probability theory) provided we are willing to alter the nature of our geographical thinking in some respect or other. Is this alteration acceptable? Are the new concepts which we evolve to match the mathematical calculus reasonable with respect to the real world? These are philosophical issues which no amount of methodology can resolve. This kind of question challenges us to stake out a new and more positive philosophy of geography on the basis of which we may create theoretical structures which, in turn, will give our discipline the identity and direction it so badly needs at the present time. Without theory we cannot hope for controlled, consistent, and rational, explanation of events. Without theory we can scarcely claim to know our own identity. It seems to me, therefore, that theory construction on a broad and imaginative scale must be our first priority in the coming decade. It will take courage and ingenuity to face up to this task. But I feel confident that it is not beyond the wit and intelligence of the current generation of geographers. Perhaps the slogan we should pin up upon our study walls for the 1970s ought to read:

'By our theories you shall know us.'

References

[Note: italicised dates of publication indicate editions of works other than the first]

Abel, T., 1948, 'The operation called verstehen', *Am. J. Sociol.* **54,** 211–18.
d'Abro, A., *1950, The evolution of scientific thought* (Dover, New York).
Achinstein, P., 1964, 'Models, analogies, and theories', *Philosophy Sci.* **31,** 328–50.
Achinstein, P., 1965, 'Theoretical models', *Br. J. Phil. Sci.* **16,** 102–20.
Ackerman, A. E., 1963, 'Where is a research frontier?' *Ann. Ass. Am. Geogr.* **53,** 429–40.
Ackerman, R., 1966, *Nondeductive inference* (London).
Ackoff, R. L., 1964, 'General systems theory and systems research: contrasting conceptions of systems science', in Mesarovic, M.D. (1964).
Ackoff, R. L., with Gupta, S. K. and Minas, J. S., 1962, *Scientific method: optimizing applied research decisions* (New York).
Adler, I., 1966, *A new look at geometry* (London).
Ahmad, Q., 1965, 'Indian cities: characteristics and correlates', *Res. Pap.* No. **102,** Dept. of Geog., Univ. of Chicago.
Aitchison, J. and Brown, J. A. C., 1957, *The lognormal distribution* (London).
Anderle, O. F., 1960, 'A plea for theoretical history', *History and Theory* **1,** 27–56.
Anscombe, F. J., 1950, 'Sampling theory of the negative binomial and logarithmic series distributions', *Biometrika* **37,** 358–82.
Anscombe, F. J., 1964, 'Some remarks on Bayesian statistics', in Shelly, M. W. and Bryan, G. L. (eds.), (1964).
Apostel, L., 1961, 'Toward the formal study of models in the non-formal sciences', in *The concept and role of the model in mathematics and natural and social sciences* (Synthese Library, Dordrecht).
Arrow, K. J., *1951, Social choice and individual values* (New York).
Arrow, K. J., 1959, 'Mathematical models in the social sciences', in Lerner, D. and Lasswell, H. D. (eds.), *The policy sciences* (Stanford).
Ashby, W. R., *1963, An introduction to cybernetics* (Wiley Science Editions, New York).
Ashby, W. R., *1966, Design for a brain* (London).
Attneave, F., 1950, 'Dimensions of similarity', *Am. J. Psychol.* **63,** 515–56.
Attneave, F., 1959, *Applications of information theory to psychology* (New York).
Bagnold, R. A., 1941, *The physics of blown sands and desert dunes* (London).
Ballabon, M. G., 1957, 'Putting the "economic" into economic geography', *Econ. Geogr.* **33,** 217–23.
Bambrough, R., 1964, 'Principia metaphysica', *Philosophy* **39,** 97–109.
Barker, S. F., 1964, *Philosophy of mathematics* (Englewood Cliffs, N.J.).
Barraclough, G., 1955, *History in a changing world* (Oxford).
Barrows, H., 1923, 'Geography as human ecology', *Ann. Ass. Am. Geogr.* **13,** 1–14.
Barry, R. G., 1967, 'Models in meteorology and climatology', in Chorley, R. J. and Haggett, P. (eds.), (1967).
Bartlett, M. S., 1955, *An introduction to stochastic processes* (Cambridge).

Bartlett, M. G., 1964, 'The spectral analysis of two-dimensional point processes', *Biometrika* **51**, 299–311.

Beach, E. F., 1957, *Economic models* (New York).

Beck, R., 1967, 'Spatial meaning and the properties of the environment', in Lowenthal, D. (ed.), (1967).

Bell, E. T., *1953, Men of mathematics* (Penguin, Harmondsworth).

Bergmann, G., 1958, *Philosophy of science* (Madison).

Berlyne, D. E., 1958, 'The influence of complexity and novelty in visual figures on orienting responses', *J. exp. Psychol.* **55**, 289–96.

Berry, B. J. L., 1958, 'A note concerning methods of classification', *Ann. Ass. Am. Geogr.* **48**, 300–3.

Berry, B. J. L., 1960, 'An inductive approach to the regionalization of economic development', in Ginsberg, N. (ed.), 'Geography and economic development', *Res. Pap.* No. **62**, Dept. of Geog., Univ. of Chicago.

Berry, B. J. L., 1961, 'A method for deriving multifactor uniform regions', *Przegl. geogr.* **33**, 263–82.

Berry, B. J. L., 1962, 'Sampling, coding, and storing flood plain data', U.S. Dept. of Ag., Farm Econ. Div., *Agric. Handbook* **237.**

Berry, B. J. L., 1963, 'Commercial structure and commercial blight', *Res. Pap.* No. **85,** Dept. of Geog., Univ. of Chicago.

Berry, B. J. L., 1964A, 'Approaches to regional analysis: a synthesis', *Ann. Ass. Am. Geogr.* **54,** 2–11.

Berry, B. J. L., 1964B., 'Cities as systems within systems of cities', *Pap. Reg. Sci. Ass.*, **13,** 147–63.

Berry, B. J. L., 1965, 'Identification of declining regions,' in Wood, W. D. and Thoman, R. S. (eds.), *Areas of economic stress in Canada* (Kingston, Ontario).

Berry, B. J. L., 1966, 'Essays on commodity flows and the spatial structure of the Indian economy', *Res. Pap.* No. **111,** Dept. of Geog., Univ. of Chicago.

Berry, B. J. L., 1967A, *Geography of market centres and retail distribution* (Englewood Cliffs, N.J.).

Berry, B. J. L., 1967B, 'Grouping and regionalizing', in Garrison, W. L. and Marble, D. (eds.), (1967).

Berry, B. J. L. and Baker, A. M., 1968, 'Geographic sampling', in Berry, B. J. L. and Marble, D. (eds.), (1968).

Berry, B. J. L. and Barnum, H. G., 1962, 'Aggregate relations and elemental components of central place systems', *J. reg. Sci.* **4,** No. 1, 35–68.

Berry, B. J. L. and Garrison, W. L., 1958, 'The functional bases of the central place hierarchy', *Econ. Geogr.* **34,** 145–54.

Berry, B. J. L. and Marble, D. (eds.), 1968, *Spatial analysis* (Englewood Cliffs, N.J.).

Berry, B. J. L. and Pred, A., 1961, 'Central place studies: a bibliography of theory and applications', *Reg. Sci. Res. Inst., Biblphy. Ser.* No. **1.**

Beth, E. W., 1965, *Mathematical thought* (Dordrecht).

Bidwell, O. W. and Hole, F. D., 1964, 'Numerical taxonomy and soil classification', *Soil Sci.* **97,** 58–62.

Blache, P. Vidal de la, *1926, Principles of human geography* (English edition, London).

Blalock, H. M., *1964, Causal inferences in non-experimental research* (Chapel Hill).

Blalock, H. M. and Blalock, A., 1959, 'Toward a clarification of system analysis in the social sciences', *Philosophy Sci.* **26,** 84–92.

Blaut, J. M., 1959, 'Microgeographic sampling', *Econ. Geogr.* **35,** 79–88.

Blaut, J. M., 1962, 'Object and relationship', *Prof. Geogr.* **14,** 1–7.

Board, C., 1967, 'Maps as models', in Chorley, R. J. and Haggett, P. (eds.), (1967)·
Boulding, K. E., 1956, *The image* (Ann Arbor).
Boulding, K. E., 1964, 'General systems as a point of view', in Mesarovic, M. D. (ed.), (1964).
Braithwaite, R. B., *1960, Scientific explanation* (Harper torchbooks, New York).
Bridgman, P. W., 1922, *Dimensional analysis* (New Haven).
Brodbeck, M., 1959, 'Models, meaning, and theories', in Gross, L. (ed.), *Symposium on sociological theory* (Evanston).
Broek, J. O. M., 1932, *The Santa Clara Valley, California: a study in landscape changes* (Utrecht).
Bromberger, S., 1963, 'A theory about the theory of theory and about the theory of theories', *Del. Semin. Phil. Sci.* **2**, 79–106.
Brookfield, H., 1964, 'Questions on the human frontiers of geography', *Econ. Geogr.* **40**, 283–303.
Bross, I. D. J., 1953, *Design for decision* (New York).
Brown, L., 1965, 'Models for spatial diffusion research', *Tech. Rep.* No. **3**, Spatial Diffusion Study, Dept. of Geog., Northwestern Univ.
Brown, R., 1963, *Explanation in social science* (London).
Brunhes, J., *1920, Human geography* (English edition: London).
Brush, J. E., 1953, 'The hierarchy of central places in southwestern Wisconsin', *Geogr. Rev.* **43**, 380–402.
Bryson, R. A. and Dutton, J. A., 1967, 'The variance spectra of certain natural series', in Garrison, W. L. and Marble, D. (eds.), 'Quantitative geography, Part II', *NWest. Univ. Stud. Geogr.* No. **14**.
Buck, R. C., 1956, 'On the logic of general behaviour systems theory', *Minn. Stud. Phil. Sci.* **1**, 223–38.
Bunge, M., *1963, Causality: the place of the causal principle in modern science* (Meridian books; New York).
Bunge, W., *1966*, 'Theoretical geography', *Lund Stud. Geogr.*, Series C No. **1** (second edition, Lund).
Burton, I., 1963, 'The quantitative revolution and theoretical geography', *Can. Geogr.* **7**, 151–62.
Campbell, N. R., 1928, *An account of the principles of measurement and calculation* (New York).
Carey, G. W., 1966, 'The regional interpretation of Manhattan population and housing patterns through factor analysis', *Geogrl Rev.* **56**, 551–69.
Carnap, R., 1942, *Introduction to semantics* (Cambridge, Mass.).
Carnap, R., 1950, *Logical foundations of probability* (Chicago).
Carnap, R., 1952, *The continuum of inductive methods* (Chicago).
Carnap, R., 1956, 'The methodological character of theoretical concepts', *Minn. Stud. Phil. Sci.* **1**, 38–76.
Carnap, R., 1958, *Introduction to symbolic logic and its applications* (New York).
Casetti, E., 1964, 'Multiple discriminant functions', and 'Classificatory and regional analysis by discriminant iterations', *Tech. Rep.* **11** and **12**, ONR Research Project, Dept. of Geog., Northwestern University.
Casetti, E., 1966, 'Analysis of spatial association by trigonometric polynomials', *Can. Geogr.* **10**, 199–204.
Cassirer, E., *1957, The philosophy of symbolic forms: vol. 3, the phenomenology of knowledge* (English edition: New Haven).
Cattell, R. B., 1965, 'Factor analysis: an introduction to essentials, Parts I and II', *Biometrics* **21**, 190–215, 405–35.

Cattell, R. B. (ed.), 1966, *Handbook of multivariate experimental psychology* (Chicago).

Caws, P., 1959, 'Definition and measurement in physics', in Churchman, C. W. and Ratoosh, P. (eds.), (1959).

Caws, P., 1965, *The philosophy of science* (Princeton, N.J.).

Chamberlin, T. C., 1897, 'The method of multiple working hypotheses', *J. Geol.* **5,** 837–48.

Chapman, J. D., 1966, 'The status of geography', *Can. Geogr.* **3,** 133–44.

Childe, V. G., 1949, 'Social worlds of knowledge', *L. T. Hobhouse Memorial Trust Lecture* No. **19.**

Chisholm, M. D. I., 1960, 'The geography of commuting', *Ann. Ass. Am. Geogr.* **50,** 187–8, 491–2.

Chisholm, M. D. I., 1962, *Rural settlement and land use* (London).

Chisholm, M. D. I., 1967, 'General systems theory and geography', *Trans. Inst. Br. Geogr.* **42,** 45–52.

Chorley, R. J., 1962, 'Geomorphology and general systems theory', *Prof. Pap., U.S. Geol. Surv.* **500–B.**

Chorley, R. J., 1964, 'Geography and analogue theory', *Ann. Ass. Am. Geogr.* **54,** 127–37.

Chorley, R. J., 1965, 'A re-evaluation of the geomorphic system of W. M. Davis', in Chorley, R. J. and Haggett, P. (eds.), (1965A).

Chorley, R. J., 1967, 'Models in geomorphology', in Chorley, R. J. and Haggett, P. (eds.), (1967).

Chorley, R. J., Dunn, A. J. and Beckinsale, R. P., 1964, *The study of landforms* Vol. I (London).

Chorley, R. J., and Haggett, P. (eds.), 1965A, *Frontiers in geographical teaching* (London).

Chorley, R. J. and Haggett, P., 1965B, 'Trend-surface mapping in geographical research', *Trans. Inst. Br. Geogr.* **37,** 47–67.

Chorley, R. J. and Haggett, P. (eds.), 1967, *Models in geography* (London).

Choynowski, M., 1959, 'Maps based on probabilities', *J. Am. Statist. Ass.* **54,** 385–8.

Christaller, W., *1966, Central places in Southern Germany* (American edition, Englewood Cliffs, N.J.).

Christian, R. R., *1965, Introduction to logic and sets* (New York).

Churchman, C. W., *1948, Theory of experimental inference* (New York).

Churchman, C. W., *1961, Prediction and optimal decision* (Englewood Cliffs, N.J.).

Churchman, C. W. and Ratoosh, P. (eds.), *1959, Measurement; definitions and theories* (New York).

Clark, A. H., 1954, 'Historical geography', in James, P. E. and Jones, C. F. (eds.), (1954).

Clark, K. G. T., 1950, 'Certain underpinnings of our arguments in human geography', *Trans. Inst. Br. Geogr.* **16,** 15–22.

Clark, P. J. and Evans, F. C., 1954, 'Distance to nearest neighbour as a measure of spatial relationships in populations', *Ecology* **35,** 445–53.

Clarkson, G. P. E., 1963, *The theory of consumer demand: a critical appraisal* (Englewood Cliffs, N.J.).

Cliff, A., 1968, 'The neighbourhood effect in the diffusion of innovations', *Trans. Inst. Br. Geogr.* **42,** 75–84.

Cline, M. G., 1949, 'Basic principles of soil classification', *Soil Sci.* **67,** 81–91.

Cochran, W. G., 1953, *Sampling techniques* (New York).

Cohen, R. and Nagel, E., 1934, *An introduction to logic* (New York).

Coleman, J. S., 1964, *Introduction to mathematical sociology* (New York).

Collingwood, R. G., 1946, *The idea of history* (Oxford).

Cooley, W. W. and Lohnes, P. R., 1964, *Multivariate procedures for the behavioral sciences* (New York).

Coombs, C. H., 1964, *A theory of data* (New York).

Coxeter, H. S. M., 1961, *Introduction to geometry* (New York).

Cramér, H., 1954, *Mathematical methods of statistics* (Princeton, N. J.).

Cramér, H., 1955, *The elements of probability theory; and some of its applications* (New York).

Curry, L., 1962A, 'The geography of service centres within towns: the elements of an operational approach', in Norborg, K. (ed.), *I.G.U. Symposium in Urban Geography* (Lund).

Curry, L., 1962B, 'Climatic change as a random series', *Ann. Ass. Am. Geogr.* **52**, 21–31.

Curry, L. 1962C, 'The climatic resources of intensive grassland farming: the Waikato, New Zealand', *Geogrl Rev.* **52**, 174–94.

Curry, L., 1964, 'The random spatial economy: an exploration in settlement theory', *Ann. Ass. Am. Geogr.* **54**, 138–46.

Curry, L., 1966A, 'Chance and Landscape', in House, J. W. (ed.), *Northern Geographical essays* (Newcastle-upon-Tyne).

Curry, L., 1966B, 'Seasonal programming and Bayesian assessment of atmospheric resources', in Sewell, W. R. D. (ed.), 'Human dimensions of weather modification', *Res. Rep.* No. **105**, Dept. of Geog., Univ. of Chicago.

Curry, L., 1967, 'Central places in the random spatial economy', *J. reg. Sci.* **7**, No. 2 (supplement), 217–38.

Dacey, M. F., 1960, 'A note on the derivation of nearest neighbor distances', *J. reg. Sci.* **2**, 81–7.

Dacey, M. F., 1962, 'Analysis of central place and point patterns by a nearest neighbor method', in Norborg, K. (ed.), *I.G.U. Symposium in Urban Geography, 1962* (Lund).

Dacey, M. F., 1963, 'Order neighbor statistics for a class of random patterns in multidimensional space', *Ann. Ass. Am. Geogr.* **53**, 505–15.

Dacey, M. F., 1964A, 'Modified Poisson probability law for point pattern more regular than random', *Ann. Ass. Am. Geogr.* **54**, 559–65.

Dacey, M. F., 1964B, 'A family of density functions of Lösch's measurements on towns distribution', *Prof. Geogr.* **16**, 5–7.

Dacey, M. F., 1964C, 'Individuation in substance and space-time languages', and 'Carnap's classification of rules of designation and positional coordinates', *Memo, GSI* (mfd) **1** and **2**, Reg. Sci., Univ. of Penn. (Philadelphia).

Dacey, M. F., 1965A, 'The geometry of central place theory', *Geogr. Annlr Series B* **47**, 111–24.

Dacey, M. F., 1965B, 'Some observations on a two-dimensional language', *Tech. Rep.* No. **7**, *ONR Task* No. 389–142, Dept. of Geog., Northwestern Univ. (Evanston).

Dacey, M. F., 1966A, 'A probability model for central place location', *Ann. Ass. Am. Geogr.* **56**, 550–68.

Dacey, M. F., 1966B, 'A county-seat model for the areal pattern of an urban system', *Geogrl Rev.* **56**, 527–42.

Dacey, M. F., 1967, 'An empirical study of the areal distribution of houses in Puerto Rico', unpublished paper, Dept. of Geog., Northwestern Univ. (Evanston).

Dacey, M. F., 1968, 'A review on measures of contiguity for two and k-color maps', in Berry, B. J. L. and Marble, D. F. (eds.), (1968).

Dacey, M. F. (undated), 'Some assumptions of geographic mapping', unpublished ms., Dept. of Geog., Northwestern Univ. (Evanston).

Dacey, M. F. and Marble, D. F., 1965, 'Some comments on certain technical aspects of geographic information systems', *Tech. Rep.* No. **2**, *ONR Task* No. 389–142, Dept. of Geog., Northwestern Univ. (Evanston).

Darby, H. C., 1953, 'On the relations of geography and history', *Trans. Inst. Br. Geogr.* **19**, 1–11.

Darby, H. C., 1962, 'The problem of geographical description', *Trans. Inst. Br. Geogr.* **30**, 1–14.

David, F. N. and Barton, D. E., 1962, *Combinatorial chance* (London).

Davis, W. M., *1954, Geographical essays* (Dover, New York).

Deutsch, K. W., *1966, The nerves of government; models of political communication and control* (New York).

Devons, E. and Gluckman, M., 1964, 'Modes and consequences of limiting a field of study', in Gluckman, M. (ed.), *Closed systems and open minds* (Edinburgh).

Dickinson, R. E., 1957, 'The geography of commuting: the Netherlands and Belgium', *Geogrl Rev.* **47**, 521–38.

Dodd, S. C., 1950, 'The interactance hypothesis: a gravity model fitting physical masses and human groups', *Am. Soc. Rev.* **15**, 245–56.

Dodd, S. C., 1953, 'Testing message diffusion in controlled experiments: charting the distance and time factors in the interactance hypothesis', *Am. Soc. Rev.* **18**, 410–16.

Dodd, S. C., 1962, 'How momental laws can be developed in sociology', *Synthese* **14**, 277–99.

Donagan, A., 1964, 'Historical explanation: the Popper-Hempel theory reconsidered', *History and Theory* **3**, 3–26.

Downs, R. M., 1967, 'Approaches to, and problems in, the measurement of geographic space perception', *Semin. Pap. Ser. A.* No. **9**, Univ. of Bristol (Bristol).

Downs, R. M., 1968, 'The role of perception in modern geography', *Semin. Pap. Ser. A* No. **11**, Univ. of Bristol (Bristol).

Dray, W. H., 1957, *Laws and explanation in history* (Oxford).

Dray, W. H., 1964, *Philosophy of history* (Englewood Cliffs, N.J.)

Duncan, O. D., Cuzzort, R. P. and Duncan, B., 1961, *Statistical geography; problems in analyzing areal data* (Glencoe, Ill.).

Dury, G. H. (ed.), 1966, *Essays in geomorphology* (London).

Edwards, A. W. F. and Cavalli-Sforza, L. L., 1965, 'A method for cluster analysis', *Biometrics* **21**, 362–75.

Einstein, A., 1923, *Sidelights of relativity* (New York).

Eisenstadt, S. N., 1949, 'The perception of time and space in a situation of culture-contact', *Jl R. anthrop. Inst.* **79**, 63–8.

Ekman, G., Lindman, R. and William-Olsson, W., 1963, 'A psychophysical study of cartographic symbols', *Geogr. Annlr* **45**, 262–71.

Ellis, B., 1966, *Basic concepts of measurement* (Cambridge).

Eyre, S. R. and Jones, G. R. J., 1966, *Geography as human ecology: methodology by example* (London).

Feller, W., *1957, An introduction to probability theory and its applications*, Vol. **1** (New York).

Fishburn, P. C., 1964, *Decision and value theory* (New York).

Fisher, R. A., *1936, Statistical methods for research workers* (Edinburgh).

Fisher, R. A., 1956, *Statistical methods and scientific inference* (Edinburgh).

Fisher, R. A., *1966, Design of experiments* (Edinburgh).

Forde, C. D., 1934, *Habitat, economy, and society* (London).

Forrester, J. W., 1961, *Industrial dynamics* (Cambridge, Mass.).

Freeman, D., 1966, 'Social anthropology and the scientific study of human behaviour', *Man*, New Series, **1**, 330–42.

Freeman, T. W., 1966, *The geographer's craft* (London).

Galtung, J., 1967, *Theory and methods of social research* (Oslo).

Garner, W. R., 1962, *Uncertainty and structure as psychological concepts* (New York).

Garrison, W. L., 1956, 'Applicability of statistical inference to geographical research', *Geogrl Rev.* **46,** 427–9.

Garrison, W. L., 1957, 'Verification of a location model', *NWest. Univ. Stud. Geogr.* No. **2,** 133–40.

Garrison, W. L. and Marble, D. F., 1957, 'The spatial structure of agricultural activity', *Ann. Ass. Am. Geogr.* **47,** 137–44.

Garrison, W. L. and Marble, D. F., 1965, 'A prolegomenon to the forecasting of transportation development', *Transportation Center Report*, Northwestern Univ. (Evanston).

Garrison, W. L. and Marble, D. F. (eds.), 1967, 'Quantitative geography, Parts I and II', *NWest. Univ. Stud. Geogr.* Nos. **13** and **14.**

George, F. H., 1967, 'The use of models of science', in Chorley, R. J. and Haggett, P. (eds.), (1967).

Getis, A., 1963, 'The determination of the location of retail activities with the use of a map transformation', *Econ. Geogr.* **39,** 14–22.

Getis, A., 1964, 'Temporal analysis of land use patterns with the use of nearest neighbor and quadrat methods', *Ann. Ass. Am. Geogr.* **54,** 391–9.

Getis, A., 1967A, 'A method for the study of sequences in geography', *Trans. Inst. Br. Geogr.* **42,** 87–92.

Getis, A., 1967B, 'Occupancy theory and map pattern analysis', *Semin. Pap. Ser. A.* No. **1,** Dept. of Geog., Univ. of Bristol.

Ghiselli, E. E., 1964, *Theory of psychological measurement* (New York).

Glacken, C. J., 1967, *Traces on the Rhodian Shore* (Berkeley and Los Angeles).

Golledge, R., 1967, 'A conceptual model of the market decision process', *J. reg. Sci.* **7,** No. 2 (supplement), 239–58.

Goodall, D. W., 1954, 'Objective methods for the classification of vegetation: III, and essay in the use of factor analysis', *Aust. J. Bot.* **2,** 304–24.

Goodman, L., 1959, 'Some alternatives to ecological correlation', *Am. J. Sociol.* **64,** 610–25.

Gould, P., 1966, 'On mental maps', *Mich. Int-Com. Math. Geogr.* No. **9.**

Gouldner, A. W., 1959, 'Reciprocity and autonomy in functional theory', in Gross, L. (ed.), *Symposium on sociological theory* (Evanston).

Gower, J. C., 1967, 'Multivariate analysis and multidimensional geometry', *The Statistician* **17,** 13–28.

Gregg, J. R. and Harris, F. T. C. (eds.), 1964, *Form and strategy in science* (Dordrecht).

Greig-Smith, P., *1964, Quantitative plant ecology* (London).

Grigg, D. B., 1965, 'The logic of regional systems', *Ann. Ass. Am. Geogr.* **55,** 465–91.

Grigg, D. B., 1967, 'Regions, models, and classes', in Chorley, R. J. and Haggett, P. (eds.), (1967).

Grünbaum, A., 1963, *Philosophical problems of space and time* (New York).

Gurvitch, G., 1964, *The spectrum of social time* (Dordrecht).

Hacking, I., 1965, *The logic of statistical inference* (Cambridge).

Hadden, J. K. and Borgatta, E. F., 1965, *American cities: their social characteristics* (Chicago).

Hagen, E., 1961, 'Analytical models in the study of social systems', *Am. J. Sociol.* **67,** 144–51.

Hägerstrand, T., 1953, *Innovationsförloppet ur korologisk synpunkt* (Lund).

Hägerstrand, T., 1957, 'Migration and area', in Hannerberg, D., Hägerstrand, T. and Odeving, B. (eds.), 'Migration in Sweden—a symposium', *Lund Stud. Geogr. Ser. B,* No. **13.**

Hägerstrand, T., 1963, 'Geographic measurements of migration', in Sutter, J. (ed.), *Human displacements: entretiens de Monaco en sciences humaines* (Monaco).

Hägerstrand, T., 1967, 'The computer and the geographer', *Trans. Inst. Br. Geogr.* **42,** 1–19.

Haggett, P., 1965A, *Locational analysis in human geography* (London).

Haggett, P., 1965B, 'Scale components in geographical problems', in Chorley, R. J. and Haggett, P. (eds.), (1965A).

Haggett, P., 1967, 'Network models in geography', in Chorley, R. J. and Haggett, P. (eds.), (1967).

Hagood, M. J., 1943, 'Statistical methods for delineation of regions applied to data on agriculture and population', *Social Forces* **21,** 288–97.

Haight, F. A., 1967, *Handbook of the Poisson distribution* (New York).

Hallowell, A. I., 1942, 'Some psychological aspects of measurement among the Saulteaux', *Am. Anthrop.* **44,** 62–77.

Hallowell, A. I., 1955, *Culture and experience* (Philadelphia).

Hansen, M. H., Hurwitz, W. N. and Madow, W. G., 1953, *Sample survey methods and theory* Vol. 1 (New York).

Hanson, N. R., *1965, Patterns of discovery* (Cambridge).

Harbaugh, J. W. and Preston, F. W., 1968, 'Fourier series analysis in geology', in Berry, B. J. L. and Marble, D. (eds.), *Spatial analysis* (Englewood cliffs, N.J.).

Harris, B., 1964, 'A note on the probability of interaction at a distance', *J. reg. Sci.* **5,** No. **2,** 31–5.

Harris, C. D., 1954, 'The market as a factor in the localization of industry in the United States', *Ann. Ass. Am. Geogr.* **44,** 315–48.

Hartshorne, R., 1939, *The nature of geography* (Chicago).

Hartshorne, R., 1955, 'Exceptionalism in geography re-examined', *Ann. Ass. Am. Geogr.* **45,** 205–44.

Hartshorne, R., 1958, 'The concept of geography as a science of space, from Kant and Humboldt to Hettner', *Ann. Ass. Am. Geogr.* **48,** 97–108.

Hartshorne, R., 1959, *Perspective on the nature of geography* (Chicago).

Harvey, D. W., 1966A, 'Theoretical concepts and the analysis of land use patterns', *Ann. Ass. Am. Geogr.* **56,** 361–74.

Harvey, D. W., 1966B, 'Geographical processes and point patterns: testing models of diffusion by quadrat sampling', *Trans. Inst. Br. Geogr.* **40,** 81–95.

Harvey, D. W., 1967A, 'Models of the evolution of spatial patterns in human geography', in Chorley, R. J. and Haggett, P. (eds.), (1967).

Harvey, D. W., 1967B, 'Behavioural postulates and the construction of theory in human geography', *Semin. Pap., Ser. A.* No. **6,** Dept. of Geog., Univ. of Bristol (Bristol). (Forthcoming in *Geographica Polonica*.)

Harvey, D. W., 1968A, 'Some methodological problems in the use of the Neyman Type A and negative binomial probability distributions in the analysis of spatial series', *Trans. Inst. Br. Geogr.* **43,** 85–95.

Harvey, D. W., 1968B, 'Pattern, process, and the scale problem in geographical research', *Trans. Inst. Br. Geogr.* **45,** 71–78.

Hempel, C. G., 1949, 'Geometry and empirical science', in Feigl, H. and Sellars, W. (eds.), *Readings in philosophical analysis* (New York).

Hempel, C. G., 1951, 'General systems theory and the unity of science', *Hum. Biol.* **23,** 313–22.

Hempel, C. G., 1959, 'The logic of functional analysis', in Gross, L. (ed.), *Symposium on sociological theory* (Evanston).

Hempel, C. G., 1965, *Aspects of scientific explanation* (New York).

Henshall, J. D. and King, L. J., 1966, 'Some structural characteristics of peasant agriculture in Barbados', *Econ. Geogr.* **42,** 74–84.

Hesse, M. B., 1963, *Models and analogies in science* (London).

Hilbert, D., *1962, Foundations of geometry* (English edition, La Salle).

Hilbert, D., and Cohn-Vossen, S., *1952, Geometry and the imagination* (English edition, New York).

Holmes, J., 1967, 'Problems in location sampling', *Ann. Ass. Am. Geogr.* **57,** 757–80.

Homans, C., 1950, *The human group* (London).

Howard, R. N., 1966, 'Classifying a population into homogeneous groups', in Lawrence, J. R. (ed.). *Operational research in the social sciences* (London).

Howard, I. P. and Templeton, W. B., 1966, *Human spatial orientation* (New York).

Hughes, H. S., 1959, *Consciousness and society* (London).

Huxley, J., 1963, 'The future of man: evolutionary aspects', in Wolsterholme, G. (ed.), *Man and his future* (London).

Imbrie, J., 1963, 'Factor and vector analysis in programs for analyzing geologic data', *Tech. Rep.* **6,** *ONR Project* 389–135, Dept. Geol., Columbia Univ., New York.

Isard, W., 1956, *Location and the space economy* (New York).

Isard, W., 1960, *Methods of regional analysis* (New York).

Isard, W. and Dacey, M. F., 1962, 'On the projection of individual behavior in regional analysis, Parts I and II', *J. reg. Sci.* **4,** No. **1,** 1–34, No. **2,** 51–83.

Iwanicka-Lyra, E., 'The delimitation of the Warsaw agglomeration', *Pol. Acad. Sci. Stud.* **17,** 163–72.

James, P. E. and Jones, C. F., 1954, *American geography: inventory and prospect* (Syracuse).

Jammer, M., 1954, *Concepts of space* (Cambridge, Mass.).

Jarvie, J. C., 'Limits to functionalism and alternatives to it in anthropology', in Martindale, D. (ed.), *Functionalism in the social sciences* (Philadelphia).

Jeffrey, R. C., 1965, *The logic of decision* (New York).

Jenks, G. F., 1963, 'Generalization in statistical mapping', *Ann. Ass. Am. Geogr.* **53,** 15–26.

Johnston, J., 1963, *Econometric methods* (New York).

Jones, E., 1956, 'Cause and effect in human geography', *Ann. Ass. Am. Geogr.* **46,** 369–77.

Joynt, C. B. and Rescher, N., 1961, 'The problem of uniqueness in history', *History and Theory* **1,** 150–62.

Kansky, K. J., 1963, 'Structure of transportation networks', *Res. Pap.* No. **84,** Dept. of Geog., Univ. of Chicago.

Kao, R., 1963, 'The use of computers in the processing and analysis of geographic information', *Geogrl Rev.* **53,** 530–47.

Kao, R., 1967, 'Geometric projections in system studies', in Garrison, W. L. and Marble, D. (eds.), 'Quantitative geography, Part II', (1967).

Kaplan, A., 1964, *The conduct of inquiry* (San Francisco).

Kates, R. W., 1962, 'Hazard and choice perception in flood plain management', *Res. Pap.* No. **78,** Dept. of Geog., Univ. of Chicago.

Katz, J. J., 1962, *The problem of induction and its solution* (Chicago).

Kemeny, J. G., 1959, *A philosopher looks at science* (Princeton, N.J.).

Kemeny, J. G., Snell, J. L. and Thompson, G. L., 1956, *Introduction to finite mathematics* (Englewood Cliffs, N.J.).

Kendall, M. G., 1939, 'The geographical distribution of crop productivity in England', *Jl R. Statist. Soc.* **102,** 21–48.

Kendall, M. G., 1957, *A course in multivariate analysis* (London).

Kendall, M. G., 1961, 'Natural law in the social sciences', *Jl R. Statist. Soc. Ser. A* **124,** 1–16.

Kendall, M. G. and Moran, P. A. P., 1963, *Geometrical probability* (London).

Kendall, M. G. and Stuart, A., 1963–7, *The advanced theory of statistics,* Vol. I (2nd edition, 1963), Vol. II (2nd edition, 1967), and Vol. III (1966), (London).

Keynes, J. M., *1962, A treatise on probability* (Harper torchbooks, New York).

King, C. A. M., 1961, *Beaches and coasts* (London).

King, C. A. M., 1966, *Techniques in geomorphology* (London).

King, L. J., 1961, 'A multivariate analysis of the spacing of urban settlements in the United States', *Ann. Ass. Am. Geogr.* **51,** 222–33.

King, L. J., 1966, 'Cross-sectional analysis of Canadian urban dimensions: 1951–1961', *Can. Geogr.* **10,** 205–24.

Kinsman, B., 1965, *Wind waves* (Englewood Cliffs, N.J.).

Kirk, W., 1951, 'Historical geography and the concept of the behavioural environment', *Indian Geogr. J.,* Silver Jubilee Edition.

Kish, L., 1965, *Survey sampling* (New York).

Klein, F., *1939, Elementary mathematics from an advanced standpoint: geometry* (English edition, New York).

Klir, J. and Valach, M., *1967, Cybernetic modelling* (English edition, London).

Kluckhohn, C., 1954, 'Culture and behaviour', in Lindzey, G. (ed.), *Handbook of social psychology, Vol. II* (Reading, Mass.).

Koestler, A., 1964, *The act of creation* (London).

Koopman, B. O., 1940, 'The axioms and algebra of intuitive probability', *Bull. Am. Math. Soc.* **46,** 763–74.

Koopmans, T. C., 1957, *Three essays on the state of economic science* (New York).

Körner, S., 1955, *Conceptual thinking* (Cambridge).

Körner, S., 1960, *The philosophy of mathematics* (London).

Körner, S., 1966, *Experience and theory* (London).

Krumbein, W. C., 1960, 'The geological population as a framework for analyzing numerical data in geology', *Lpool Manch Geol. J.* **2,** 341–68.

Krumbein, W. C. and Graybill, F. A., 1965, *An introduction to statistical models in geology* (New York).

Kruskal, J. B., 1964, 'Multidimensional scaling by optimizing goodness of fit to a non-metric hypothesis', *Psychometrika* **29,** 1–28.

Kuhn, A., *1966, The study of society: a multidisciplinary approach* (London).
Kuhn, T. S., 1962, *The structure of scientific revolutions* (Chicago).
Kulldorff, G., 1955, 'Migration probabilities', *Lund Stud. Geogr.*, Ser. B, No. **14.**
Kyburg, H. E. and Smokler, H. E. (eds.), 1964, *Studies in subjective probability* (New York).
Langhaar, H. L., 1951, *Dimensional analysis and theory of models* (New York).
de Laplace, *1951, A philosophical essay on probabilities* (English edition, Dover, New York).
Leach, E. R., 1957, 'The epistemological background to Malinowski's empiricism', in Firth, R. (ed.), *Man and culture: an evaluation of the work of Bronislaw Malinowski* (London).
Lee, T. R., 1963–4, 'Psychology and living space', *Trans. Bartlett Soc.* **2,** 9–36.
Lehman, H., 1965, 'Functional explanation in biology', *Philosophy Sci.* **32,** 1–20.
Lenneberg, E. H., 1962, 'The relationship of language to the formation of concepts', *Synthese* **14,** 103–9.
Leopold, L. B. and Langbein, W. B., 1962, 'The concept of entropy in landscape evolution', *Prof. Pap.* **500A,** U.S. Geol. Survey.
Leopold, L. B. and Wolman, M. G., 1957, 'River channel patterns: braided, meandering, and straight', *Prof. Pap.* **282-B,** U.S. Geol. Survey.
Leopold, L. B., Wolman, M. G. and Miller, J. P., 1964, *Fluvial processes in geomorphology* (San Francisco).
Levi-Strauss, C., *1963, Structural anthropology* (English edition, New York).
Lewis, P. W., 1965, 'Three related problems in the formulation of laws in geography', *Prof. Geogr.* **17,** No. 5, 24–7.
Lewis, G. M., 1966, 'Regional ideas and reality in the Cis-Rocky Mountain West', *Trans. Inst. Br. Geogr.* **38,** 135–50.
Lindley, D. V., 1965, *Introduction to probability and statistics*, Vols. I and II (Cambridge).
Lösch, A., *1954, The economics of location* (English edition, New Haven).
Lovell, K., 1961, *The growth of basic mathematical and scientific concepts in children* (London).
Lowenthal, D., 1961, 'Geography, experience, and imagination: towards a geographical epistemology', *Ann. Ass. Am. Geogr.* **51,** 241–60.
Lowenthal, D. (ed.), 1967, 'Environmental perception and behavior', *Res. Pap.* No. **109,** Dept. of Geog., Univ. of Chicago.
Lowenthal, D. and Prince, H. C., 1964, 'The English landscape', *Geogrl Rev.* **54,** 304–46.
Lowry, I., 1964, 'Model of metropolis', *Rand Corporation Memo* RM–4035–RC.
Lowry, I. S., 1965, 'A short course in model design', *Jl Am. Inst. Planners* **31,** No. 2, 158–66.
Luce, R. D. and Raiffa, H., 1957, *Games and decisions* (New York).
Lukermann, F., 1965, 'The "calcul des probabilités" and the École française de Géographie', *Can. Geogr.* **9,** 128–37.
Mahalanobis, P. C., 1927, 'Analysis of race mixture in Bengal', *Jnl Ass. Soc. Bengal* **23,** 301–33.
Mahalanobis, P. C., 1936, 'On the generalized distance in statistics', *Proc. Natn. Inst. Sci.* (India), **12,** 49.
Mandelbaum, M., 1961, 'Historical explanation: the problem of covering laws', *History and Theory* **1,** 229–42.
Martin, A. F., 1951, 'The necessity for determinism', *Trans. Inst. Br. Geogr.* **17,** 1–12.

Martindale, D., 1965, *Functionalism in the social sciences* (Philadelphia).

Massarik, F., 1965, 'Magic, models, man, and the cultures of mathematics', in Massarik, F. and Ratoosh, P. (eds.), *Mathematical explorations in behavioral science* (Homewood, Illinois).

Matalas, N. C. and Reiher, B. J., 1967, 'Some comments on the use of factor analysis', *Wat. Resour. Res.* **3**, 213–23.

Matérn, B., 1960, 'Spatial variation', *Meddelanden från Statens Skogsforsknings-institut* **5**, No. 3 (Stockholm).

McCarty, H. H., 1954, 'An approach to the theory of economic geography', *Econ. Geogr.* **30**, 95–101.

McCarty, H. H., Hook, J. C. and Knos, D. S., 1956, *The measurement of association in industrial geography* (Dept. of Geog., Univ. of Iowa).

McConnell, M., 1966, 'Quadrat methods in map analysis', *Discuss. Pap.* No. **3**, Dept. of Geog., Univ. of Iowa.

McCord, J. R. and Moroney, R. M., 1964, *An introduction to probability theory* (New York).

Medvedkov, Y. V., 1967, 'The concept of entropy in settlement pattern analysis', *Pap. Reg. Sci. Ass.* **18**, 165–8.

Meier, R. L., 1962, *A communications theory of urban growth* (Cambridge, Mass.).

Melluish, R. K., 1931, *An introduction to the mathematics of map projections* (Cambridge).

Merton, R. K., *1957, Social theory and social structure* (Glencoe, Ill.).

Mesarovic, M. D. (ed.), 1964, *Views on general systems theory* (New York).

Meyerhoff, H., *1960, Time in literature* (Berkeley and Los Angeles).

Mill, J. S., *1950, Philosophy of scientific method* (edited, Hafner editions, New York).

Miller, J. G., 1965, 'Living systems: basic concepts', *Behavl Sci.* **10**, 193–237.

Miller, R. L. and Kahn, J. S., 1962, *Statistical analysis in the geological sciences* (New York).

Monkhouse, F. J., 1965, *A dictionary of geography* (London).

Montefiore, A. C. and Williams, W. W., 1955, 'Determinism and possibilism', *Geogrl Stud.* **2**, 1–11.

Moore, W. G., *1967, A dictionary of geography* (London).

Morgenstern, O., *1965, On the accuracy of economic observations* (Princeton, N.J.).

Moser, C. A., 1958, *Survey methods in social investigation* (London).

Moser, C. A. and Scott, W., 1961, *British towns: a statistical study of their social and economic differences* (London).

Nagel, E., 1939, 'Principles of the theory of probability', *Internat. Encyclopedia of Unified Sci.* **1**, No. 6.

Nagel, E., 1949, 'The meaning of reduction in the natural sciences', in Stouffer, R. C. (ed.), *Science and civilisation* (Madison).

Nagel, E., 1961, *The structure of science* (New York).

N.A.S. (National Academy of Sciences), 1965, 'The science of geography', *Natn. Res. Coun.—ad hoc* committee on geography (Washington).

Neyman, J., 1939, 'On a new class of "contagious" distributions, applicable in entomology and bacteriology', *Ann. Math. Statist.* **10**, 35–57.

Neyman, J., 1950, *First course in probability and statistics* (New York).

Neyman, J., 1960, 'Indeterminism in science and new demands on statisticians', *J. Am. Statist. Ass.* **55**, 625–39.

Nordbeck, S., 1965, 'The Law of allometric growth', *Mich. Int-Com. Math. Geogr.* No. **7**.

Nowak, S., 1960, 'Some problems of causal interpretation of statistical relation-ships', *Philosophy Sci.* **27**, 23–38.

Nunnally, J. C., 1967, *Psychometric theory* (New York).

Nye, J. F., 1952, 'The mechanics of glacier flow', *Jnl Glaciology* **2**, 82–93.

Nystuen, J. D., 1963, 'Identification of some fundamental spatial concepts', *Pap. Mich. Acad. Sci., Arts and Letters*, **48**, 373–84.

Oakeshott, M., 1933, *Experience and its modes* (London).

Obeyesekere, G., 1966, 'Methodological and philosophical relativism', *Man*, New Series, **1**, 368–74.

Olsson, G., 1965A, 'Distance and human interaction: a bibliography and review', *Reg. Sci. Res. Inst., Biblphy Ser.* No. **2**.

Olsson, G., 1965B, 'Distance and human interaction: a migration study', *Geogr. Annlr Ser. B.* **47**, 3–43.

Olsson, G. 1967, 'Geography 1984', *Semin. Pap., Ser. A.* No. **8**, Dept. of Geog., Univ. of Bristol.

Olsson, G. and Persson, A., 'The spacing of central places in Sweden', *Pap. Reg. Sci. Ass.* **12**, 87–93.

Orcutt, G. H., Greenberger, M., Korbel, J. and Rivlin, A., 1961; *Microanalysis of socioeconomic systems* (New York).

Pahl, R. E., 1967, 'Sociological models in geography', in Chorley, R. J. and Haggett, P. (eds.), (1967).

Parsons, T., 1949, *The structure of social action* (Glencoe).

Parsons, T., 1951, *The social system* (London).

Parzen, E., 1960, *Modern probability theory and its applications* (New York).

Pearson, E. S., 1962, 'Contribution', in Savage, L. J., *et al.*, (1954).

Philbrick, A. K., 1957, 'Principles of areal functional organization in regional human geography', *Econ. Geogr.* **33**, 299–336.

Piaget, J.,*1930*, *The child's conception of physical causality* (English edition, London).

Piaget, J. and Inhelder, B., 1956, *The child's conception of space* (English edition, London).

Plackett, R. L., 1966, 'Current trends in statistical inference', *Jl R. statist. Soc. Ser. A* **129**, 249–67.

Platt, R. S., 1948, 'Determinism in geography', *Ann. Ass. Am. Geogr.* **38**, 126–32.

Poincaré, H., *1952*, *Science and hypothesis* (Dover, New York).

Popper, K., 1952, *The open society and its enemies*, Vol. II (London).

Popper, K., 1957, *The poverty of historicism* (London).

Popper, K., 1963, *Conjectures and refutations* (London).

Popper, K., *1965*, *The logic of scientific discovery* (Harper torchbooks, New York).

Pred, A. R., 1966, *The spatial dynamics of U.S. urban-industrial growth* (Cambridge, Mass.).

Pred, A., 1967, *'Behavior and location, Part I'*, Lund Stud. Geogr. Ser. B, No. 27.

Quastler, H., 1965, 'General principles of systems analysis', in Waterman, T. H. and Morowitz, H. J. (eds.), *Theoretical and mathematical biology* (New York).

Radcliffe-Brown, A. R., 1952, *Structure and function in primitive society* (London).

Radcliffe-Brown, A. R., 1957, *A natural science of society* (Glencoe, Ill.).

Raisz, E., 1948, *General cartography* (New York).

Ramsey, F. P., *1960*, *The foundations of mathematics, and other logical essays* (edited by R. B. Braithwaite, Paterson, N.J.).

Ramsey, I. T., 1964, *Models and mystery* (London).

Rao, C. R., 1948, 'The utilisation of multiple measurement in problems of biological classification', *Jl R. statist. Soc., Ser. B*, 159–203.

Rao, C. R., 1952, *Advanced statistical methods in biometric research* (New York).

Rapoport, A., 1953, *Operational philosophy* (New York).

Rapoport, A., 1960, *Fights, games, and debates* (Ann Arbor).

Reichenbach, H., 1949, *The theory of probability* (Berkeley and Los Angeles).

Reichenbach, H., 1956, *The direction of time* (Berkeley and Los Angeles).

Reichenbach, H., *1958*, *The philosophy of space and time* (Dover, New York).

Rescher, N., 1964, 'The stochastic revolution and the nature of scientific explanation', *Synthese* **14**, 200–15.

Reynolds, R. B., 1956, 'Statistical methods in geographical research', *Geogrl Rev.* **46**, 129–32.

Richardson, L. F., 1939, 'Generalized foreign policy', *Br. J. Psychol.*, *Monographs Supplements* **23**.

Robbins, L., 1932, *An essay on the nature and significance of economic science* (London).

Roberts, F. S. and Suppes, P., 1967, 'Some problems in the geometry of visual perception', *Synthese* **17**, 173–201.

Robinson, A. H., 1952, *The look of maps* (Madison).

Robinson, A. H., 1956, 'The necessity of weighting values in correlation of area data', *Ann. Ass. Am. Geogr.* **46**, 233–6.

Robinson, A. H., *1960*, *Elements of cartography* (New York).

Robinson, A. H., 1965, 'The potential contribution of cartography in liberal education', in *Ass. Am. Geogr.*, Commission on College Geography, *Geography in undergraduate education* (Washington).

Rogers, A., 1965, 'A stochastic analysis of the spatial clustering of retail establishments', *J. Am. Statist. Ass.* **60**, 1094–103.

Rogers, E. M., 1962, *Diffusion of innovations* (New York).

Rosen, R., 1967, *Optimality principles in biology* (London).

Rostow, W. W., 1960, *The stages of economic growth* (Cambridge).

Rudner, R. S., 1966, *Philosophy of social science* (Englewood Cliffs, N.J.).

Russell, B., 1914, *Our knowledge of the external world* (London).

Russell, B., 1948, *Human knowledge: its scope and limits* (New York).

Ryle, G., 1949, *The concept of mind* (London).

Saarinen, T. F., 1966, 'Perception of the drought hazard on the Great Plains', *Res. Pap.* No. **106**, Dept. of Geog., Univ. of Chicago.

Sauer, C. O., *1963*, *Land and life* (edited by Leighley, J. B., Berkeley).

Savage, L. J., 1954, *The foundations of statistics* (New York).

Sawyer, W. W., 1955, *Prelude to mathematics* (Harmondsworth, Middx).

Schaefer, F. K., 1953, 'Exceptionalism in geography: a methodological examination', *Ann. Ass. Am. Geogr.* **43**, 226–49.

Scheidigger, A. E., 1961, *Theoretical geomorphology* (Heidelberg).

Schneider, M., 1959, 'Gravity models and trip distribution theory', *Pap. Reg. Sci. Ass.* **5**, 51–8.

Schnore, L. F., 1961, 'Geography and human ecology', *Econ. Geogr.* **37**, 207–217.

Segall, M. H., Campbell, D. T. and Herskovits, M. J., 1966, *The influence of culture on visual perception* (Indianapolis).

Shelly, M. W. and Bryan, G. L., 1964, *Human judgments and optimality* (New York).

Siegel, S., 1956, *Nonparametric statistics for the behavioral sciences* (New York).

Simon, H. A., 1953, 'Causal ordering and identifiability', in Hood, W. C. and Koopmans, T. C. (eds.), *Studies in econometric method* (New York).

Simon, H. A., 1957, *Models of man* (New York).

Simpson, G. G., 1961, *Principles of animal taxonomy* (New York).

Simpson, G. G., 1963, 'Historical science', in Albritton, C. C. (ed.), *The fabric of geology* (Reading, Mass.).

Sinnhuber, K. A., 1954, 'Central Europe—Mitteleuropa—Europe Centrale', *Trans. Inst. Br. Geogr.* **20,** 15–39.

Skilling, H., 1964, 'An operational view', *Am. Scient.* **52,** 388A–96A.

Smart, J. J., 1959, 'Can biology be an exact science', *Synthese* **11,** 359–68.

Smith, C. T., 1965, 'Historical geography: current trends and prospects', in Chorley, R. J. and Haggett, P. (eds.), (1965A).

Smith, R. H. T., 1965A, 'Method and purpose in functional town classification', *Ann. Ass. Am. Geogr.* **55,** 539–48.

Smith, R. H. T., 1965B, 'The functions of Australian towns', *Tijdschr. econ. soc. Geogr.* **56,** 81–92.

Sokal, R. R. and Sneath, P. H. A., 1963, *Principles of numerical taxonomy* (San Francisco).

Sorre, M., 1962, 'The role of historical explanation in human geography', in Wagner, P. L. and Mikesell, M. W. (eds.), *Readings in cultural geography* (Chicago).

Spate, O. H. K., 1952, 'Toynbee and Huntington: a study in determinism', *Geogrl J.* **118,** 406–24.

Spate, O. H. K., 1957, 'How determined is possibilism', *Geogrl Stud.* **4,** 3–12.

Spate, O. H. K., 1960, 'Quantity and quality in geography', *Ann. Ass. Am. Geogr.* **50,** 377–94.

Spector, M., 1965, 'Models and theories', *Br. J. Phil. Sci.* **16,** 121–42.

Sprout, H. and Sprout, M., 1965, *The ecological perspective on human affairs* (Princeton, N.J.).

Stamp, L. D., 1961, *A glossary of geographical terms* (London).

Stebbing, L. S., 1961, *A modern elementary logic* (London).

Stevens, S. S., 1935, 'The operational basis of psychology', *Am. J. Psychol.* **47,** 323–30.

Stevens, S. S., 1959, 'Measurement, psychophysics, and utility', in Churchman, C. W. and Ratoosh, P. (eds.), (1959).

Stewart, J. Q., 1948, 'Demographic gravitation: evidence and applications', *Sociometry* **11,** 31–57.

Stoddart, D. R., 1965, 'Geography and the ecological approach', *Geography* **50,** 242–51.

Stoddart, D. R., 1966, 'Darwin's impact on geography', *Ann. Ass. Am. Geogr.* **56,** 683–98.

Stoddart, D. R., 1967A, 'Organism and ecosystem as geographical models', in Chorley, R. J. and Haggett, P. (1967).

Stoddart, D. R., 1967B, 'Growth and structure of geography', *Trans. Inst. Br. Geogr.* **41,** 1–19.

Stoll, R. R., 1961, *Sets, logic, and axiomatic theories* (San Francisco).

Stone, R., 1960, 'A comparison of the economic structure of regions based on the concept of distance', *J. reg. Sci.* **2,** No. 1, 1–20.

Strotz, R. H. and Wold, H., 1960, 'A triptych on causal systems', *Econometrica* **28,** 417–63.

Stuart, A., 1962, *Basic ideas of scientific sampling* (London).

Suppes, P., 1961, 'A comparison of the meaning and uses of models in mathematics and the empirical sciences', in *The concept and role of the model in mathematics and natural and social sciences* (Synthese library, Dordrecht).

Swanson, J. W., 1967, 'On models', *Br. J. Phil. Sci.* **17,** 297–311.

Tarski, A., *1965, Introduction to logic* (New York).

Taylor, G., 1937, *Environment, race and nation* (Toronto).

Taylor, G., 1951, *Geography in the twentieth century* (London).

Theil, H., 1967, *Economics and information theory* (Amsterdam).

Theodorsen, G. A., 1961, *Studies in human ecology* (New York).

Thomas, E. N., 1962, 'The stability of distance-population-size relationships for Iowa towns from 1900–1950', in Norborg, K. (ed.), *I.G.U. Symposium in urban geography* (Lund).

Thomas, E. N. and Anderson, D. L., 1965, 'Additional comments on weighting values in correlation analysis of areal data', *Ann. Ass. Am. Geogr.* **55**, 492–505.

Thompson, d'Arcy W., *1961*, *On growth and form* (abridged edition, Cambridge).

Tissot, M. A., 1881, *Mémoire sur la représentation des surfaces* (Paris).

Tobler, W., 1961, 'Map transformations of geographic space', *Doctoral dissertation*, Dept. of Geog., Univ. of Washington.

Tobler, W., 1963, 'Geographic area and map projections', *Geogrl Rev.* **53**, 59–78.

Tobler, W., 1964, 'Automation in the preparation of thematic maps', *Jl Brit. Cartographic Soc.*, 1–7.

Tobler, W., 1966A, 'On geography and geometry', *Unpublished ms.* Dept. of Geog., Univ. of Michigan (Ann Arbor).

Tobler, W., 1966B, 'Numerical map generalization', *Mich. Int-Com. Math. Geogr.* No. **8**.

Torgerson, W. S., 1958, *Theory and methods of scaling* (New York).

Torgerson, W. S., 1965, 'Multidimensional scaling of similarity', *Psychometrika* **30**, 379–93.

Toulmin, S., 1960A, *Reason in ethics* (Cambridge).

Toulmin, S., 1960B, *The philosophy of science* (New York).

Toulmin, S. and Goodfield, J., 1965, *The discovery of time* (London).

Tukey, J. W., 1962, 'The future of data analysis', *Ann Math. Statists* **33**, 1–67.

Tuller, A., 1966, *An introduction to geometries* (Princeton, N.J.).

Van Paassen, C., 1957, *The classical tradition of geography* (Groningen).

Vining, R., 1955, 'A description of certain spatial aspects of an economic system', *Econ. Devt. and Cult. Change* **3**, 147–95.

Von Bertalanffy, L., 1951, 'An outline of genera systems theory', *Br. J. Phil. Sci.* **1**, 134–65.

Von Bertalanffy, L., 1962, 'General systems theory: a critical review', *Gen. Syst.* **7**, 1–20.

Von Neumann, J. and Morgenstern, O., *1964*, *Theory of games and economic behavior* (Wiley science editions, New York).

Wallis, J. R., 1965, 'Multivariate statistical methods in hydrology—a comparison using data of known functional relationship', *Wat. Resour. Res. Pap.* **1**, 447–61.

Ward, J. H., 1963, 'Hierarchical grouping to optimize an objective function', *J. Am. Statist. Ass.* **58**, 236–44.

Warntz, W., 1965, 'A note on surfaces and paths and applications to geographical problems', *Mich. Int-Com. Math. Geogr.* No. **6**.

Warntz, W., 1967, 'Global science and the tyranny of space', *Pap. Reg. Sci. Soc.* **19**, 7–19.

Watkins, J. W. N., 1952, 'Ideal types and historical explanation', *Br. J. Phil. Sci.* **3**, 22–43.

Watson, J. W., 1955, 'Geography: a discipline in distance', *Scott. Geogr. Mag.* **71**, 1–13.

Watson, R. A., 1966, 'Is geology different?' *Philosophy Sci.* **33**, 172–85.

Webb, E. J., Campbell, D. T., Schwartz, R. D. and Sechrest, L., 1966, *Unobtrusive measures: non-reactive research in the social sciences* (Chicago).

Weber, M., *1949*, *The methodology of the social sciences* (English edition, Glencoe, Ill.).

Whitehead, A. N. and Russell, B., 1908–11, *Principia mathematica* (Three volumes, Oxford).

Whittlesey, D., 1929, 'Sequent occupance', *Ann. Ass. Am. Geogr.* **19**, 162–5.

Wiener, N., *1961*, *Cybernetics* (Cambridge, Mass.).

Wilson, A. G., 1967, 'A statistical theory of spatial distribution models', *Transpn. Res.* **1**, 253–269.

Wilson, E. B., 1955, 'Review of scientific explanation', *J. Am. Statist. Ass.* **50**, 1354–7.

Wilson, N. L., 1955, 'Space, time, and individuals', *J. Phil.* **52**, 589–98.

Winch, P., 1958, *The idea of a social science* (London).

Wisdom, J. O., 1952, *Foundations of inference in natural science* (London).

Wold, H., 1954, 'Causality and econometrics', *Econometrica* **22**, 162–77.

Wold, H. and Jureen, L., 1953, *Demand analysis* (New York).

Woldenberg, M. J. and Berry, B. J. L., 1967, 'Rivers and central places: analogous systems?' *J. reg. Sci.* **7**, No. 2, 129–39.

Wolpert, J., 1964, 'The decision process in spatial context', *Ann. Ass. Am. Geogr.* **54**, 537–58.

Wolpert, J., 1965, 'Behavioral aspects of the decision to migrate', *Pap. Reg. Sci. Ass.* **15**, 159–69.

Woodger, J. H., 1937, *The axiomatic method in biology* (Cambridge).

Wooldridge, S. W., 1951, 'The progress of geomorphology', in Taylor, G. (ed.), (1951).

Wooldridge, S. W., 1956, *The geographer as scientist* (London).

Wooldridge, S. W. and East, W. G., 1951, *The spirit and purpose of geography* (London).

Workman, R. W., 1964, 'What makes an explanation', *Philosophy Sci.* **31**, 241–54.

Wright, J. K., 1966, *Human nature in geography* (Cambridge, Mass.).

Wrigley, E. A., 1965 'Changes in the philosophy of geography', in Chorley. R. J. and Haggett, P. (eds.), (1965A).

Yates, F., *1960*, *Sampling methods for censuses and surveys* (London).

Zetterberg, H., *1965*, *On theory and verification in sociology* (Totawa, N.J.).

Zipf, G. K., *1949*, *Human behavior and the principle of least effort* (Cambridge, Mass.)

Author Index

Subject Index

[Bold numerals indicate where an important concept is defined or where an important idea is examined in depth]